NORTH-HOLLAND MATHEMATICS STUDIES 196

(Continuation of the Notas de Matemática)

Editor: Saul LUBKIN
University of Rochester
New York, U.S.A.

ELSEVIER

2004

Amsterdam – Boston – Heidelberg – London – New York – Oxford
Paris – San Diego – San Francisco – Singapore – Sydney – Tokyo

COMPUTATIONAL COMPLEXITY: A QUANTITATIVE PERSPECTIVE

COMPUTATIONAL COMPLEXITY: A QUANTITATIVE PERSPECTIVE

Marius ZIMAND
Department of Computer and Information Sciences
Towson University
Towson, U.S.A.

ELSEVIER

2004

Amsterdam – Boston – Heidelberg – London – New York – Oxford
Paris – San Diego – San Francisco – Singapore – Sydney – Tokyo

ELSEVIER B.V.
Sara Burgerhartstraat 25
P.O. Box 211, 1000 AE Amsterdam
The Netherlands

ELSEVIER Inc.
525 B Street, Suite 1900
San Diego, CA 92101-4495
USA

ELSEVIER Ltd
The Boulevard, Langford Lane
Kidlington, Oxford OX5 1GB
UK

ELSEVIER Ltd
84 Theobalds Road
London WC1X 8RR
UK

First edition 2004

Library of Congress Cataloging in Publication Data
A catalog record is available from the Library of Congress.

British Library Cataloguing in Publication Data
A catalogue record is available from the British Library.

ISBN: 0-444-82841-9
ISSN: 0304-0208

♾ The paper used in this publication meets the requirements of ANSI/NISO Z39.48-1992 (Permanence of Paper).
Printed in The Netherlands.

To my family

Preface

About "quantitative perspective." The subtitle of the book seems to be redundant and requires an explanation. The main purpose of computational complexity is to measure the amount of time, or of space, or of some other resource, that is necessary to solve a computational problem. Thus, by its very nature, computational complexity is a quantitative theory. However, a look at some of the best-known results in complexity (e.g., results asserting the absolute or the conditional separation of complexity classes, or the hardness of certain computational tasks) reveals that the quantitative component is, in many aspects, quite weak. For instance, from the deterministic time hierarchy theorem, we know that there exists a problem that is solvable in exponential time but not in polynomial time. This result, important as it is, raises several questions of a quantitative nature. We would like to know (a) something about the abundance of such problems, (b) if the hardness of the problem manifests itself for just a few rare and accidental inputs, or, on the contrary, for most inputs, and (c) if there is perhaps some approximation of the problem, in some natural sense, that is solvable in polynomial time.

This book analyzes such quantitative aspects of some of the most important results in computational complexity.

From a certain point of view, most theorems in computational complexity can be divided into two types. Type one consists of those theorems that state a complexity-related attribute of one individual function. Type two consists of those theorems that involve an entire class, or several classes, of functions. This taxonomy is important because it indicates the technical tools that we can use for the quantitative analysis. For type one results, the quantitative attributes of interest have a concrete numerical formulation expressed as a function of the input length. For example, if a function is hard (in some sense), we can ask on how many inputs of length n it is hard. The quantitative analysis of type two results is not so straightforward. A generic formulation of many theorems in this category is that there exists a function f in some class $\mathcal{C}$ that has property Q (for example: "There is a function in EXP that is not polynomial-time computable," or, "There exists a computable function that is speedable"). The obvious quantitative question, "How many functions f in $\mathcal{C}$ have property Q," is usually not too meaningful. Indeed, for most classes $\mathcal{C}$ and properties Q, it holds that if a function f in $\mathcal{C}$ has property Q, then almost every finite variation of f is in $\mathcal{C}$ and has property Q as well, and, therefore, the answer is trivially "An infinity." Fortunately, mathematicians have already developed tools and theories to handle this type of situations. The solution is to rephrase the question as "How large is the subset of function in C that have property Q," and to seek answers such as "small," or "large," or several nuances inbetween by using concepts from topology and measure theory. The theoretical foundations of this approach are presented in Chapter 1, Section 1.2.

A result enhanced with relevant quantitative attributes is more informative and more convincing and, therefore, clearly has theoretical merit. Besides that, a quantitative result can have, especially for type one results, practical value as well. At some point, there has been a common perception that computational complexity is a theory of "bad news," because common results, such as showing that a problem is NP-complete, assert that real-world and innocent-looking tasks are not feasible (in general, if we assume some reasonable hypothesis). In fact, "bad news" is a relative term, and, indeed, in some situations, we want an adversary to not be able to perform a task, e.g., to not be able to break our cryptography protocol. However, a "bad news" result does not automatically become useful in such a scenario. For this to happen, its hardness features have to be quantitatively evaluated and shown to manifest extensively.

Audience. My intention has been to write the book so that it appeals to a large audience. Experts in computational complexity may be interested in the special "quantitative" angle from which most results are presented. However, my primary target audience is elsewhere. In my intention, the book should benefit the most a reader who knows already the basic tenets of complexity, enjoys a rigorous mathematical treatment of a subject, wants to find out more about complexity than what is covered in a standard course, and, in particular, is interested in the novel major developments in complexity. The book is self-contained and can serve as a textbook for a course in advanced computational complexity. In general, the book should appeal to graduate computer science students and postdocs, and to researchers who have an interest in theory and need a good understanding of computational complexity, e.g., researchers in algorithms, AI, logic, and other disciplines.

Topics. The intended audience has influenced the choice of topics. Most of them are relevant outside the immediate scope of computational complexity. Also, most of them go beyond the material that is covered in a standard first course in complexity theory. One chapter is dedicated to abstract complexity theory, an older field which, however, deserves attention because it lays out the foundations of complexity. The other chapters, on the other hand, focus on recent and important developments in complexity. The book presents in a fairly detailed manner concepts that have been at the center of the main research lines in complexity in the last decade or so, such as: average-complexity, quantum computation, hardness amplification, resource-bounded measure, the relation between one-way functions and pseudo-random generators, the relation between hard predicates and pseudo-random generators, extractors, derandomization of bounded-error probabilistic algorithms, probabilistically checkable proofs, non-approximability of optimization problems, and others. In some cases, it has not been possible, given the book's scope, to present the ultimate results regarding some of these concepts. However, I have included a presentation of the proof techniques that are required to obtain such results.

Chapter 1 presents basic facts from the theory of computation, computational complexity, topology, and measure theory that are used throughout the book. Depending on the reader's familiarity with these matters, this chapter should be read first, or just browsed and used as a reference.

The other chapters are independent and can be read in any order.

Chapter 2 presents the most important results in abstract complexity theory. These are classical results, which are displayed here from a novel angle that emphasizes some important quantitative facets.

Chapter 3 explores quantitative issues regarding the most important complexity classes, namely P, NP, E, and EXP. It includes a section on average-case complexity.

Chapter 4 is dedicated to quantum computation. The discussion concentrates on the potential of quantum computation to vastly outperform classical computation.

Chapter 5 focuses on some of the basic primitive objects that are used in cryptography: One-way functions, pseudo-random generators, and hard functions and predicates. The presentation emphasizes the quantitative attributes of these primitives, an aspect that is essential for their utilization in cryptography.

Chapter 6 is dedicated to NP optimization problems. The chapter concentrates on the issue of whether individual problems from this category admit polynomial-time good approximation algorithms.

I will maintain a website for this book (accessible from my web page at `http://triton.towson.edu/~mzimand`). It will contain a list of comments and updates for the topics presented in the book, and a list of errata. Please send me your comments and any error that you find in the book.

Acknowledgments. I would like to thank Saul Lubkin who gave me the idea to write this book and encouraged me over the years to complete this undertaking. I am grateful to Richard Chang, William Gasarch, Sanjay Gupta, Omer Horvitz, and Jon Squire who read parts of the book and gave me useful suggestions. I thank Mirko Janc for his expert advice on typesetting issues and for the many hours he spent struggling with my poorly Latex-edited manuscript.

Special thanks to my wife Iliana and my son Paul. This book would not exist without their patience, support, and love.

Baltimore, February 18, 2004. *Marius Zimand*

Contents

Chapter 1

Preliminaries

1.1 Short guide to computability and computational complexity

The purpose of this section is to fix some terminology and to briefly overview a few basic facts from computability theory and computational complexity theory that will be used in the book. Of course, this being a book on computational complexity, many other concepts will be introduced throughout the book.

1.1.1 Computability

Computability theory explores the capabilities and the limits of algorithms. Intuitively, an algorithm is a mechanical procedure given by a finite description. An algorithm's objective is to solve a computational problem. A computational problem consists of a (usually infinite) set of inputs and of associated tasks. For example, an input can be a positive integer and the associated task is to determine if the number is prime or not. An algorithm that solves a computational problem takes an arbitrary input and performs the associated task. One of the questions addressed by computability theory, perhaps the most basic one, is whether, for a given computational problem P, there exists or not an algorithm that solves it. According to the *Church-Turing thesis*, the informal notion of an algorithm is fully captured by the formal concept of a *Turing machine.* A Turing machine models a device (e.g., computer) that is able to perform mechanical computations. For this reason, a Turing machine is said to be a computational model. Many other computational models have been considered: Random-access machines, Markov algorithms, λ-calculus, and others. They all turned out to have identical computational power. For instance, for any Turing machine M, there is a random-access machine R that simulates the computations performed by M, and vice-versa, for any random-access machine R, there is a Turing machine M that simulates the computations performed by R.

In this book, we adopt the Turing machine as the standard model of computation. Among other things, this implies that the term *machine*, unless specified otherwise, signifies a Turing machine.

A Turing machine (TM), in its basic variant, is given by seven components denoted $Q, \Gamma, \Sigma, \delta, q_{\text{start}}, q_{\text{acc}}, q_{\text{rej}}$, whose meaning we elucidate next. A Turing machine consists of (a) a finite control, (b) one tape, representing the memory, that has a left margin and is divided into an infinite number of cells, and (c) a moving read/write head. The finite control can be in any one of a finite set Q of *states*. Each tape cell can hold one symbol from the *tape alphabet* Γ, which includes the *input alphabet* Σ and also contains a special character B, called the *blank character* ($B \notin \Sigma$). At each moment, the head is positioned on one cell of the tape (we also say, that the head is scanning the cell). In one computation step, the finite control changes its state and the head reads the symbol from the cell that it currently scans, writes a symbol into that cell, and moves one cell to the left or right (the head cannot move to the left of the leftmost cell). Initially, the head is positioned on the leftmost cell, the finite control is in a designated special state q_{start}, called the *start state*, and the tape contains the input string $x \in \Sigma^*$ written on consecutive cells starting from the leftmost cell; all the other cells contain the B (blank) symbol. There are two designated special states q_{acc} and q_{rej}, called the *accepting state* and the *rejecting state*. The computation terminates when the finite control goes into the state q_{acc} or into the state q_{rej}. The computation of a Turing machine is stipulated by the *transition function* δ. Turing machines can be *deterministic* or *nondeterministic* depending on the type of the function δ.

For a deterministic Turing machine, δ takes the form $\delta : Q \times \Gamma \to Q \times \Gamma \times \{L, R\}$. Suppose that $\delta(q, a) = (q', b, R)$, where $q, q' \in Q$ and $a, b \in \Gamma$ and R is a special symbol meaning "move right" (L is another special symbol meaning "move left"). The effect is that if the machine is in state q and the scanned cell contains the symbol a, then the machine goes to state q', writes symbol b into the scanned cell, and moves to the right cell. (similarly to R, L tells the head to move to the left cell, unless the currently scanned cell is the leftmost one, in which case the head does not move). If the machine goes into the state q_{acc}, then the input string is accepted, and if the machine goes into the state q_{rej}, then the input string is rejected. There is the possibility that the machine never goes into q_{acc} or q_{rej}, in which case we say that the machine does not halt, or that it loops. For a Turing machine M, $L(M)$ is the set of input strings $x \in \Sigma^*$ that are accepted by M. $L(M)$ is called the language accepted by M. The status of a Turing machine M at a given time t is completely given by the content of the tape, the state of the finite control, and the position of the head, all at time t. These elements taken together define a *configuration*. A standard way to represent a configuration is as follows. Suppose that at time t, M is in state q, the tape contains the symbols $b_0 \ldots b_{k-1}$ (b_0 is the symbol in the leftmost cell and b_{k-1} is the rightmost symbol other than blank), and the head is scanning cell j. Then the configuration C_t of M at time t is represented by the string $b_0 \ldots b_{j-1}(b_j, q)b_{j+1} \ldots b_{k-1}$. The transition function δ defines a derivation rule $\vdash$ between configurations: $C_t \vdash C_{t+1}$ if configuration

C_{t+1} follows after configuration C_t according to a computation step given by δ. A halting computation of M on an arbitrary input $x \in \Sigma^*$ is described by a finite sequence of configurations $C_0, C_1, \ldots, C_k$, where C_0 is the initial configuration, C_k is the (halting) configuration in which the state is q_{acc} or q_{rej}, and $C_{i-1} \vdash C_i$, for each $1 \leq i \leq k$. The running time of M on x is k, if there is a sequence of configurations as above, and it is ∞ if M does not halt on input x. The *running time*, or the *time complexity* of M is the function $f : \mathbb{N} \rightarrow \mathbb{N}$, where, for all $n \in \mathbb{N}$, $f(n)$ is the maximum running time of M on input x taken over all input strings x of length n. The space utilized by M on input x is the positive integer that gives the index[1] of the rightmost tape cell reached by the head during the computation of M on x. The *space complexity* of M is the function $f : \mathbb{N} \rightarrow \mathbb{N}$, where $f(n)$ is the maximum space utilized by M on x taken over all input strings $x \in \Sigma^*$ of length n.

For a *nondeterministic Turing machine*, function f has the form $f : Q \times \Gamma \rightarrow \mathcal{P}(Q \times \Gamma \times \{L, R\})$, where $\mathcal{P}(A)$ denotes the power set of the set A. This implies that from one given configuration, there are finitely many possible next configurations and the machine has to choose (or guess, as we sometimes say) the configuration into which it is going. The computation of a Turing machine N on an input x is described by a tree whose branches correspond to the different possible computation paths of the machine. This tree is called the *computation tree* of N on input x. As in the case of deterministic machines, a computation path may terminate in a configuration containing the state q_{acc} (called an accepting configuration), or in a configuration containing the state q_{rej} (called an rejecting configuration), or may not terminate. The machine accepts the input if and only if there exists a computation path in the computation tree that leads to an accepting configuration. The language accepted by a nondeterministic Turing machine N, denoted $L(N)$, is the set of strings accepted by N. The running time of a nondeterministic machine N on an arbitrary input $x \in L(N)$ is the minimum number of steps along a computation path, where the minimum is taken over all the accepting computation paths in the computation tree.[2] The running time of N, also called the time complexity of N, is the function $f : \mathbb{N} \rightarrow \mathbb{N} \cup \{\infty\}$, where, for all $n \in \mathbb{N}$, $f(n)$ is the maximum running time of N on input x taken over all input strings $x \in L(N)$ of length n, if there are such strings, and it is ∞, otherwise. The space complexity of N is defined in the analogous way.

A nondeterministic Turing machine N can be converted into an equivalent deterministic Turing machine M (i.e., $L(N) = L(M)$).

It is sometimes helpful to consider variants of the Turing machine model that have enhanced "hardware." In particular, in Chapter 6, we will use Turing machines that have k read/write tapes, for some $k \in \mathbb{N}$. All the concepts that have

---

[1]The leftmost cell has index 1, the cell to its right has index 2, and so on.

[2]Note that the running time is not defined for inputs that are not accepted. Typically, for natural nondeterministic algorithms, we can impose that all the computation paths have the same number of steps. In such a case, the running time is defined for all inputs to be the common length of all computation paths in the tree.

effective operators (i.e., operators that are both total and effective) is governed by the Kreisel-Lacombe-Shoenfield Theorem. We state it in a particular form that is related to total effective operators (for a proof, see [Cal88, p. 192]):

If F is a total effective operator, then there exists a computable function $g\colon \mathbb{N} \to \mathbb{N}$ such that, for every computable φ_i,

$$\mathrm{Graph}(F(\varphi_i)) = \bigcup_{j \in C_i} \mathrm{Graph}(\alpha_{g(j)}),$$

where $C_i = \{j \in \mathbb{N} \mid \mathrm{Graph}(\alpha_j) \subseteq \mathrm{Graph}(\varphi_i)\}$.

1.1.2 Computational complexity

Once we have determined that a computational problem is solvable by algorithms, the next goal is to assess the necessary computational resources. This question is in the charter of computational complexity theory. At the core of this theory is the notion of a *complexity class*. A complexity class can be defined in an abstract way and this approach will be undertaken in Chapter 2. However, it is more common to consider concrete complexity classes that are defined by (a) a model of computation, (b) a computational resource, and (c) a function that bounds the allowed amount of the resource. Typically, a complexity class consists of the decision problems solved by computational devices in a given model of computation using an amount of a given resource that is bounded by a given function. The models of computation used in this book are provided primarily by different types of Turing machines. Sometimes we will also consider circuit-based models, but we defer for the moment the discussion of such models. The most important computational resources are (a) the running time of an algorithm, and (b) the space used by an algorithm. For the case of deterministic and nondeterministic Turing machines, these two resources have already been introduced (see our earlier discussion of time complexity and space complexity). The basic time and space classes are defined as follows.

Definition 1.1.2 *Let $f\colon \mathbb{N} \to \mathbb{N}$ be a function.*

(a) DTIME$[f(n)]$ *is the class of languages accepted by deterministic Turing machines of time complexity bounded by $f(n)$,*

(b) NTIME$[f(n)]$ *is the class of languages accepted by nondeterministic Turing machines of time complexity bounded by $f(n)$,*

(c) DSPACE$[f(n)]$ *is the class of languages accepted by deterministic Turing machines of space complexity bounded by $f(n)$,*

(d) NSPACE$[f(n)]$ *is the class of languages accepted by nondeterministic Turing machines of space complexity bounded by $f(n)$.*

The utilization of Turing machines is not essential modulo a polynomial factor in the bounding function, because the common models of computation can simulate each other in polynomial time and polynomial space. In principle any function that maps natural numbers into natural numbers can be used in the role of f in the above definition. However, some functions lead to abnormal phenomena. For example, by the Gap Theorem (see Theorem 2.4.1), for any computable functions r and a, there exists a computable function f such that $f(n) \geq a(n)$ and $\mathrm{DTIME}[f(n)] = \mathrm{DTIME}[r(f(n))]$. For instance, if $r(n) = 2^{2^n}$, this implies that the incrementation of the running time from $f(n)$ to $2^{2^{f(n)}}$ does not allow the computation of any "new" function. The typical natural bounding functions (i.e., functions such as 1, n, $n \log n$, n^3, 2^n, etc.) do not exhibit such phenomena and there is a formal way to delimit such functions. A function $t\colon \mathbb{N} \to \mathbb{N}$ is *fully time-constructible* if there exists a deterministic Turing machine that halts after exactly $t(n)$ steps on every input of length n. A function $s\colon \mathbb{N} \to \mathbb{N}$ is *fully space-constructible* if there exists a deterministic Turing machine that uses exactly $s(n)$ space on every input of length n. All the above natural functions, as well as many other nice functions, are fully time-constructible and fully space-constructible. Moreover, if $t_1(n)$ and $t_2(n)$ are fully time-constructible, then $t_1 + t_2$, $t_1 \cdot t_2$, and $t_1^{t_2}$ are fully time-constructible. The same fact holds for fully space-constructible functions. The following hierarchy theorems are known for fully time-constructible functions and for fully space-constructible functions.

Theorem 1.1.3 (Hierarchy theorems, see [DK00]) *Let t_1 and t_2 be fully time-constructible functions, and let s_1 and s_2 be fully space-constructible functions with $s_1(n) \geq \log n$ and $s_2(n) \geq \log n$, for all n.*

(a) *If* $t_1(n) \log t_1(n) = o(t_2(n))$, *then* $\mathrm{DTIME}[t_1] \subsetneqq \mathrm{DTIME}[t_2]$.

(b) *If* $t_1(n+1) = o(t_2(n))$, *then* $\mathrm{NTIME}[t_1] \subsetneqq \mathrm{NTIME}[t_2]$.

(c) *If* $s_1(n) = o(s_2(n))$, *then* $\mathrm{DSPACE}[s_1] \subsetneqq \mathrm{DSPACE}[s_2]$.

(d) *If* $s_1(n) = o(s_2(n))$, *then* $\mathrm{NSPACE}[s_1] \subsetneqq \mathrm{NSPACE}[s_2]$.

The following relations are known between deterministic and nondeterministic complexity classes.

Theorem 1.1.4 (Deterministic classes vs. nondeterministic Classes, see [DK00]) *Let $f_1\colon \mathbb{N} \to \mathbb{N}$ be a fully space-constructible function, $f_1(n) \geq n$, for all $n \in \mathbb{N}$, and let $f_2\colon \mathbb{N} \to \mathbb{N}$ be a fully space-constructible function, $f_2(n) \geq \log n$, for all $n \in \mathbb{N}$. Then,*

(a) $\mathrm{NTIME}[f_1(n)] \subseteq \bigcup_{c \geq 0} \mathrm{DTIME}[2^{c f_1(n)}]$.

(b) $\mathrm{NSPACE}[f_2(n)] \subseteq \bigcup_{c \geq 0} \mathrm{DTIME}[2^{c f_2(n)}]$.

(c) $\mathrm{NTIME}[f_1(n)] \subseteq \mathrm{DSPACE}[f_1(n)]$.

(d) $\mathrm{NSPACE}[f_2(n)] \subseteq \mathrm{DSPACE}[(f_2(n))^2]$.

There are numerous complexity classes and some are defined by different mechanisms (in Chapter 6, for instance, we will see complexity classes defined by syntactically restricted formulas in some logical systems). However, it is universally accepted that the following classes are the most important.

Definition 1.1.5
- L = DSPACE[$\log n$] *(deterministic space);*
- NL = NSPACE[$\log n$] *(nondeterministic space);*
- P = $\bigcup_{k\geq 1}$ DTIME[n^k] *(polynomial time);*
- NP = $\bigcup_{k\geq 1}$ NTIME[n^k] *(nondeterministic polynomial time);*
- PSPACE = $\bigcup_{k\geq 1}$ DSPACE[n^k] *(polynomial space, equal to* $\bigcup_{k\geq 1}$ NSPACE[n^k]*);*
- E = $\bigcup_{c\geq 0}$ DTIME[2^{cn}]*;*
- NE = $\bigcup_{c\geq 0}$ NTIME[2^{cn}]*;*
- EXP = $\bigcup_{k\geq 1}$ DTIME[2^{n^k}] *(exponential time);*
- NEXP = $\bigcup_{k\geq 1}$ NTIME[2^{n^k}] *(nondeterministic exponential time);*

Nondeterministic computation can be viewed in a different way. We can assume without loss of generality that a nondeterministic Turing machine has exactly two choices at each non-final step (i.e., each non-halting configuration has exactly two ordered successor configurations). In this case, at each step, a guess takes the form of a bit b, such that $b = 0$ ($b = 1$) means the machine will go in the first successor configuration (respectively, the second successor configuration). In fact, we can consider that the machine on input x makes all the guesses upfront in the form of a binary string y, written perhaps on a separate tape, after which the rest of the computation runs in a deterministic fashion. Via these observations, the following alternative definition of the class NP can be shown.

Theorem 1.1.6 *For any language A, $A \in$ NP if and only if there is a predicate Q computable in polynomial time and a polynomial p such that, for any input x,*

$$x \in A \Leftrightarrow \exists y \in \Sigma^* \ (|y| \leq p(|x|) \text{ and } Q(x,y) = 1).^3$$

The same mechanism can be used to define probabilistic complexity classes (used to capture the hardness of probabilistic algorithms). We will assume again that the Turing machine, which we call, this time, a *probabilistic Turing machine*, has exactly two choices at each step. Intuitively, each choice is taken according to a coin flip. As above, we can consider that all the necessary coin flips are made upfront before the start of the actual computation and are written on a separate tape in the form of a binary string, usually called the *random string*. We will

---

[3] We denote by $|x|$ the length of the string x.

only consider polynomial-time probabilistic Turing machines. This means that, for each such machine M, there is a polynomial p such that, for all input x, all the computation paths in the computation tree of M on x have length $p(|x|)$. The most important probabilistic complexity classes are given in the following definition. First we need a notational convention: If M is a probabilistic Turing machine, $M(x, y) = 1$ denotes the fact that M on input x and with random string y halts in an accepting configuration.

Definition 1.1.7 (a) (PP) *A language A is in* PP *if there is a polynomial-time probabilistic Turing machine M such that*

$$x \in L \Leftrightarrow \text{Prob}_y(M(x, y) = 1) > 1/2.$$

(b) (BPP) *A language A is in* BPP *if there is a constant $\epsilon > 0$ and a polynomial-time probabilistic Turing machine M such that*

$$x \in L \Leftrightarrow \text{Prob}_y(M(x, y) = 1) > 1/2 + \epsilon,$$
$$x \notin L \Leftrightarrow \text{Prob}_y(M(x, y) = 1) < 1/2 - \epsilon.$$

(c) (RP) *A language A is in* RP *if there is a polynomial-time probabilistic Turing machine M such that*

$$x \in L \Leftrightarrow \text{Prob}_y(M(x, y) = 1) > 1/2,$$
$$x \notin L \Leftrightarrow \text{Prob}_y(M(x, y) = 1) = 0.$$

By repeating the computation several times, the error probabilities for BPP-computation and RP-computation can be made very small. We state this result for the case of BPP-computation.

Theorem 1.1.8 *Let $L \in$ BPP. Then, for every polynomial q, there exists a polynomial-time probabilistic machine M such that*

$$x \in L \Leftrightarrow \text{Prob}_y(M(x, y) = 1) > 1 + \frac{1}{2^{q(|x|)}},$$
$$x \notin L \Leftrightarrow \text{Prob}_y(M(x, y) = 1) < \frac{1}{2^{q(|x|)}}.$$

We turn to circuit complexity. We will limit to boolean circuits and, unless specified otherwise, a circuit in this book is a boolean circuit. Boolean circuits have been introduced to model electronic circuits whose gates perform logical bit operations. Formally, a boolean circuit is an acyclic directed graph whose nodes, also called *gates*, are classified in the following three categories:

(a) *Input gates* are nodes that have no incoming edge and one outgoing edge; the input nodes are labeled by distinct input variables $x_1, x_2, \ldots, x_n$ or by the boolean constants 0 and 1.

(b) *Inner gates* that are labeled with one of the boolean operators: AND, OR, and NOT. Unless specified otherwise, AND gates and OR gates have two incoming edges and one outgoing edge, and NOT gates have one incoming node and one outgoing edge.

(c) There is one *output gate* labeled AND, OR, or NOT. The output gate has two incoming edges if it is labeled AND or OR, and one incoming edge if it is labeled NOT; the output gate has no outgoing edges.

A circuit computes a predicate function in the following way. Assume that the input gates are labeled $x_1, x_2, \ldots, x_n$. Then the circuit calculates a function that maps $\{0,1\}^n$ to $\{0,1\}$. The n-bit input string gives a boolean assignment to the variables in the obvious way: The first bit of the input is assigned to x_1, the second bit of the output is assigned to x_2, and so on. Then each gate reads the values from its incoming edges, applies the boolean operator with which it is labeled, and (except for the output gate) further sends the result through the outgoing edge. The bit calculated by the output gate is the value calculated by the circuit.

We can consider circuits that have an ordered set of output gates. The value calculated by such a circuit is obtained by concatenating in order the bits calculated by the output gates.

Circuits can be probabilistic as well. Such circuits have additional input gates that are assigned random bits. The value calculated by the circuit is a random variable that depends on these random bits.

Note that a circuit, unlike a Turing machine, only takes inputs of a fixed length stipulated by the number of input gates that are labeled with variables. Therefore, it is often helpful to consider a *family of circuits* $C = \{C_1, \ldots, C_n, \ldots\}$, where, for each $n \in \mathbb{N}$, C_n is a circuit that admits inputs of length n.

The *size* of a circuit C, denoted size(C), is the number of gates in C. This is the main complexity measure for circuits. In the case of a family of circuits $(C_n)_{n\in\mathbb{N}}$, the size complexity of the family is a function g of n given by $g(n) = \text{size}(C_n)$. Since each gate performs an operation, the size is similar to the time complexity of Turing machines. It can be shown that any deterministic Turing machine of time complexity $t(n)$, where t is a fully time-constructible function, can be simulated by a family of circuits $(C_n)_{n\in\mathbb{N}}$ of size $t(n)\log t(n)$. Moreover, the family of circuits is *uniform*, in the sense that there is an efficient algorithm that on input 1^n produces the circuit C_n. In some situations, we will ignore the logarithmic factor and we will simply assume that an algorithm that performs $t(n)$ elementary operations can be implemented by a family of circuits (not necessarily uniform) of size $O(t(n))$.

An important aspect, due to the non-uniformity of the general model, is that circuits can calculate even non-computable functions. For instance, consider the notoriously non-computable set

$$K = \{i \in \mathbb{N} \mid \text{ the } i\text{-th Turing machine halts on input } i\}.$$

(K represents the famous halting problem.) Let $A = \{x \mid |x| \in K\}$. The language A is non-computable as well. However, for each $n \in \mathbb{N}$, we can easily construct a small circuit C_n that either (a) accepts all inputs of length n in case $n \in K$ (for example, C_n calculates x_1 OR $\overline{x_1}$),[4] or (b) rejects all inputs of length n in case $n \notin K$ (for example, C_n calculates x_1 AND $\overline{x_1}$). In fact, it can be shown that any function $f\colon \{0,1\}^n \to \{0,1\}$ can be calculated by a boolean circuit of size at most $(1+O(\frac{1}{\sqrt{n}}))\frac{2^n}{n} = O(2^n)$ (in brief, the circuit stores the truth table of the function; see, e.g., [Sav98, page 80]). Circuits will be used in Chapter 5, dedicated to cryptographical primitives, to model adversaries that want to compromise certain cryptographical protocols. In such circumstances, circuits are more meaningful than uniform models of computation in proving lower bounds: The fact that a certain task cannot be done by a circuit of size S shows that any adversary has to perform at least $\approx S$ elementary operations to accomplish the task.

We can bound the number of circuits having size t, where t is an arbitrary natural number, in the following way. An arbitrary gate of a circuit is described by its type, which can be AND, OR, NOT, or input gate, and by the numerical identification (number ID) of its at most two predecessors gates (i.e., the gates that provide the inputs of the current gate). Let us convene that the ID 0 means "no predecessor," which is needed for the input gates and for the non-existing predecessor of a NO gate. We need two bits to represent the type and at most $2\lceil \log t \rceil$ bits to represent via their number ID the at most two predecessor gates. Thus one gate can be described by a binary string of length at most $2+2\lceil \log t \rceil) \leq 2(\log t+2) = 2\log(4t)$. A circuit with t gates can be represented by a binary string made with t blocks of bits, each block of length at most $2\log(4t)$ describing a gate (the numeric ID of each gate is given by the rank of the corresponding block in the entire string). Therefore a circuit of size t is completely described by a binary string of length at most $t \cdot 2(\log(4t))$. Therefore the number of circuits of size t is bounded by the number of such binary strings, which is $2^{2t\cdot\log(4t)} = 2^{O(t\log t)}$.

1.2 Short guide to topology and measure theory

This section is intended to be an easily available reference for the topological and measure-theoretical notions utilized in this book.

In brief, we need concepts that enable us to declare that a certain class of sets is to a certain degree "small" or "large." The idea of discreteness is the base upon which the concept of a "small" class is defined in both the topological approach and the measure-theoretical approach. The primitive concepts that model this intuition are the *nowhere dense sets* (for topology) and the *measure zero sets* (for measure). Informally, a nowhere dense set is a set "full of holes." A measure zero set is a set on the real line that can be covered by intervals whose lengths total an arbitrary small positive value. The following sections present the technical realization of these ideas.

---

[4] We denote by $\overline{x}$ the negation of the boolean variable x.

1.2.1 Topology

A *topological space* is a pair $(X, \mathcal{O})$, where X is a set and $\mathcal{O}$ is a class of subsets of X, called the open sets of X, containing $\emptyset$ and X, and closed under finite intersections and arbitrary union. A *neighborhood* of a point $x \in X$ is an open set containing x. A *base* is a class $\mathcal{B}$ of open sets such that for every $x \in X$ and every neighborhood V of x, there exists a set $B \in \mathcal{B}$ such that $x \in B \subseteq V$. One can build a topological space starting from a base $\mathcal{B}$. Namely, suppose $\mathcal{B}$ is a class of subsets of X satisfying the properties (1) for every x, $x \in U \cap V$, for some $U, V \in \mathcal{B}$, implies that there exists $W \in \mathcal{B}$ such that $x \in W \subseteq U \cap V$, and (2) for every $x \in X$ there exists $B \in \mathcal{B}$ with $x \in B$; then there exists a unique topological space $(X, \mathcal{O})$ having $\mathcal{B}$ as a base. ($\mathcal{O}$ is the closure under arbitrary unions of sets in $\mathcal{B}$.) This is called the topological space *generated* by $\mathcal{B}$.

Definition 1.2.1 (Baire classification) *Let $(X, \mathcal{O})$ be a topological space.*

(1) *A set $A \subseteq X$ is* nowhere dense *if for every non-empty open set U_1 there exists a non-empty open set U_2 included in U_1 such that $A \cap U_2 = \emptyset$. In case $(X, \mathcal{O})$ is the topological space generated by base $\mathcal{B}$, the above is equivalent to saying that for every $U_1 \in \mathcal{B}$ there exists $U_2 \in \mathcal{B}$, U_2 included in U_1, such that $A \cap U_2 = \emptyset$.*

(2) *A set $A \subseteq X$ is of* first Baire category *(or* first category, *or* meagre*) if A can be represented as the countable union of nowhere dense sets.*

(3) *A set $A \subseteq X$ is of* second Baire category *if it is not of first Baire category.*

(4) *A set $A \subseteq X$ is* co-meagre *if its complement, $X - A$, is meagre.*

(5) *A set $A \subseteq X$ is* co-nowhere dense *if its complement, $X - A$, is nowhere dense.*

As mentioned, intuitively, a nowhere dense set A is a set "full of holes," because no matter how small an open set U_1 one may believe to be included in A, there is an entire open subset $U_2 \subseteq U_1$ that lies completely outside A (i.e., $A \cap U_2 = \emptyset$.)

The subsets of X can be classified with respect to the following taxonomy of sets of increasing size: Nowhere dense, first category, second category, co-meagre, and co-nowhere dense. Meagre sets are considered to be small sets while the sets situated above second category (inclusively) in the above hierarchy are considered to be large.

The sets of first category form a σ-ideal. This means that the class of such sets is closed under countable union and arbitrary subsets. In "reasonable" topological spaces $(X, \mathcal{O})$, the universe X is of second category. This is usually called the Baire Category Theorem for $(X, \mathcal{O})$.

In this book, we are interested in classifying classes of computable languages (or, equivalently, classes of computable predicates), and classes of computable functions. Let us consider here the former type (the other one will be used rarely and the approach follows the same pattern). In order to analyze classes of computable languages, we have to build relevant topological spaces $(X, \mathcal{O})$. X is defined as follows. Let $\Sigma = \{0, 1\}$ be the binary alphabet, and Σ^* the set of finite binary strings. Σ^∞ is the set of infinite binary strings. The set Σ^* is considered to be ordered in the lexicographical order: $\lambda < 0 < 1 < 00 \ldots$, where λ is the empty word. Let $s_i, i \geq 1$, be the i-th string in Σ^* according to this ordering and $\text{pos}(x) \in \mathbf{N} - \{0\}$ be the rank of string x in this ordering (i.e., $\text{pos}(x) = i \Leftrightarrow x = s_i$). For $x \in \Sigma^*$, $|x|$ denotes the length of x. The cardinality of a set A is denoted by $\|A\|$. For $x \in \Sigma^* \cup \Sigma^\infty$, $x(i) \in \{0, 1\}$ is the i-th bit of x, and $x(i : j)$ is the string $x(i)x(i+1)\ldots x(j)$ (defined for i and j at most $|x|$, in case $x \in \Sigma^*$). We identify a language $A \subseteq \Sigma^*$ with its characteristic sequence $A(s_1)A(s_2)\ldots A(s_n)\ldots$, where for each positive integer i, $A(s_i) = 1$ if $s_i \in A$, and $A(s_i) = 0$ if $s_i \notin A$. By this codification, $A \in \Sigma^\infty$.

Therefore, classes of computable sets are subsets of Σ^∞ and, thus, X is taken to be Σ^∞. In this book, we consider two bases $\mathcal{B}^C$ and $\mathcal{B}^S$, generating the Cantor topology and, respectively, the superset topology. Both these bases are formed by sets indexed by finite binary strings. Thus, $\mathcal{B}^C = (U_v^C)_{v \in \Sigma^*}$ and $\mathcal{B}^S = (U_v^S)_{v \in \Sigma^*}$.

For $v \in \Sigma^*$, U_v^C (the basic open set defined by v in the Cantor topology) is defined by

$$U_v^C = \{w \in \Sigma^\infty \mid \forall i \ (1 \leq i \leq |v| \Rightarrow (v(i) = w(i)))\}.$$

For $v \in \Sigma^*$, U_v^S (the basic open set defined by v in the superset topology) is defined by

$$U_v^S = \{w \in \Sigma^\infty \mid \forall i \ ((1 \leq i \leq |v| \text{ and } v(i) = 1) \Rightarrow (v(i) = w(i) = 1))\}.$$

It is readily checked that for every v_1 and v_2 in Σ^*, if $U_{v_1}^C \cap U_{v_2}^C \neq \emptyset$, then there exits v_3 in Σ^* such that $U_{v_1}^C \cap U_{v_2}^C = U_{v_3}^C$, and that for every w in Σ^∞, there exists v in Σ^* such that $w \in U_v^C$. The same properties hold for the sets in $\mathcal{B}^S$, which implies that $\mathcal{B}^C$ and $\mathcal{B}^S$ are indeed valid basis for Σ^∞. Let $\mathcal{O}^C$ be the set of open sets generated by the base $\mathcal{B}^C$, and $\mathcal{O}^S$ be the set of open sets generated by the base $\mathcal{B}^S$. The Baire Category Theorem holds for the topological spaces $(X, \mathcal{O}^C)$ and $(X, \mathcal{O}^S)$.

The Cantor topology is arguably the most natural topology on Σ^∞. It can also be defined as the infinite product of the discrete topology on Σ. The Cantor topology corresponds to extensions of finite initial binary segments (i.e., predicates with domain of the form $\{0, 1, \ldots, n\}$ for some n), an operation which is extensively used in computable function theory and in computational complexity theory. Indeed, U_v^C can be regarded as the class of predicates that extend the finite initial predicate encoded by v. The superset topology is the "next natural" topology on Σ^∞. It corresponds to extension of finite sets, an operation which is also widely used. Indeed, U_v^S can be regarded as the class of sets that are supersets of the finite set encoded by v.

A construction, similar to the one leading to the Cantor topology on the space of binary languages, can be carried out for the class of computable functions. The only difference is that the binary alphabet Σ is replaced by the infinite alphabet $\mathbb{N}$, the set of natural numbers. Such a topology is considered in Chapter 2.

Often the topology that is considered will be stated upfront, and in this case the superscripts C and S will be omitted.

Unfortunately, the classical setting is not good enough for our purposes. It can be seen that any countable class of subsets of Σ^∞ is meagre relative to the Cantor topology. The same holds relative to the superset topology for any countable class that does not contain infinite binary strings in which almost every bit is 1. Indeed, to consider just the case of the Cantor topology, if $Y = (Y_i)_{i\in\mathbb{N}}$ is a countable class of subsets of Σ^∞, then $Y = \bigcup_{i\in\mathbb{N}}\{Y_i\}$ and each class $\{Y_i\}$ is nowhere dense, because for each $U^C_{v_1} \in \mathcal{B}^C$ one can easily find $U^C_{v_2} \in \mathcal{B}^C$ such that $U^C_{v_2} \subseteq U^C_{v_1}$ and $U^C_{v_2} \cap Y_i = \emptyset$. The string v_2 is an extension of the string v_1 (this guarantees that $U^C_{v_2} \subseteq U^C_{v_1}$) and is taken such that for some j with $j > |v_1|$, $v_2(j) \neq Y_i(s_j)$. Of course, this is not surprising. By viewing Y as a set of real points in the interval $[0,1]$ (obtained by associating to each $Y_i \in \Sigma^\infty$, the real number $0.Y_i(1)Y_i(2)\ldots$, written here in base 2), we see that Y, being a countable set, is "full of holes."

As we deal with algorithmic objects, it is natural to overcome this difficulty by considering an effective or even resource-bounded version of Definition 1.2.1. Namely, returning to the previous example, we demand that, given v_1, v_2 should be found in a computable way or even by an algorithm acting within predetermined resource bounds. In other words, a set A is nowhere dense in the effective sense, if the holes can be effectively constructed. Thus, in the effective analogues of Definition 1.2.1, part (1) that we will use in the forthcoming chapters, we require the existence of computable or resource-bounded witness functions that compute U_2 starting from U_1. Also, the effective analogues of Definition 1.2.1, part (2), require that A can be represented as a *uniform* countable union of effective nowhere dense sets, in the sense that the witness function for each class in the union should be found in a uniform way given the respective class. Clearly, if a set is nowhere dense (or meagre) in the effective sense, it is also nowhere dense (meagre) in the classical sense. The converse, usually, is not true: It can happen that a set which is nowhere dense classically is not so in the effective sense because the "holes" cannot be effectively constructed. The relevant formal definitions will be given taking into account the specific features of the objects that we analyze; however, the above guidelines will be followed in all situations.

1.2.2 Measure theory

A set $A \subseteq \Sigma^*$ is represented by the infinite binary sequence $A(s_1)A(s_2)\ldots \in \Sigma^\infty$.[5] As seen earlier, such a representation can be associated with a real number in the interval $[0,1]$. (From the point of view of measure theory, it does not matter that up to two distinct languages may be mapped to the same real value.) In what follows,

[5] Σ and the strings s_i are as defined in Section 1.2.1.

we identify $[0,1]$ with Σ^∞. Consequently, in order to study measure-theoretic aspects of classes of languages, we need to introduce the standard Lebesgue measure on the interval $[0,1]$.

The basic idea of Lebesgue measure is simple and natural: To measure the size of a set A of real numbers (whose shape may be quite complicated), try to approximate A by using as measuring sticks sets of the form (a,b), that is intervals whose structure and size should not cause any controversy. For every pair of real numbers $a, b, a \leq b$, the length of the interval $I = (a,b)$, denoted $|I|$, is equal to $b - a$. A sequence of intervals (I_i) covers a set of real numbers C if C is contained in $\bigcup_i I_i$. The greatest lower bound of the sums $\sum_i |I_i|$ over all sequences of intervals (I_i) that cover C is called the outer Lebesgue measure of C and is denoted by $\mu^*(C)$. Unfortunately, some sets behave strangely and it is not true that all $C \subseteq [0,1]$ have the desirable property that $\mu^*(C) + \mu^*([0,1] - C) = 1$. To have this property, C must be Lebesgue measurable. We say that C is *Lebesgue measurable* if, for each $\epsilon > 0$, there exist a closed set F and an open set G (in the standard topology on the real line) such that $F \subseteq C \subseteq G$ and $\mu^*(G - F) < \epsilon$. It can be shown that a set C is Lebesgue measurable if

$$\mu^*(B) = \mu^*(B \cap C) + \mu^*(B \cap \overline{C}), \qquad \text{for all } B \subseteq [0,1],$$

where $\overline{C} = [0,1] - C$. Also, the class of Lebesgue measurable sets is closed under the operations of countable union and difference of sets, and that the empty set is a Lebesgue measurable set.

Definition 1.2.2 (σ-field) *A collection of sets that contains the empty set and is closed under countable union and difference of sets is called a σ-field.*

Thus the Lebesgue measurable sets form a σ-field.

Definition 1.2.3 (Lebesgue measure) *The restriction of μ^* on the σ-field of Lebesgue measurable sets in $[0,1]$ is denoted by μ and is called the Lebesgue measure on the interval $[0,1]$.*

The function μ is real-valued, non-negative, countably additive (i.e., if $(C_i)_{i\in\mathbb{N}}$ is a sequence of Lebesgue measurable pairwise disjoint sets, then $\mu(\bigcup_{i\in\mathbb{N}} C_i) = \sum_{i\in\mathbb{N}} \mu(C_i)$) and $\mu(\emptyset) = 0$ (these are the properties of a measure in the general setting). For any class of subsets of a set X of real numbers, there is a smallest σ-field of subsets of X that contains it. This is called the *σ-field generated by the class.* The members of the σ-field generated by the class of intervals included in X are called the *Borel sets* of X. It turns out that we can restrict the class of measuring sticks even more and consider only intervals of the form $[0.x_1x_2\ldots x_n, 0.x_1x_2\ldots x_n11\ldots]$. We denote this interval by B_x, where $x = x_1x_2\ldots x_n \in \Sigma^*$. Note that the length of B_x is $2^{-|x|}$ and that the sets B_x are just the basic open sets in the Cantor topology. The key point is that the σ-field generated by the collection of sets $(B_x)_{x\in\Sigma^*}$ coincides with the Borel sets of $[0,1]$ (taking into account the association between subsets of Σ^* and real numbers in $[0,1]$). It can be shown that

that is arbitrarily close to zero, this may not be possible to do effectively or within bounded computational resources.

The now standard method of defining effective and resource-bounded measure is due to Lutz [Lut92] building on earlier work of Schnorr [Sch73], and is based on effective and, respectively, resource-bounded martingales (which, roughly speaking, are betting strategies). We first sketch the method at an intuitive level. Suppose A is a set of infinite binary strings (equivalently, A is a set of real numbers in $[0,1]$) and we want to build a sequence of intervals $(I_i)_{i\in\mathrm{N}}$ covering A such that $\sum_{i\in\mathrm{N}} |I_i| \leq 2^{-k}$ (i.e., 2^{-k} is an arbitrarily small value). The procedure runs in stages. We start with 2^{-k} dollars invested in the whole interval $[0,1]$ and 0 dollars on all others subintervals of $[0,1]$. Thus initially, $\text{invest}([0,1]) = 2^{-k}$ and $\text{invest}(I) = 0$, for all other subintervals I. We can imagine playing a game as follows: At each stage, the investment on each interval I doubles its value and this new amount is reinvested on I_ℓ and I_r, the left half and the right half intervals of I, the amount invested on each of these half intervals being decided by a computable or resource-bounded betting strategy. The goal is to concentrate the investition on the intervals that give a sufficiently precise cover of A. The procedure continues for a number of stages decided by us. Summarizing, at each stage the intervals on which we invest have half the length of their father and the available amount of money doubles. By induction on the stage number, it follows that

$$\sum_{I \text{ subinterval of } [0,1]} |I| \cdot \text{invest}(I) = 2^{-k}$$

at each stage. Let us say that I is a winning interval if the final amount invested in I is greater or equal than 1 dollar. We win the game if A is covered by winning intervals. If we win the game, then we effectively found a covering of A of total length $\leq 2^{-k}$, because it follows that

$$\sum_{I \text{ winning subinterval of } [0,1]} |I| \leq 2^{-k}.$$

To show that a set A has measure zero, we would like to win the game for all $k \geq 0$, so that we find coverings of arbitrarily small total length. Equivalenly, by scaling, we can assume that we play a sequence of games, $game_1, game_2, \ldots, game_n, \ldots$, and in each game we start with the same amount, say 1 dollar. In $game_n$, a winning interval is an interval on which, at the end of the game, we have invested n dollars. If, in all the games, we manage to cover A by winning intervals, then we can infer that A has measure 0 (this argument is made rigorous in Lemma 1.2.9).

The intervals are given by the basic sets B_x. Thus, $[0,1] = B_\lambda$, $[0,1/2] = B_0$, $[1/2,1] = B_1$, and so on. Let $d(x)$ be the amount invested on B_x. The rule of the game can be written as

$$d(x0) + d(x1) = 2d(x),$$

for each $x \in \Sigma^*$. Such a function is called a martingale. The above ideas are formalized in the following definition.

Definition 1.2.6 (Martingale)

(i) *A martingale is a function* $d\colon \Sigma^* \to [0,\infty)$ *satisfying*

$$d(x) = \frac{d(x0) + d(x1)}{2},$$

for all strings in Σ^*.

(ii) *The global value of the martingale* d *is* $d(\lambda)$, *where* λ *is the empty word.*

(iii) *The set covered by a martingale* d *is*

$$S[d] = \bigcup_{w \in \Sigma^*, d(w) \geq 1} B_w.$$

(iv) *A martingale* d *covers a set* $X \subseteq \Sigma^\infty$ *if* $X \subseteq S[d]$.

(v) *For each* $n \in \mathbb{N}$, *the set* n*-covered by a martingale* d *is*

$$S^n[d] = \bigcup_{w \in \Sigma^*, d(w) \geq n} B_w.$$

(vi) *A martingale* d n*-covers a set* $X \subseteq \Sigma^\infty$ *if* $X \subseteq S^n[d]$.

(vii) *A martingale* d *succeeds on a set* $X \subseteq \Sigma^\infty$ *if, for all* $n \in \mathbb{N}$, $X \subseteq S^n[d]$.

Let us play this betting game on a few classes of sets.

Example 1.2.7 Consider a class C_1 containing a single set A and let $k \in \mathbb{N}$. We start with 2^{-k} dollars, i.e., $d(\lambda) = 2^{-k}$. We bet

$$d(x0) = \begin{cases} 2d(x), & \text{if } A(s_{|x|+1}) = 0, \\ 0, & \text{if } A(s_{|x|+1}) = 1, \end{cases}$$

and

$$d(x1) = \begin{cases} 0, & \text{if } A(s_{|x|+1}) = 0, \\ 2d(x), & \text{if } A(s_{|x|+1}) = 1. \end{cases}$$

Let $w = A(s_1)A(s_2)\ldots A(s_k)$. Observe that $d(w) = 1$ and, since $A \in B_w$, it follows that d covers the set C_1. It is easy to see that in fact d succeeds on C_1. ∎

Example 1.2.8 A set A is *sparse* if there is a polynomial p such that $\|A^{\leq n}\| \leq p(n)$, for all $n \in \mathbb{N}$. Consider C_2 the class of sparse sets and let us build a martingale that covers C_2. Let $k \in \mathbb{N}$. We start with $d(\lambda) = 2^{-k}$ and, recursively, we define $d(x0) = (3/2) \cdot d(x)$, and $d(x1) = (1/2) \cdot d(x)$. Let us look at the value of $d(A^{\leq n})$, when A is a sparse set. For all the strings of length at most n that are not in A (let $\mathrm{out}(n)$ be the number of such strings), we have increased the investment by a factor of $3/2$. For all the strings of length at most n that are take values in the set of natural numbers, or in the set of strings over some alphabet. In our context, we want them to be martingales which in general are defined to take values that are nonnegative real numbers. We will assume that effective martingales take values in the set of nonnegative dyadic rational numbers, i.e., in the set $D = \{m2^{-n} \mid m, n \in \mathbb{N}\}$. Thus, the values taken by these martingales have a finite representation and, therefore, we can talk about effectively computable martingales.

It is important to note that, depending on Γ, a class C containing a single set A may not have Γ-measure zero (the reader can look at Example 1.2.7 to see why this is so). The second basic property of measure zero sets still holds: If $C \subseteq D$ and D has Γ-measure zero, then C has Γ-measure zero. The third basic property of measure zero sets (if $C = \bigcup_{n>0} C_n$ and all C_n have measure zero, then C has measure zero) in general fails. The reason is that in order to build the martingale that succeeds on C, we need a universal function able to simulate all the martingales d_n that succeed on C_n, and such a universal function may not be in Γ. However, this difficulty can be overcome if Γ has a few nice closure properties and if there is a certain uniformity among the martingales which show that the classes in the union have Γ-measure zero. Thus, we need several definitions.

A function $d\colon \mathbb{N} \times \Sigma^* \to [0,1]$ is a *martingale system* if, for each $i \in \mathbb{N}$, the function $d_i\colon \Sigma^* \to [0,1]$, defined by $d_i(x) = d(i,x)$ is a martingale.

A class C is a *Γ-uniform union of Γ-measure zero sets* if there is a countable family $(C_n)_{n\in\mathbb{N}}$ and d a martingale system such that (a) $d \in \Gamma$, (b) $C = \bigcup_{n\in\mathbb{N}} C_n$, and (c), for all $n \in \mathbb{N}$, d_n succeeds on C_n.

We say that a class of functions Γ is *closed under bounded sum* if, for any martingale system d, with $d \in \Gamma$, the function $d'\colon \Sigma^* \to [0,1]$, defined by

$$d'(x) = \sum_{i\leq|x|} d_i(x)$$

is in Γ as well.

We say that a class of functions Γ is *closed under finite variation* in case the following property holds: If f is any function in Γ and f' is obtained from f by modifying it on a finite set of inputs, then f' is in Γ as well.

Proposition 1.2.12 *Let Γ be a class of functions closed under bounded sum and under finite variation. Let C be a Γ-uniform union of Γ-measure zero sets. Then C has Γ-measure zero.*

Proof. For each $n \in \mathbb{N}$, let $\delta_n(x)$ be defined such that for each $x \neq \lambda$, $d_n(x) = \delta_n(x) \cdot d_n(\text{pred}(x))$, where $\text{pred}(x)$ is the prefix of x of length $|x| - 1$. Also let

$$\hat{d}_n(x) = \begin{cases} 2^{-n}, & \text{if } |x| < n, \\ \delta_n(x) \cdot \hat{d}_n(\text{pred}(x)), & \text{if } |x| \geq n. \end{cases}$$

It is easy to check that $\hat{d}_k$ is a martingale and that

$$\hat{d}_n(x) \geq \left(\frac{1}{2^{2n} \cdot d_n(\lambda)}\right) \cdot d_n(x).$$

Thus $\hat{d}_n$ succeeds on C_n, because d_n and $\hat{d}_n$ differ by at most a constant factor. Also, it is easy to see that the functions $(\hat{d}_n)_{n\in\mathbb{N}}$ form a martingale system. Let us define

$$\begin{aligned} d(x) &= \sum_{n=0}^{\infty} \hat{d}_n(x) \\ &= \sum_{n=0}^{|x|} \hat{d}_n(x) + \sum_{n=|x|+1}^{\infty} 2^{-n} \\ &= \sum_{n=0}^{|x|} \hat{d}_n(x) + 2^{-|x|}. \end{aligned}$$

Then d is a martingale and d is in Γ. Also, observe that for all $n \in \mathbb{N}$, for x sufficiently large (i.e., for x with $|x| \geq n$), $d(x) > \hat{d}_n(x)$. It follows that d succeeds on C_n. This holds for every n, and, therefore, we conclude that d succeds on C. ∎

Example 1.2.13 Let C be the class of sets A such that for almost every $n \in \mathbb{N}$, $0^n \in A$. We show that C has Γ-measure zero, where Γ is the class of computable functions. It is clear that this class of functions is closed under bounded sum and finite variation. Let

$$C_i = \{A \mid \text{for all } n \geq i, 0^n \in A\}.$$

We define a martingale d_i by (i) $d_i(\lambda) = \alpha$ (α is an arbitrary strictly positive real value), (ii) $d_i(x) = d_i(\text{pred}(x))$, if $s_{|x|} \neq 0^n$, and $s_{|x|} \neq 0^{n-1}1, n \geq i$, and (iii) $d_i(x) = 2 \cdot d_i(\text{pred}(x))$, if $s_{|x|} = 0^n, n \geq i$ and $d_i(x) = 0$ if $s_{|x|} = 0^{n-1}1, n \geq i$.

Since $0^n = s_{2^n}$, for all n, it follows that

$$d_i(A(1:2^{n+1})) = 2 \cdot d_i(A(1:2^n)),$$

for all $n \geq i$. Therefore d_i succeeds on C_i. Thus, the function $d(i,x) = d_i(x)$ verifies the conditions in Proposition 1.2.12. ∎

Chapter 2

Abstract complexity theory

2.1 Chapter overview and basic definitions

Abstract complexity theory studies fundamental quantitative aspects of computing. The theory reveals complexity related properties that are machine-independent in the sense that they are valid regardless of the particularities of a specific computational model. For instance, abstract complexity theory addresses fundamental questions such as:

(1) Is there an absolute upper bound for the complexity of all (or of most) computable function?

(2) Is there a best way (resource-wise) to calculate a computable function?

(3) If we spend more computational resources, can we compute more functions?

At the center of the theory lies the concept of an *abstract complexity measure*. This notion models almost any realistic computational resource that one may be interested in, in particular the running time and the amount of memory used in a computation. One could envision other computational resources as well such as the consumed energy, the number of accesses to registers and memory, etc.

In this chapter, we take an abstract approach. We first consider $\varphi_0, \varphi_1, \ldots, \varphi_n, \ldots$, an acceptable gödelization of the set of p.c. functions (the reader may find useful to review Section 1.1). The *abstract complexity measure* (or just the complexity measure) of a p.c. function φ_i is given by a partial function Φ_i attached to φ_i such that the system $(\varphi_i, \Phi_i)_{i\in\mathbb{N}}$ satisfies the following two axioms, called the *Blum axioms*:

(BA 1) For each $i \in \mathbb{N}$, $\Phi_i(x)$ is defined if and only if $\varphi_i(x)$ is defined, or, in more formal notation,

$$\Phi_i(x) < \infty \Longleftrightarrow \varphi_i(x) < \infty.$$

One can easily observe that the extensions $\sqsubseteq$ in the above definition can be taken to be proper ($\sqsubset$), and this will be the case in all our further considerations.

As mentioned, a topology over the set of p.c. predicates can be introduced similarly. The above definition can be stated in terms of the relativized topology of p.c. predicates (i.e., $\{0,1\}$-valued functions), by simply considering that $(\alpha_n)_{n\in\mathbb{N}}$ enumerates FPRED, the set of all p.c. predicates having the domain equal to a finite initial segment of $\mathbb{N}$, (i.e., each α_n can be considered to be a binary string). In this case, the topology is generated by the basic open sets $(U_t)_{t\in\text{FPRED}}$, where

$$U_t = \{f \mid t \sqsubseteq f, f \text{ is a p.c. predicate}\}.$$

This abuse of notation (i.e., we use the same notation for the basic open sets of both PC and PRED) will always be clarified by context.

2.2 Complexity classes

IN BRIEF: Any complexity class is topologically small.

The central concept in computational complexity is that of a complexity class.

Definition 2.2.1 (Complexity class) *Let $\Phi = ((\varphi_i, \Phi_i))_{i\in\mathbb{N}}$ be a Blum space and let g be a computable function. The complexity class defined by g in Φ is*

$$C_g^\Phi = \{f \in \text{COMP} \mid \textit{there exists } i \textit{ such that } \varphi_i = f \textit{ and } \Phi_i(x) < g(x) \textit{ a.e. } x\}.$$

Sometimes, when the Blum space Φ is clear from the context, the superscript Φ will be omitted.

We address here the following very basic questions:

(1) How large can a complexity class be? Is it true that if we can afford a vast amount of computational resources then we can solve any computable problem or at least most of them?

(2) (Counterpart of question (1)) Given a bound $g(n)$, how large is the class of functions for which each algorithm requires at least $g(n)$ computational resources for almost every input n?

The answer to the first question is, perhaps disappointingly, that however large g is, the class of problems solvable within resources bounded from above by g is of the first Baire category. The answer to the second question is that for every increasing and computable g the class of functions that are almost everywhere g-hard is of the second Baire category. The underlying topology is the Cantor topology defined on the space of p.c. functions (see Definition 2.1.2).

Theorem 2.2.2 INFORMAL STATEMENT: *Any complexity class is topologically small.*

FORMAL STATEMENT: *For any Blum space* Φ *and for any computable function* g, C_g^Φ *is effectively of the first Baire category.*

Proof. For any pair i, j of natural numbers, let

$$C_{\langle i,j\rangle} = \begin{cases} \{\varphi_i\}, & \text{if } \Phi_i(x) < g(x) \text{ for all } x > j \\ \emptyset, & \text{otherwise.} \end{cases}$$

One can easily check that $C_g^\Phi = \bigcup_{i,j\in\mathbb{N}} C_{\langle i,j\rangle}$, and thus, it suffices to show that $C_{\langle i,j\rangle}$ is effectively nowhere dense via some function f. The function f is specified by the following description of $\alpha_{f(\langle i,j\rangle,n)}$ for arbitrary i, j, and $n \in \mathbb{N}$:

- for $x \in \text{dom}(\alpha_n)$,
$$\alpha_{f(\langle i,j\rangle,n)}(x) = \alpha_n(x);$$
- for $x = |\alpha_n|$,
$$\alpha_{f(\langle i,j\rangle,n)}(x) = \begin{cases} \max(1-\varphi_i(x), 0), & \text{if } \Phi_i(x) < g(x) \\ 0, & \text{otherwise.} \end{cases} \tag{2.1}$$

The function f is computable. Indeed, the condition $\Phi_i(x) < g(x)$ can be effectively tested (because of the second Blum axiom) and in case this relation holds then $\varphi_i(x)$, and, consequently, $\max(1-\varphi_i(x), 0)$ can be computed.

We only have to show that

$$C_{\langle i,j\rangle} \cap U_{\alpha_{f(\langle i,j\rangle),n)}} = \emptyset, \tag{2.2}$$

for all sufficiently large n.

If $C_{\langle i,j\rangle}$ is empty, the above relation clearly holds. If $C_{\langle i,j\rangle}$ is not empty, then, for $|\alpha_n| > j$,

$$\alpha_{f(\langle i,j\rangle),n)}(|\alpha_n|) \neq \varphi_i(|\alpha_n|),$$

and therefore,

$$\varphi_i \notin U_{\alpha_{f(\langle i,j\rangle),n)}}.$$

Since in this case $C_{\langle i,j\rangle} = \{\varphi_i\}$, the relation (2.2) does indeed hold. ∎

Thus any complexity class is topologically small. We now turn to question (2). Formally, let g be an increasing computable function and define the set

$$\text{HARD}(\Phi, g) = \{f \in \text{COMP} \mid \text{ for every } \varphi_i = f,\ \Phi_i(x) > g(x) \text{ a.e. } x\} \tag{2.3}$$

of all computable functions requiring at least $g(x)$ complexity. The class $\text{HARD}(\Phi, g)$ is in a sense dual to C_g^Φ but it is not the complement of C_g^Φ. A classic result, due to Rabin [Rab60], asserts that $\text{HARD}(\Phi, g)$ is non-empty. We will show a stronger fact: For any increasing g and for every Blum space Φ, $\text{HARD}(\Phi, g)$ is topologically not small (i.e., it is of second Baire category). This result is an easy consequence of the main theorem of the next section and, therefore, will be presented there.

2.3 The speed-up phenomenon

IN BRIEF: There are computable functions that do not have a most-efficient algorithm. Moreover, there are computable functions f that are "speedable": Given any algorithm A for f, there exists an algorithm B for f that is more efficient than A. In fact, the set of speedable functions is topologically not small.

No matter how large a threshold function g is, the set of computable functions requiring algorithms that need more than $g(x)$ computational resources is topologically not small.

When someone seeks an algorithm for a problem, he or she has in mind two objectives: (a) correctness (i.e., the algorithm should indeed provide the right solution) and (b) optimal efficiency (the algorithm should work using minimum resources). The fact that there are problems for which objective (a) is not attainable came as a surprise when it has been discovered in the pioneering years of the theory of computation, but is now more or less common knowledge. It is less known that even when objective (a) is achievable, it can happen that objective (b) is not: There are computable problems (that is problems for which objective (a) is reachable) that do not admit a most efficient algorithm. This result, called the *Speed-Up Theorem*, is due to Blum [Blu67]. It states that there exists a computable function f such that if φ_i is any algorithm computing it, i.e., $\varphi_i = f$, then there exists another algorithm φ_j with $\varphi_j = f$ and whose complexity measure is smaller by an arbitrarily large amount than the complexity measure of the initial algorithm φ_i. We will prove a strong variant of this result. First, we will show that speed-up holds even if we restrict to computable predicates, demonstrating that the speed-up is not some fluke effect of functions which take unreasonably large values. Consequently, we consider the topology generated by $(U_t)_{t \in \text{FPRED}}$. Secondly, the rate of speed-up is given by an arbitrary total effective operator. Thus, for a total effective operator F, let SPEED(Φ, F) denote the class of computable functions having F-speedup almost everywhere. More precisely,

$$\text{SPEED}(\Phi, F) = \{f \in \text{COMP} \cap \text{PRED} \mid \text{ for all } \varphi_i = f, \text{ there exists } \varphi_j = f \text{ such that } F(\Phi_j)(x) < \Phi_i(x) \text{ a.e. } x\}. \tag{2.4}$$

For example, if, for all x,

$$F(g(x)) = 2^{g(x)}$$

then $f \in \text{SPEED}(\Phi, F)$ means that if $\varphi_i = f$, then there is an index j such that $\varphi_j = f$, and $\Phi_i(x) > 2^{\Phi_j(x)}$, a.e. x. Thus, φ_j achieves exponential speed-up when compared to φ_i.

Finally, the strong version of the Speed-Up Theorem shows that for any total effective operator F, from a topological point of view, there are *many* F-speedable functions.

Theorem 2.3.1 (Speed-Up theorem) INFORMAL STATEMENT: *The set of computable predicates for which there exists a more efficient algorithm than any given one is topologically not small.*

FORMAL STATEMENT: *For every Blum space* Φ *and for every total effective operator* F*, the set* SPEED(Φ, F) *is effectively of second Baire category.*

Proof. Fix a total effective operator F; for conciseness, let SPEED denote the set SPEED(Φ, F). Assume that SPEED is meagre. This means that there exists a decomposition

$$\text{SPEED} = \bigcup_{j \geq 0} \text{SPEED}_j$$

and a computable function f such that for every $j \geq 0$,

$$\alpha_n \sqsubset \alpha_{f(\langle j,n \rangle)} \text{ for all } n, \text{ and}$$

$$\text{SPEED}_j \cap U_{\alpha_{f(\langle j,n \rangle)}} = \emptyset \text{ a.e. } n.$$

We construct a computable function g satisfying for every $i \in \mathbb{N}$ the following two requirements:

$$R(i) : \alpha_{f(\langle i,m \rangle)} \sqsubset g \text{ i.o. } m,$$
$$Q(i) : \text{if } \Phi_i(x) < p_i(x) \text{ i.o. } x, \text{ then } g \neq \varphi_i,$$

where $(p_i)_{i \in \mathbb{N}}$ is a sequence of functions which will be defined later. The sequence has the property that for all i there is j with $\varphi_j = g$ and

$$p_i(x) \geq F(\Phi_j(x)) \text{ a.e. } x. \tag{2.5}$$

Let us first observe that this is enough for obtaining the desired conclusion. Indeed, conditions $Q(i)$ in conjunction with the property exhibited in Equation (2.5) guarantee that $g \in$ SPEED. By the initial assumption, $g \in \text{SPEED}_i$, for some i. Condition $R(i)$ implies that $g \in U_{\alpha_{f(\langle i,m \rangle)}}$ i.o. m, which contradicts the fact that $\text{SPEED}_i \cap U_{\alpha_{f(\langle i,m \rangle)}} = \emptyset$ a.e. m. The proof proceeds by defining, in a construction by stages, a family of p.c. functions $(z_s)_{s \geq 0}$, $z_s : \mathbb{N}^4 \xrightarrow{o} \mathbb{N}$. More precisely, at step s we construct the function z_s which is defined on a finite subset of $\mathbb{N}^4$. The function g we are interested in will be obtained in two steps:

- First we take the function z to be the limit (as s goes to ∞) of this construction (we assure that the limit exists);
- in the second step, we fix to convenient values the first three input variables of z (denoted by n, w and v) and finally obtain the desired function g.

More precisely, n will be the fixed point derived by using the Recursion Theorem, w will be selected in such a way as to achieve the desired rate of speed-up, and v will be chosen such that α_v patches the initial segment of the faster program so that it exactly computes the desired function.

Henceforth we will be using the notation $\lambda x.f(n,w,v,x)$ to represent the fact that n,w, and v are fixed parameters and we view the sequence of values $f(n,w,v,x)$ as a function of x only.

We present an overview of the construction. At stage s we construct the p.c. function z_s, such that $\lambda x.z_s(n,w,v,x)$ *tries* to properly extend $\lambda x.z_{s-1}(n,w,v,x)$. It may be the case that some subcomputations in stage s cannot be performed (more precisely, the fourth condition in what is denoted *Test* in the algorithm, see Figure (2.1)). In this case, the computation loops forever at some point in stage s, and, naturally, $\lambda x.z_t(n,w,v,x)$ is undefined for all $t \geq s$. However, for the value n_0 obtained through the use of the Recursion Theorem, this will not happen and, consequently, for all w and v and for all s, $\lambda x.z_s(n_0,w,v,x)$ properly extends $\lambda x.z_{s-1}(n_0,w,v,x)$. The particular feature of the construction is that, for all w and v, the extended part of $\lambda x.z_s(n,w,v,x)$ uses from the previous stages information related to the construction of $\lambda x.z_{s-1}(n,0,0,x)$ (and not $\lambda x.z_{s-1}(n,w,v,x)$, as one might expect). In this way, at all stages s, we extend the domain of $\lambda x.z_s(n,w,v,x)$ for all w and v by the same set. This allows us to define at each stage s the integer value $Lh_s(n)$ such that for every $w,v \geq 0$,

$$\mathrm{dom}(\lambda x.z_s(n,w,v,x)) = \{0,\ldots,Lh_s(n)-1\}.$$

We also use the sets $\mathrm{DIAG}_s(n,w,v)$ having the property that $i \in \mathrm{DIAG}_s(n,w,v)$ implies that by stage s we have insured that $\lambda x.z_s(n,w,v,x) \neq \varphi_i(x)$ for some $x < Lh_s(n)$. In the computation of $\lambda x.z_s(n,w,v,x)$ we focus at stage s on a pair of natural numbers (j,k), denoted $\mathrm{ACTIVE}_s(n,w,v)$ and called the *active pair*, with the intention to fulfill $R(j)$. Satisfying $R(j)$ is easy. Indeed, assume that the function $\lambda x.z_{s-1}(n,w,v,x)$ constructed up to stage s gives the initial segment α_m i.e.,

$$\alpha_m = z_{s-1}(n,w,v,0)\ldots z_{s-1}(n,w,v,Lh_{s-1}(n)-1).$$

Then, because $\alpha_{f(\langle j,m\rangle)}$ properly extends α_m, it is enough to define $\lambda x.z_s(n,w,v,x)$ as an extension of $\lambda x.z_{s-1}(n,w,v,x)$, so that

$$\alpha_{f(\langle j,m\rangle)} = z_s(n,w,v,0)\ldots z_s(n,w,v,Lh_s(n)-1).$$

However, if at some stage we discover an index i with i less than $\langle j,k\rangle$ (and also having some additional properties) such that $Q(i)$ can be satisfied, we prefer to do this. When $R(j)$ is satisfied, the next pair of natural numbers in a standard ordering of $\mathbb{N}^2$ becomes the new *active pair*. We will use a function called *next* that on input a pair of natural numbers produces the successor pair. The construction is given in Figure (2.1).

We denote $z(n,w,v,x) = \lim_{s\to\infty} z_s(n,w,v,x)$ (the limit exists by the way the function z_s extends z_{s-1} for each s). Let t be a computable function such that $\varphi_{t(n,w,v)}(x) = z(n,w,v,x)$. Now we turn to the definition of the functions $(p_i)_{i\geq 0}$. First, let ψ be the p.c. function defined by the following clauses:

- $\psi(n,\langle i,x\rangle) = 0$, if $x \leq i$ or if there exists $m \leq i$ with $\Phi_n(\langle 0,m\rangle) \geq x$;

The construction of $z(n,w,v,x)$: (n, w, v should be viewed as parameters; at each stage the function is defined for additional values of x)

Stage $s = -1$: Put $z_{-1}(n,w,v,x) = \infty$, for all $n, w, v, x \geq 0$, $Lh_{-1}(n) = 0$, for every $n \geq 0$, $\text{DIAG}_{-1}(n,w,v) = \emptyset$, for all $n, w, v \geq 0$, $\text{ACTIVE}_{-1}(n,w,v) = (0,0)$.

Stage $s \geq 0$: Take m such that $\alpha_m = \lambda x. z_{s-1}(n,0,0,x)$ and let

$$(j,k) = \text{ACTIVE}_{s-1}(n,0,0).$$

We define $z_s(n,w,v,x)$ for all $x < |\alpha_{f(\langle j,m\rangle)}|$ in two parts.
Part 1: For x with $0 \leq x < Lh_{s-1}(n)$, define $z_s(n,w,v,x) = z_{s-1}(n,w,v,x)$.
Part 2:

for each x such that $Lh_{s-1}(n) \leq x < |\alpha_{f(\langle j,m\rangle)}|$ **do**
    if $x \in \text{dom}(\alpha_v)$, **then** $z_s(n,w,v,x) = \alpha_v(x)$
    else if there exists i such that:
        (1) $i < \langle j,k\rangle$,
        (2) $i \notin \text{DIAG}_{s-1}(n,0,0)$,
        (3) $w \leq i < x$,
        (4) $\Phi_i(x) \leq \varphi_n(\langle i,x\rangle)$
    [Conditions 1) - 4) will be further on denoted as the *Test*.
    The *Test* is checked in increasing order of i in the range
    $w, w+1, \ldots, \min(x, \langle j,k\rangle)$. If for some i as above the
    Test cannot be evaluated (because $\varphi_n(\langle i,x\rangle)$ is not defined)
    then, of course, $z_s(n,w,v,x)$ is not defined and the procedure loops forever.]
        then (Satisfy $Q(i)$):
            choose the least such i,
            $z_s(n,w,v,x) = \max(1 - \varphi_i(x), 0)$
            $\text{DIAG}_{s-1}(n,w,v) = \text{DIAG}_{s-1}(n,w,v) \cup \{i\}$
        else (Satisfy $R(j)$)
            $z_s(n,w,v,x) = \alpha_{f(\langle j,m\rangle)}(x)$
        end if
    end if
end for
if for all x with $Lh_{s-1}(n) \leq x < |\alpha_{f(\langle j,m\rangle)}|$ it holds that
no i satisfies the *Test* or $x \in \text{dom}(\alpha_v)$ **then**
    $\text{ACTIVE}_s(n,w,v) = next(\text{ACTIVE}_{s-1}(n,w,v))$
else
    $\text{ACTIVE}_s(n,w,v) = \text{ACTIVE}_{s-1}(n,w,v)$
end if
$\text{DIAG}_s(n,w,v) = \text{DIAG}_{s-1}(n,w,v)$
$Lh_s(n) = |\alpha_{f(\langle j,m\rangle)}|$

End of construction

Figure 2.1: Construction of $z(n,w,v,x)$.

- $\psi(n, \langle i, x\rangle) = \max\{F(\Phi_{t(n,i+1,v)})(x) \mid v \leq x\}$, if the first condition fails to hold, but $\Phi_n(\langle j, y\rangle)$ is defined for all $y \leq x$ and all $i < j \leq x$;
- $\psi(n, \langle i, x\rangle) = \infty$, in the remaining situations.

By the Recursion Theorem, there exists n_0 such that $\varphi_{n_0}(u) = \psi(n_0, u)$, for all u. Fix such an n_0.

We are finally ready to define the function g mentioned in the introductory paragraphs of this proof. We set

$$g(x) = \varphi_{t(n_0,0,0)}(x).$$

We also define the sequence of functions $p_i, i \in \mathbb{N}$. For each i and x, let $p_i(x) = \varphi_{n_0}(\langle i, x\rangle)$; the complexity of $p_i(x)$ is then $\Phi_{n_0}(\langle i, x\rangle)$. One has:

$$p_i(x) = \begin{cases} 0, & \text{if } x \leq i \text{ or there exists } m \leq i \text{ with } \Phi_{n_0}(\langle 0, m\rangle) \geq x, \\ \max\{F(\Phi_{t(n_0,i+1,v)})(x) : v \leq x\}, & \\ & \text{if } p_j(y) \text{ is defined for all } y \leq x \text{ and all } j, i < j \leq x, \\ \infty, & \text{otherwise.} \end{cases} \tag{2.6}$$

(The clauses that give the three ways in which p_i is calculated should be evaluated in the order in which they are presented, i.e., top-down.)

We continue our proof with a series of intermediate results.

Claim 2.3.2 *For all natural numbers i and x, $\varphi_{n_0}(\langle i, x\rangle)$ is defined, i.e., p_i is computable for all i.*

Proof. We show first that $\varphi_{n_0}(\langle 0, x\rangle)$ is defined for all x, i.e., p_0 is computable. Suppose there exists w such that $p_0(w)$ is undefined. Hence $\Phi_{n_0}(\langle 0, w\rangle)$ is undefined as well from which we derive that

$$\Phi_{n_0}(\langle 0, w\rangle) \geq x, \text{ for all } x.$$

Then for all $i \geq w$,

$$p_i(x) = 0 \text{ for all } x,$$

by the first clause in the definition of p_i. We retain that p_i is computable for all $i \geq w$. It follows that for all v, $\lambda x.z(n_0, w, v, x)$, and thus $\varphi_{t(n_0,w,v)}$, are computable. Indeed, this is a consequence of the fact that condition (4) in the *Test* can be effectively checked (because $\varphi_{n_0}(\langle i, x\rangle) = p_i(x)$ is defined for all $i \geq w$) and, if needed, $\varphi_i(x)$ can be computed in order to fulfill $Q(i)$. From Equation (2.6) we notice that $p_{w-1}(x)$ is defined for all x because in the definition of p_{w-1} provided in Equation (2.6) one of the first two clauses applies and the values $F(\Phi_{t(n_0,w,v)})$ are defined for all v. In the same manner, we deduce that p_{w-2} is computable , and then p_{w-3} is computable, ..., p_0 is computable, which contradicts the assumption. So, in fact, p_0 is computable. Next we observe that $p_0(x)$ is defined by the second clause in Equation (2.6) for almost every x, which implies that for all j, p_j is computable. ∎

Claim 2.3.3 *For all v, $p_i(x) \geq F(\Phi_{t(n_0,i+1,v)})(x)$, a.e. x.*

Proof. Since p_i is computable, it follows that p_i is defined by the second clause in equation (2.6) for almost every x. The conclusion follows. ∎

Claim 2.3.4 *For all w, there exists v such that the functions $\varphi_{t(n_0,w,v)}$ and $\varphi_{t(n_0,0,0)}$ are equal.*

Proof. There exists a stage s when

$$\mathrm{DIAG}_s(n_0, 0, 0) \cap \{0, 1, \ldots, w-1\} = \bigcup_{t \geq 0} \mathrm{DIAG}_t(n_0, 0, 0) \cap \{0, 1, \ldots, w-1\},$$

i.e., all the indices less than $w - 1$ that ever enter in the set $\mathrm{DIAG}(n_0, 0, 0)$ must be there by some stage s (where $\mathrm{DIAG}(n_0, 0, 0)$ is the limit of the sets $\mathrm{DIAG}_t(n_0, 0, 0)$).

Take v such that $\alpha_v = \lambda x.z_s(n_0, 0, 0, x)$. For $x < |\alpha_v|$, one has

$$\varphi_{t(n_0,w,v)}(x) = \alpha_v(x) = z_s(n_0, 0, 0, x) = \varphi_{t(n_0,0,0)}(x).$$

If $x \geq |\alpha_v|$, then $\varphi_{t(n_0,w,v)}(x) = \varphi_{t(n_0,0,0)}(x)$, by the construction of z_s. Indeed, $\varphi_{t(n_0,w,v)}$ and $\varphi_{t(n_0,0,0)}$ could differ only because at some stage t the procedures computing them may satisfy the *Test* for i and respectively i' with $i \neq i'$, or it may happen that only $\varphi_{t(n_0,0,0)}$ satisfies the *Test* while $\varphi_{t(n_0,w,v)}$ does not. However, after stage s, the procedure computing $\varphi_{t(n_0,0,0)}$ does not satisfy any $Q(i')$ with $i' < w$ (by the definition of s). Recall our strategy according to which stage s of the construction of $\lambda x.z_s(n, w, v, x)$ uses information related to $\lambda x.z_{s-1}(n, 0, 0, x)$, namely the information given by $\mathrm{ACTIVE}_{s-1}(n, 0, 0)$ and $\mathrm{DIAG}_{s-1}(n, 0, 0)$. It follows that at each stage $t > s$, whenever a condition $Q(\cdot)$ is satisfied, the computations of $\varphi_{t(n_0,w,v)}$ and $\varphi_{t(n_0,0,0)}$ select the same value i for satisfying $Q(i)$. Thus, after stage s, the computations of $\varphi_{t(n_0,w,v)}$ and $\varphi_{t(n_0,0,0)}$ produce the same output, in spite of the fact that the computation of $\varphi_{t(n_0,0,0)}$ spends exceedingly more resources (in fact, this is the source of speed-up). ∎

Claim 2.3.5 *For every $j \in \mathbb{N}$, there exist infinitely many $m \in \mathbb{N}$ such that $\alpha_{f(\langle j,m \rangle)} \sqsubseteq g$, i.e., $R(j)$ is satisfied for every j.*

Proof. We show first that every pair of natural numbers (j, k) becomes eventually the *active pair* in the algorithm computing $\varphi_{t(n_0,0,0)} = \lambda x.z(n_0, 0, 0, x)$. We proceed by induction on $\langle j, k \rangle$. For $\langle j, k \rangle = 0$ the statement clearly holds (see stage $s = -1$). There are only finitely many $i \in \mathbb{N}$ less than $\langle j, k \rangle$ that can satisfy the *Test* and, once such an i has satisfied the *Test*, it will not attempt doing this any more, since i is inserted in $\mathrm{DIAG}(n_0, 0, 0)$ (see requierement (2) in *Test*). Eventually no such i is found, so the construction satisfies $R(j)$ and takes the next pair of natural numbers as the *active pair*. Consider now the stage s when $R(j)$ is satisfied and let m be as defined at this stage. From Part 1 of the algorithm (see Figure (2.1)) we see that

$$z_s(n_0,0,0,x) = z_{s-1}(n_0,0,0,x) = \alpha_m(x) = \alpha_{f(\langle j,m\rangle)}(x),$$

for all x with $0 \le x \le Lh_{s-1}(n_0)$. Since at stage s, $R(j)$ is satisfied, we conclude that

$$g = \varphi_{t(n_0,0,0)}(\cdot) = z(n_0,0,0,\cdot) \sqsupseteq z_s(n_0,0,0,\cdot) = \alpha_{f(\langle j,m\rangle)}.$$

It can be seen that the value of the parameter m is different each time $R(j)$ is satisfied, because for n_0 each step s of the construction provides a proper extension of the function $\lambda x.z(n_0,0,0,x)$. It follows that $\alpha_{f(\langle j,m\rangle)} \sqsubseteq g$ i.o. m. ∎

Claim 2.3.6 *If, for some i, $g = \varphi_i$, then $\Phi_i(x) > p_i(x)$ a.e. x. In other words, all the requierements $Q(i)$ are satisfied.*

Proof. There is a stage $s \ge 0$ in the computation of $z(n_0,0,0,x)$ such that from that moment on all active pairs (j,k) satisfy $\langle j,k\rangle > i$ and all $Q(i')$ for $i' < i$ that are ever to be satisfied have been already satisfied. In case $i \in \mathrm{DIAG}_{s-1}(n_0,0,0)$, φ_i was already diagonalized, i.e.,

$$\varphi_i \ne z(n_0,0,0,\cdot) = g,$$

contradicting the assumption. So, $i \notin \mathrm{DIAG}_{s-1}(n_0,0,0)$. Suppose now that

$$\Phi_i(x) \le p_i(x) = \varphi_{n_0}(\langle i,x\rangle), \text{ for some } x \ge \max(Lh_{s-1}(n),i).$$

There exists a stage $t \ge s$ such that $Lh_t(n) \ge x \ge Lh_{s-1}(n)$. It follows that at stage t the index i satisfies the *Test*, and thus the relation $z_t(n_0,0,0,\cdot) \ne \varphi_i(x)$ is realized. Hence

$$g = \varphi_{t(n_0,0,0,\cdot)} = z(n_0,0,0,\cdot) \ne \varphi_i,$$

yielding again a contradiction. ∎

We finish the proof of the theorem by showing that $g \in$ SPEED. Indeed, let $\varphi_i = g$. Then

$$\Phi_i(x) > p_i(x) \ge F(\Phi_{t(n_0,i+1,v)})(x) \text{ a.e. } x,$$

for all $v \in \mathbb{N}$. The first inequality comes from Claim 2.3.6 and the second one from Claim 2.3.3. Using Claim 2.3.4 take v such that $\varphi_{t(n_0,i+1,v)} = \varphi_{t(n_0,0,0)}$ and let $j = t(n_0,i+1,v)$. Then $\varphi_j = g$ (by the definition of g) and $\Phi_i(x) > F(\Phi_j)(x)$ a.e. x, which concludes the proof. ∎

We now return to the class HARD(Φ,g) defined in (2.3). As noted, this class of functions is topologically not small.

Theorem 2.3.7 INFORMAL STATEMENT: *The set of functions whose computation require an amount of resources superior to any given threshold function (no matter how large) is topologically not small.*

FORMAL STATEMENT: *For every Blum space Φ and for every increasing computable function g,* HARD(Φ,g) *is effectively of the second Baire category.*

Proof. Consider the total effective operator F defined by $F(\varphi_i)(x) = g(x)$, for all $i, x \in \mathbb{N}$, and apply Theorem 2.3.1 to F. This works because $\text{SPEED}(\Phi, F) \subseteq \text{HARD}(\Phi, g)$. ∎

It would be desirable to use the Speed-Up Theorem effectively, in the sense that, given a speedable function f and an algorithm that computes it, one would like to construct a faster algorithm for f. This is not possible. Blum [Blu71] has shown that for no speedable function f can one find algorithmically a faster program than an arbitrary given program for f. Although for some speedable functions it is possible to effectively bound the size of the index of the faster program (see [MF72, HY71]), Schnorr [Sch72] has shown that it is not possible for any speedable function to simultaneously bound the size of the faster program and the threshold input value starting from which the faster program is indeed faster. The topological analysis of the Speed-Up Theorem easily yields another facet of the non-effectiveness that shrouds the speed-up phenomenon. Given any sound deductive system, it is not possible to prove, except for a tiny set of functions, that a function is speedable. Such a phenomenon is called logical independence. Intuitively, a *deductive system* $\mathcal{T}$ consists of a system of axioms and a set of deduction rules. The axioms are some particular expressions in a logical language. Starting with the axioms and using the deduction rules, one can generate some other expressions called *theorems*. Encoding logical expressions by natural numbers in a standard and bijective way, we will just assume that the set of theorems derivable in $\mathcal{T}$ is computably enumerable. In other words, there is a computable function f whose image is the set Th of all theorems derivable in $\mathcal{T}$. A deductive system is *sound* if all the derivable theorems are true in a standard interpretation of arithmetic. For example, if "*the function computed by the i-th Turing machine is F-speedable*" is a theorem derivable in the sound deductive system, then indeed the function computed by the i-th Turing machine belongs to $\text{SPEED}(\Phi, F)$ (we are assuming that the Blum space Φ, the operator F, and a standard enumeration of all Turing machines have been fixed).

Definition 2.3.8 *A set of computable functions* $\mathcal{F}$ *is* effectively enumerable *if there is a computable function* $h\colon \mathbb{N} \times \mathbb{N} \to \mathbb{N}$ *such that* $g \in \mathcal{F}$ *if and only if there exists* $i \in \mathbb{N}$ *with* $g(x) = h(i, x)$*, for all* $x \in \mathbb{N}$. *We say that the function* h *enumerates* $\mathcal{F}$.

Proposition 2.3.9 *Let* $\mathcal{F}$ *be an effectively enumerable set of computable functions. Then* $\mathcal{F}$ *is effectively of the first Baire category.*

Proof. Let h be the function that enumerates $\mathcal{F}$. Then $\mathcal{F} = \bigcup \mathcal{F}_i$, where $\mathcal{F}_i = \{h(i, \cdot)\}$. For each i, consider the function f_i defined by $\alpha_{f_i(v)} = vy$, where $y = \max\{1 - h(i, |v|), 0\}$. For each i, $\mathcal{F}_i$ is effectivelly nowhere dense because $\mathcal{F}_i \cap U_{\alpha_{f_i(v)}} = \emptyset$. It follows easily that $\mathcal{F}$ is effectively of the first Baire category. ∎

Proposition 2.3.10 *Consider $\mathcal{P}$ a property of computable functions. Suppose that there is a sound deductive system $\mathcal{T}$ such that, for each computable function having property $\mathcal{P}$, there is a Turing machine M for which the sentence "The function computed by M has property $\mathcal{P}$" is a theorem of $\mathcal{T}$. Then the set of functions having property $\mathcal{P}$ is effectively of the first Baire category.*

Proof. Let $\mathcal{A} = \{f \mid f$ is computable and has property $\mathcal{P}\}$. If $\mathcal{A}$ is finite, then clearly, it is of the first Baire category. Therefore, let us assume that $\mathcal{A}$ is infinite. By Proposition 2.3.9, it is sufficient to show that $\mathcal{A}$ is effectively enumerable. Since the set of theorems in $\mathcal{T}$ is computably enumerable, we can obtain a second enumeration containing the subset of theorems of the form "The function computed by M has property $\mathcal{P}$." We can now define a function $h\colon \mathbb{N} \times \mathbb{N} \to \mathbb{N}$ by taking $h(j, \cdot)$ to be the function calculated by the machine M that appears in the j-th theorem in the second enumeration. The hypothesis implies that h enumerates $\mathcal{A}$. ∎

Theorem 2.3.11 *Let $\mathcal{T}$ be any sound deductive system and F a total effective operator. There exists a function $h \in$ SPEED $=$ SPEED(Φ, F) such that, for each machine M computing h, the sentence* "The function computed by M belongs to SPEED" *is not a theorem of $\mathcal{T}$. Moreover the set of such functions h is effectively of the second Baire category.*

Proof. Assume that the set of functions h having the above property is of the first Baire category. By Proposition 2.3.10 the set of functions computed by some machine M for which it is provable in $\mathcal{T}$ that M computes a function in SPEED is also of the first category. Since the union of two first Baire category sets is of first Baire category as well, it would follow that SPEED is of the first Baire category. This contradicts Theorem 2.3.1. ∎

2.4 Gap and compression

IN BRIEF: There are situations when the augmentation of computational resources does not yield more computational power.

Is it true that if we increase the amount of computational resources we can solve more problems? The intuition says yes, and most people, when buying a more powerful computer, expect to be able to do more. In fact, surprisingly, the answer is sometimes, and in a (topological) sense not rarely, negative.

To formalize the above question, we consider an arbitrary total effective operator F such that for any computable function f, $F(f)$ is much bigger than f. The question becomes: If f is a computable function, is it true that

$$C_f^\Phi \subsetneqq C_{F(f)}^\Phi?$$

(We use $\subsetneqq$ to denote proper inclusion.) The following theorem, called the Gap Theorem, shows that the answer is negative: For any increasing total effective operator F, there are functions f such that $C_f^\Phi = C_{F(f)}^\Phi$. In fact, as we have done before, we will prove a strong topological version of this theorem, which shows that there are *many* such functions.

Theorem 2.4.1 (Gap theorem) INFORMAL STATEMENT: *For each operator F, there is a computable function t such that increasing the amount of resources from t to $F(t)$ does not add any computational power. Moreover the set of such t is topologically not small.*

FORMAL STATEMENT: *Let F be a total effective operator such that, for all i and x, $F(\varphi_i)(x) \geq \varphi_i(x)$. Then the set*

$$\mathrm{GAP}(\Phi, F) = \{t \in \mathrm{COMP} \mid C_t^\Phi = C_{F(t)}^\Phi\}$$

is effectively of the second category.

Proof. To reach a contradiction, suppose that $\mathrm{GAP}(\Phi, F) = \bigcup_{j\geq 0} \mathrm{GAP}_j$ and there exists a computable function f such that for all natural j:

(i) $\alpha_n \sqsubset \alpha_{f(\langle j,n\rangle)}$,

(ii) $\mathrm{GAP}_j \cap U_{\alpha_{f(\langle j,n\rangle)}} = \emptyset$,

for sufficiently large n. We construct a function $t \in \mathrm{GAP}(\Phi, F)$ such that for all j there are infinitely many n with $\alpha_{f(\langle j,n\rangle)} \sqsubseteq t$. It will follow that for some i,

$$t \in \mathrm{GAP}_i \cap U_{\alpha_{f(\langle i,n\rangle)}} \text{ i.o. } n,$$

a contradiction.

By the Kreisel-Lacombe-Shoenfield Theorem there exists a computable function $g\colon \mathbb{N} \to \mathbb{N}$ such that for every computable φ_i,

$$\mathrm{Graph}(F(\varphi_i)) = \bigcup_{j\in C_i} \mathrm{Graph}(\alpha_{g(j)}),$$

where $C_i = \{j \in \mathbb{N} \mid \mathrm{Graph}(\alpha_j) \subseteq \mathrm{Graph}(\varphi_i)\}$.

The function t will be defined in stages. At stage s we construct a finite initial segment t_s of t and keep track of the value Lh_s such that $\mathrm{dom}(t_s) = \{0, 1, \ldots, Lh_s - 1\}$.

Construction of t.

Stage $s = 0$: Put $t_0(0) \equiv 0$, $Lh_0 \equiv 1$.

Stage $s > 0$: Let $s \equiv \langle j, k\rangle$. (The pair (j, k) acts like the *active pair* in the previous proofs.) Let m be such that $t_{s-1} = \alpha_m$. We define

$$t_s(x) \stackrel{\mathrm{def}}{=} t_{s-1}(x), \text{ for } 0 \leq x < Lh_{s-1},$$

and

$$t_s(x) \stackrel{\mathrm{def}}{=} \alpha_{f(\langle j,m\rangle)}(x), \text{ for } Lh_{s-1} \leq x < |\alpha_{f(\langle j,m\rangle)}|.$$

Note that we have insured that $\alpha_{f(\langle j,m\rangle)} \sqsubseteq t_s$. Next we define $s+1$ computable extensions of the function t_s constructed so far, namely $t^{(s)}, t^{(s-1)}, \ldots, t^{(0)} : \mathbb{N} \to \mathbb{N}$ in the following way:

$$t^{(s)}(z) = \begin{cases} t_s(z), & \text{if } z < |\alpha_{f(\langle j,m\rangle)}|, \\ 0, & \text{otherwise,} \end{cases}$$

and for $i = s-1, s-2, \ldots, 0$ (in this order)

$$t^{(i)}(z) = \begin{cases} t_s(z), & \text{if } z < |\alpha_{f(\langle j,m\rangle)}|, \\ F(t^{(i+1)})(z), & \text{otherwise.} \end{cases}$$

We truncate the functions $t^{(s)}, t^{(s-1)}, \ldots, t^{(0)}$ to some finite initial segment functions $u^{(s)}, u^{(s-1)}, \ldots, u^{(0)}$ as will be described below. It is useful to keep in mind that

$$F(t^{(i)})(z) = t^{(i-1)}(z), \text{ for every } z \geq |\alpha_{f(\langle j,m\rangle)}|.$$

The idea in defining $u^{(i)}$ from $t^{(i)}$ is to retain in the graph of $u^{(i)}$ enough elements from the graph of $t^{(i)}$ so that if $t' : \mathbb{N} \to \mathbb{N}$ is an extension of $u^{(i)}$ then

$$F(t')(z) = u^{(i-1)}(z),$$

for any z in the domain of $u^{(i-1)}$ that is at least $|\alpha_{\langle j,m\rangle}|$.

To this end, let

$$B_{z,i} = \{y \in \mathbb{N} \mid \alpha_y(n) = t^{(i)}(n), \text{ for every } n \leq z\}.$$

We first define $u^{(0)}$ as follows. Let

$$A_0 = |\alpha_{f(\langle j,m\rangle)}|$$

and

$$u^{(0)}(z) = \begin{cases} t^{(0)}(z), & \text{if } z \leq A_0, \\ \infty, & \text{otherwise,} \end{cases}$$

Then, inductively for $i = 0, \ldots, s-1$, we define

$$A_{i+1} = \min\left\{ z \mid z \geq A_i \text{ and } \mathrm{Graph}(u^{(i)}) \subseteq \bigcup_{y \in B_{z,i+1}} \mathrm{Graph}(\alpha_{g(y)}) \right\},$$

and

$$u^{(i+1)}(z) = \begin{cases} t^{(i+1)}(z), & \text{if } z \leq A_{i+1}, \\ \infty, & \text{otherwise.} \end{cases}$$

Recall that g is the function corresponding to the operator F according to the Kreisel-Lacombe-Shoenfield Theorem. Since for $z \geq A_0$ and for any $i \in \{0, \ldots, s-1\}$,

$$F(t^{(i+1)}) = t^{(i)}(z),$$

it follows that

$$\{(z, t^{(i)}(z)) \mid z \geq A_0\} \subseteq \bigcup_{z \geq A_0} \bigcup_{y \in B_{z,i+1}} \text{Graph}(\alpha_{g(y)}).$$

Because $u^{(i)}$ is a restriction of $t^{(i)}$, it follows that the value A_i is defined and furthermore can be calculated effectively. We also conclude that, as claimed above, if t' is an extension of $u^{(i+1)}$, then $F(t')(z) = u^{(i)}(z)$, for all $z \leq A_i$.

For all $h \in \{0, 1, \ldots, s-1\}$, and $i \in \{0, 1, \ldots, s\}$, we say that $u^{(i)}$ is *unsafe* for h if

(i) $u^{(i)}(z) < \Phi_h(z)$, for some z with $Lh_{s-1} \leq z \leq A_i$, and

(ii) $i = 0$, or $(\Phi_h(z) \leq u^{(i-1)}(z), i > 0)$, for all z with $Lh_{s-1} \leq z \leq A_{i-1}$.

Keep in mind that, for all $i \in \{1, 2, \ldots, s\}$,

$$Lh_{s-1} \leq A_0 \leq A_1 \leq \ldots \leq A_s,$$

and $u^{(i)}(z) \leq u^{(i-1)}(z)$, for all z with $Lh_{s-1} \leq z \leq A_{i-1}$. Hence, for every $h \in \{0, 1, \ldots, s-1\}$, at most one $u^{(i)}$ is *unsafe* for h. Since there are $s+1$ such extensions $u^{(i)}$, at least one of them is *safe* for all $h < s$, and such an $u^{(i)}$ can be found effectively. We extend t_s by setting it to be equal to a certain $u^{(i_0)}$ which is *safe* for every $h < s$ and set $Lh_s \equiv A_{i_0} + 1$.

End of stage s.

Finally, let $t \equiv \lim_{s \to \infty} t_s$.

End of construction of t.

For each i there are infinitely many k such that $\alpha_{f(\langle i,k \rangle)} \sqsubseteq t$, since at each step $s = \langle i, k \rangle$, we make $\alpha_{f(\langle i,k \rangle)} \sqsubseteq t_s$ and $t_s \sqsubseteq t$. It remains to show that $t \in \text{GAP}(\Phi, F)$. Suppose there exists j such that

$$\Phi_j(x) \leq F(t)(x) \text{ a.e. } x, \text{ and}$$
$$\Phi_j(x) > t(x) \text{ i.o. } x.$$

There exists a stage $s > j$ such that $\Phi_j(x) \leq F(t)(x)$ for all x with $Lh_{s-1} \leq x < Lh_s$ and $\Phi_j(x) > t(x)$ for some x with $Lh_{s-1} \leq x < Lh_s$. We claim that this contradicts the choice of a *safe* initial segment at stage s. Let $u^{(i_0)}$ be the *safe* segment selected at stage s. Note that if $Lh_{s-1} \leq x < Lh_s$, then for all $x \leq A_{i_0 - 1}$,

$$t(x) = u^{(i_0)}(x),$$

and, in case $i_0 > 0$,

$$F(t)(x) = u^{(i_0 - 1)}(x)$$

If $i_0 = 0$, then $u^{(i_0)}$ is a *safe* segment if and only if

$$u^{(i_0)}(x) \geq \Phi_j(x),$$

Proof. We have $G = \bigcup_{i \in \mathbb{N}} G_i$ where $G_i = \{g_i\}$. The set G is of the first Baire category via the function $f(i,n)$ that gives $\alpha_{f(i,n)}$ defined as follows.

For $x \in \text{dom}(\alpha_n)$, $\alpha_{f(i,n)}(x) = \alpha_n(x)$. For $x = |\alpha_n|$,

$$\alpha_{f(i,n)}(x) = \begin{cases} 1+y, & \text{if } g_i(x) \leq y \\ y, & \textit{otherwise.} \end{cases}$$

Clearly, for all i and for all n, $U_{\alpha_n} \subset U_{\alpha_{f(i,n)}}$, and $G_i \cap U_{\alpha_{f(i,n)}} = \emptyset$. ∎

Our next result shows that the meagerness of a measured set is essential because it gives the necessary property promised above. If we fix the operator F that gives the increasing factor and consider a second category set A of computable functions, then the compression property does not hold for all functions f in A.

Proposition 2.4.5 *Let F be a total effective operator. The set*

$$A_F = \{s \in \text{COMP}\} \mid \textit{there exists } i \textit{ with } \Phi_i(x) \leq F(s)(x) \textit{ a.e. } x \\ \textit{and for all } j \textit{ with } \varphi_j = \varphi_i,\ \Phi_j(x) > s(x) \textit{ a.e. } x\}$$

is effectively of the first Baire category.

Proof. We start by noticing that $A_F \subseteq \bigcup_{i,m \geq 0} A_{\langle i,m \rangle}$, where

$$A_{\langle i,m \rangle} = \{s \in \text{COMP} \mid \Phi_i(x) \leq F(s)(x) \text{ and } \Phi_i(x) > s(x) \text{ for all } x \geq m\}.$$

Let g be the function that corresponds to the operator F according to the Kreisel-Lacombe-Shoenfield Theorem, i.e., for every computable function s,

$$\text{Graph}(F(s)) = \bigcup_{j \in C_i} \text{Graph}(\alpha_{g(j)}), \tag{2.7}$$

where $C_i = \{j \in \mathbb{N} \mid \text{Graph}(\alpha_j) \subseteq \text{Graph}(s)\}$. The sets $A_{\langle i,m \rangle}$ are nowhere dense via the following uniform family of witness functions $f_{(\langle i,m \rangle,n)}$. We fix i, m and n. To present the function $f_{(\langle i,m \rangle,n)}$, we will give the function $\alpha_{f_{(\langle i,m \rangle,n)}}$ defined on an finite initial segment of $\mathbb{N}$ and which is actually required in the Definition 2.1.2. Let $k = 1 + \max(m, |\alpha_n|)$.

(1) For $x \in \text{dom}(\alpha_n)$, $\alpha_{f(\langle i,m \rangle,n)}(x) = \alpha_n(x)$. So, $U_{\alpha_n} \subseteq U_{\alpha_{f(\langle i,m \rangle,n)}}$.

(2) For $x \geq |\alpha_n|$ and $x < k$, $\alpha_{f(\langle i,m \rangle,n)}(x) = 0$.

(3) For $x = k$: Let β be the function given by the values of $\alpha_{f(\langle i,m \rangle,n)}$ defined so far, i.e.,

$$\beta(x) = \alpha_{f(\langle i,m \rangle,n)}(x), \quad x = 0, 1, \ldots, k-1.$$

Take α_j an extension of β such that $k \in \text{dom}(\alpha_{g(j)})$. Such an extension must exist (by relation 2.7), and j can be calculated effectively. There are two cases.

Case 1. $\Phi_i(k) \leq \alpha_{g(j)}(k)$. Note that by the second Blum axiom this condition can be checked. In this case, $\Phi_i(k)$ is defined and we let

$$\alpha_{f(\langle i,m \rangle,n)}(k) = \Phi_i(k) + 1.$$

It follows that, if s is an extension of $\alpha_{f(\langle i,m\rangle,n)}$, then $\Phi_i(k) < \Phi_i(k) + 1 = s(k)$.

Case 2. $\Phi_i(k) > \alpha_{g(j)}(k)$. Then we set for all $x \in \text{dom}(\alpha_j)$,

$$\alpha_{f(\langle i,m\rangle,n)}(x) = \alpha_j(x).$$

Note that since α_j is an extension of β, this does not conflict the previous assignments. Now, if s is an extension of $\alpha_{f(\langle i,m\rangle,n)}$, then $\text{Graph}(\alpha_j) \subseteq \text{Graph}(s)$, which, by the Kreisel-Lacombe-Shoenfield Theorem, implies $\text{Graph}(\alpha_{g(j)}) \subseteq \text{Graph}(F(s))$. Therefore, in this case, $\Phi_i(k) > \alpha_{g(j)}(k) = F(s)(k)$. In both cases, we obtain $U_{\alpha_{f(\langle i,m\rangle,n)}} \cap A_{\langle i,m\rangle} = \emptyset$, which finishes the proof. ∎

2.5 Union theorem

IN BRIEF: The union of a set of increasing complexity classes is a complexity class.

Let $\{f_1, \ldots, f_k, \ldots\}$ be a set of computable functions such that for each i and n, $f_i(n) < f_{i+1}(n)$. We also assume that the set of functions is computably enumerable. For example, we may think at the case when $f_i(n) = n^i$. Consider a Blum space Φ. Is the union of the complexity classes $C^{\Phi}_{f_i}$ a complexity class? The next theorem, shows that the answer is positive. (We consider a fixed but arbitrary Blum space Φ and henceforth we drop the superscript Φ.)

Theorem 2.5.1 (Union theorem) *Let $\{f_i \mid i \in \mathbb{N}\}$ be a computably enumerable set of computable functions such that for all i and all n, $f_i(n) < f_{i+1}(n)$. Then there exists a computable function t such that*

$$C_t = \bigcup_{i \in \mathbb{N}} C_{f_i}.$$

Proof. The plan for the construction of t is as follows. For each j, we maintain during the construction an integer value i_j having the meaning that we are guessing that φ_j is in $C_{f_{i_j}}$. If we find out that $\Phi_j(n) > f_{i_j}(n)$ for some n (and thus our guess is false) then we assign to $t(n)$ a value that is smaller than $\Phi_j(n)$, placing ϕ_j (at least momentarily) out of C_t. We also modify our guess for ϕ_j and assume that Φ_j is bounded by some larger f_i. If φ_j is in $\bigcup C_{f_i}$, after some point our guess will be correct and t will be larger than Φ_j. This will place φ_j in C_t. On the other hand, if φ_j is not in $\bigcup C_{f_i}$, then t will be less than Φ_j infinitely many times and thus φ_j is not in C_t.

To implement this plan, the construction uses a list of integers $i_1, i_2, i_3, \ldots$ which play the role described above. At step n, $i_n = n$ is entered in the list, so that at the limit we have guesses for all partial computable functions. At step n, we also check if our current guesses are correct, that is we verify if for all integers

i_j in the list, $\Phi_j(n) \leq f_{i_j}(n)$. If some indices j do not pass this test, we pick j to be the smallest of them and make $t(n) = f_{i_j}(n)$. In this case, we also update the value of i_j to n.

Construction of t.

Step -1: The list is empty and the function t is not defined in any point.

Step n, $n \geq 0$: Insert $i_n = n$ in the list. For all $j \in \{1, \ldots, n\}$ check if

$$\Phi_j(n) > f_{i_j}(n). \tag{2.8}$$

(a) If there is no j that verifies the Equation (2.8), make $t(n) = f_n(n)$ and go to Step $n+1$.

(b) If there are indices j that verify the Equation (2.8), let j be the one with the smallest value i_j. In case of a tie take the smallest j. Set $t(n) = f_{i_j}(n)$ and $i_j = n$.

End of construction.

We show now that

$$C_t = \bigcup_{i \in \mathbb{N}} C_{f_i}.$$

If ϕ_k is in $\bigcup_{i \in \mathbb{N}} C_{f_i}$, then there is some ℓ such that $\Phi_k(n) \leq f_\ell(n)$ a.e. n. We claim that $f_\ell(n) \leq t(n)$ a.e. n, which is what we need to conclude that $\phi_k \in C_t$. Note that there is n_0, such that at all steps n with $n \geq n_0$, the index i_j that is picked at Step n (b) is at least ℓ. Indeed, only $i_1, i_2, \ldots, i_{\ell-1}$ start with values that are less than ℓ, and each time one of these indices is changed after step ℓ, it gets a value larger than ℓ. Since $t(n)$ is set to either $f_n(n)$ or $f_{i_j}(n)$ (for the j picked at step n, if some index j is picked), it follows that, for all $n \geq n_0$, $t(n) \geq f_\ell(n)$ and, therefore, $\Phi_k(n) \leq t(n)$ for all $n \geq n_0$.

Let now j be such that $\varphi_j \in C_t$, which means $\Phi_j(n) \leq t(n)$ a.e. n. The key observation is that j cannot be the index chosen at (*b*) but for finitely many times, because otherwise $t(n) = f_{i_j}(n) < \Phi_j(n)$ for infinitely many n. This implies that there must be some index j_0 such that $\Phi_j(n) \leq f_{j_0}(n)$ a.e. n, because otherwise j would be infinitely many often the smallest index for which $\Phi_j(n) \leq f_{i_j}(n)$, and j would be picked at (*b*) and this, as we have just noted, is not possible. Therefore, $\varphi_j \in C_{f_{j_0}} \subseteq \bigcup_{i \in \mathbb{N}} C_{f_i}$. ∎

2.6 Effective measure

IN BRIEF: Each complexity class has effective measure zero. The class of hard problems has effective measure one.

The size of a class of p.c. functions can be investigated from a measure-theoretical point of view as well. In this section we present the measure-theoretical counterpart of some of the topological results that we have seen earlier. More precisely, we show that any complexity class C_g^Φ has effective measure 0, and that, on the other hand, for any computable function g, HARD(Φ, g) (see Equation (2.3)) has effective measure 1.

We will be using the technical apparatus developed in Section 1.2.2. One first observation is that effective measure is defined only for predicates (i.e., $\{0,1\}$-valued functions). The class of computable predicates will be denoted by CPRED. If f is a predicate then it can be associated to the infinite binary string $f(0)f(1)\ldots f(n)\ldots$. Via this association, a class of predicates will be viewed as a subset of Σ^∞, where $\Sigma = \{0,1\}$. Effective measure is obtained by taking $\Gamma = \text{COMP}$ in Definition 1.2.11.

Definition 2.6.1 *A class $C \subseteq \Sigma^\infty$ has effective measure zero if there is a computable martingale that succeeds on $C \cap \text{CPRED}$.*

A class $C \subseteq \Sigma^\infty$ has effective measure one, if the complement of C has effective measure zero.

Theorem 2.6.2 *Let Φ be a Blum space. For every computable function g, $\text{HARD}(\Phi, g) \cap \text{CPRED}$ has computable measure one.*

Proof. We need to show that the set

$$\text{HARD}^c(g) = \{f \in \text{CPRED} \mid \text{ there exists } \varphi_i = f, \Phi_i(x) \leq g(x) \text{ i.o. } x\}$$

has effective measure zero. It is immediate that $\text{HARD}^c(g) \subseteq \bigcup H_i$, where

$$H_i = \begin{cases} \{f \in \text{CPRED} \mid \varphi_i = f\}, & \text{if there exist infinitely many } x \\ & \quad \text{such that } \Phi_i(x) \leq g(x), \\ \emptyset, & \text{otherwise.} \end{cases}$$

Our intention is to use Proposition 1.2.12 to show that $\bigcup H_i$ has effective measure zero, because this would imply that $\text{HARD}^c(g)$ has effective measure zero. Thus we need to build a computable function $d\colon \mathbb{N} \times \Sigma^* \to [0, \infty)$ such that, for each i, $d(i, \cdot)$ is a martingale that succeeds on H_i (the other conditions in Proposition 1.2.12 are easily seen to hold in the context here). The function d is defined as follows:

$$d(i, \lambda) = 1,$$

and, inductively,

$$d(i, x0) = \begin{cases} 0, & \text{if } \Phi_i(|x|) \leq g(|x|) \text{ and } \varphi_i(|x|) = 1, \\ 2d(i, x), & \text{if } \Phi_i(|x|) \leq g(|x|) \text{ and } \varphi_i(|x|) = 0, \\ d(i, x), & \text{otherwise,} \end{cases}$$

$$d(i, x1) = \begin{cases} 2d(i, x), & \text{if } \Phi_i(|x|) \leq g(|x|) \text{ and } \varphi_i(|x|) = 1, \\ 0, & \text{if } \Phi_i(|x|) \leq g(|x|) \text{ and } \varphi_i(|x|) = 0, \\ d(i, x), & \text{otherwise.} \end{cases}$$

It can readily be checked that $d(i, \cdot)$ is a martingale. Also, by induction on h, one can see that $d(i, x) \geq 2^h$ if and only if there exist at least h integers y with $y \leq |x| - 1$ and $\Phi_i(y) < g(y)$. Therefore, it follows that for all n, $H_i \subseteq S^n[d(i, \cdot)]$. As noted, this implies that $\mathrm{HARD}^c(g)$ has effective measure zero, and hence $\mathrm{HARD}(\Phi, g)$ has effective measure one. ∎

From the above proof we also derive that all complexity classes are small in the sense of effective measure theory. Indeed, if $((\varphi_i)_{i\in\mathbb{N}}, (\Phi_i)_{i\in\mathbb{N}})$ is a Blum space of $\{0,1\}$-valued p.c. functions, and g is a computable function, then the complexity class defined by g, C_g^{Φ}, is included in $\mathrm{HARD}^c(g)$, the set defined in the above proof and shown to have effective measure zero in CPRED. Therefore:

Theorem 2.6.3 *Let $((\varphi_i)_{i\in\mathbb{N}}, (\Phi_i)_{i\in\mathbb{N}})$ be a Blum space of $\{0,1\}$-valued p.c. functions and g a computable function. Then C_g^{Φ} has effective measure zero.*

Measured sets (see Definition 2.4.2) of computable predicates can also be easily analyzed: As shown in the next theorem any such set has effective measure zero. Many natural classes of computable predicates form measured sets, e.g., the class of primitive computable predicates, every computably enumerable complexity class of predicates, and, in fact, any computably enumerable set of computable predicates. It follows that all these classes (and all their subclasses, like all levels in Grzegorczyk's hierarchy, see, for example, [Cal88]) have effective measure zero.

Theorem 2.6.4 *If $\Gamma = (\gamma_i)_{i\in\mathbb{N}}$ is a measured set of computable predicates, then Γ has effective measure zero.*

Proof. We decompose $\Gamma = \bigcup \Gamma_i$, where $\Gamma_i = \{\gamma_i\}$. It is immediate to build a computable function $d \colon \mathbb{N} \times \Sigma^* \to [0,1]$ such that each restriction $d(i, \cdot)$ is a martingale that succeeds on Γ_i. According to Proposition 1.2.12 this is enough for our purpose. Namely,

$$d(i, \lambda) = 1,$$

and inductively,

$$d(i, x0) = \begin{cases} 0, & \text{if } \gamma_i(|x|) = 1, \\ 2d(i, x), & \text{if } \gamma_i(|x|) = 0, \end{cases}$$

$$d(i, x1) = \begin{cases} 2d(i, x), & \text{if } \gamma_i(|x|) = 1, \\ 0, & \text{if } \gamma_i(|x|) = 0. \end{cases}$$

∎

It is noteworthy that both topology and measure theory agree in their view of complexity classes and classes of hard functions: From both standpoints, complexity classes are small, and the classes of hard functions are large (they also agree on measured sets). Such an agreement is by no means mandatory, as the two approaches use quite different yardsticks for classification. The fact that both classification schemas have reached a common conclusion should enforce the intuition that only a tiny part of problems can be algorithmically solved with a bounded amount of computational resources.

2.7 Comments and bibliographical notes

The heyday of abstract complexity theory was in the late 1960s and the early 1970s. The first attempt of an axiomatic approach has been made, however, a few years earlier by Rabin [Rab60, Rab59]. He showed the existence of functions that are almost everywhere hard to compute. Blum, influenced by the works of Rabin, Hartmanis, Lewis and Stearns, really triggered abstract complexity theory (as a matter of fact, also called Blum complexity theory) in his paper [Blu67]. This paper contains the formulation of the two Blum axioms. Moreover, besides setting the bedrock of the theory, the paper also contains some of its most important results, such as the Speed-Up Theorem (in a more restrictive variant that does not involve operators) and the Compression Theorem. The Gap Theorem has been discovered independently by Trakhtenbrot [Tra67] and by Borodin [Bor72]. Levin [Lev73, Lev74] (paper [Lev74] is partially translated in [Lev95]) and independently Meyer and Winklman [MW79] have established conditions under which a set of p.c. functions can be the set of complexities of all algorithms that calculate a p.c. function. From this result both the Speed-Up Theorem and the Compression Theorem are easily derivable. An improved version of this result has been obtained by Seiferas and Meyer [SM95]. The operator version of the Speed-Up Theorem is due to Meyer and Fischer [MF72], and the operator version of the Gap Theorem has been discovered by Constable [Con72]. Different and somewhat easier proofs for the operator versions of the Speed-Up Theorem and of the Gap Theorem have been given by Young [You73].

The Union Theorem is due to McCreight and Meyer [MM65] This paper contains another important theorem, the Naming Theorem, which states that one can name complexity classes with functions from a measured set (i.e., there is a

Table 2.1: Effective topology and measure theory in abstract complexity. Notes: (1) Measure refers to effective measure. (2) Category refers to effective Baire category. (3) "?" indicates an open problem.

Object	Category	Measure	Where
complexity class	I	0	Category: [Meh73] Measure: [CZ96]
a.e. hard functions	II	1	Category: [CZ96] Measure: [CZ96]
i.o. speedable functions	II	?	[CIZ92]
a.e. speedable functions	II	?	[CZ92]
operator speedable functions	II	?	[CZ96]
functions yielding gaps	II	?	[CZ92]
functions yielding operator gaps	II	?	[CZ96]
measured set of functions	I	0	Category: [Cal82] Measure: [CZ96]

measured set G such that each complexity class coincides with C_g^Φ for some g in G).

Mehlhorn [Meh73] has been the first to utilize topological tools in abstract complexity theory. Theorem 2.2.2 is due to him. The topological analysis of the Speed-Up Theorem has been first undertaken by Calude, Istrate, and Zimand [CIZ92], who have obtained a partial result for functions that are speedable infinitely often and for the simpler case in which the speed-up factor is given by a computable function rather than an operator. The full result from Theorem 2.3.1 has been established by Calude and Zimand [CZ96]. The paper [CZ96] also contains the topological analysis of the Gap Theorem and Compression Theorem covered in Section 2.4. The measure-theoretical results from Section 2.6 are from the same paper [CZ96].

A summary of the results utilizing the topological or the measure-theoretical approach is given in Table 2.7 including references to the original papers.

Presentations of different aspects of abstract complexity theory containing additional results and references can be found in Brainerd and Landweber [BL74], Calude [Cal88], Machtey and Young [MY78], Seiferas [Sei90]. The classical paper of Hartmanis and Hopcroft [HH71] is an excellent, comprehensive survey of the area.

Chapter 3

Polynomial time, nondeterministic polynomial time, and exponential time

3.1 Chapter overview and basic definitions

Programmers, even those who had little exposure to complexity theory, use *polynomial time*, *nondeterministic polynomial time*, and *exponential time* as a rough and basic efficiency-related taxonomy for classifying algorithms. These three types of running time define the classes P, NP, and E (or EXP, depending on the definition of exponential time), on which we focus in this chapter.

The class P contains all computational problems that can be solved in "reasonable" time, and, therefore, it is considered to be the class of feasible problems. P also contains problems solvable by algorithms with huge time complexity, such as, say, $n^{2^{100}}$, which are in fact unusable. However, such problems do not arise in practice, and, on the other hand, there are many theoretical advantages in the equivalation "P = Class of Feasible Problems." First, the solvability of a problem in polynomial time seems to be an intrinsic feature of the problem, independent of computability issues. Problems in P seem to have an underlying "nice" structure, which can be captured by some mathematical theory, and which polynomial-time algorithms can exploit. Secondly, the class P is robust to changes in the computational model, because all reasonable ("classical" [1]) computational models can simulate each other in polynomial time.

[1]Note however that this argument has been recently challenged by some new computational models based on quantum theory. See Chapter 4.

The class NP contains all the problems that can be solved in polynomial time by nondeterministic machines. There exists an alternative characterization of NP that is more intuitive and has also the merit of being independent of a particular computational model: A set A is in NP if and only if input instances in A admit *membership proofs* whose validity can be verified in (deterministic) polynomial time (see Theorem 1.1.6 for a formal statement). For example, let us consider the NP set SAT. SAT consists of the set of satisfiable boolean formulas. We recall (see also Section 3.2) that a boolean formula ϕ is given by a well-formed expression consisting of variables and the logical operators $\vee$ (and), $\wedge$ (or), and negation ($\overline{x}$ denotes the negation of the variable x). A truth assignment for ϕ is an assignment of the boolean values T (*true*) and F (*false*) to each variable in ϕ. The formula ϕ is satisfiable if there exists a truth assignment that makes it true. For instance, the formula $(x \vee \overline{y}) \wedge (\overline{x} \vee y)$ is satisfiable. One truth assignment that satisfies it (which can serve as a membership proof, also sometimes called a *solution*) is the truth assignment $x = \mathrm{T}$ and $y = \mathrm{T}$. The formula $x \wedge y \wedge (\overline{x} \vee \overline{y})$ is not satisfiable. The set SAT is in NP, because each $\phi \in$ SAT has a truth assignment that makes it true and given a truth assignment one can check in polynomial time (in the length of ϕ) whether it makes ϕ true or not. Finding a satisfying truth assignment is another issue. In short, for P problems, it is feasible to *find* solutions, while, for NP problems, it is feasible to *check* solutions. Clearly, P $\subseteq$ NP, and even though it is very reasonable to believe that the inclusion is proper, the issue of whether P is properly included in NP is, as everyone knows, open.

The class NP is important also because numerous important computational problems are NP-complete. Intuitively, a problem A is NP-complete if $A \in$ NP and any other problem in NP can be solved in deterministic polynomial time with the "help" of A. Therefore, an NP-complete problem cannot be solved in deterministic polynomial time, unless all problems in NP are in P. The "help" is formalized by the concept of a polynomial-time reduction. There are more types of reductions and, correspondingly, there are more notions of NP-completeness. The most general type of a polynomial-time reduction is obtained via *oracle Turing machines.* Intuitively, an oracle Turing machine operates like a normal Turing machine except that it is allowed to ask whether a given string belongs to a given set A, called the *oracle set*, and the correct answer is assumed to be automatically obtained in one computation step. More precisely, an oracle Turing machine M is obtained by adding to a normal machine an extra tape, called the *oracle query tape*, and three distinguished states q_{query}, q_{y}, and q_{n}. The oracle set A can be any language over the tape alphabet Γ. The machine, in addition to the normal operations, can write a string $u \in \Gamma^*$ on the oracle query tape. From time to time, as dictated by the transition function, the finite control goes in the state q_{query} with some string $u \in \Gamma^*$ on the oracle query tape. We say that M *queries* the oracle A whether $u \in A$ or not. When this happens, the machine goes in the next step in state q_{y} if $u \in A$ and it goes in state q_{n} if $u \notin A$. An oracle Turing machine M working with oracle set A is denoted M^A and $L(M^A)$ is the language accepted by M with oracle A. We are now prepared to define the most general type of polynomial-time reducibility, the *polynomial-time Turing reducibility*, also called *Cook reducibility.* A problem B is polynomial-time Turing reducible to a problem A (notation $B \leq_T^p A$) if there is a polynomial-time oracle machine M such that $B = L(M^A)$. In other words, $B \leq_T^p A$ if B is polynomial-time computable given unrestricted access to A. The $\leq_m^p$ reducibility (called *polynomial-time many-one reducibility* and also known as *Karp reducibility*) is basically the most restrictive type of reducibility. A problem B is $\leq_m^p$ reducible to a problem A (notation $B \leq_m^p A$) if there is a polynomial-time computable function f such that, for all x, $x \in B$ if and only if $f(x) \in A$ (i.e., if A is viewed as an oracle, the reduction algorithm on input x constructs on the oracle query tape a single string $f(x)$ and the answer of the oracle to the query "$f(x) \in A$?" directly determines whether $x \in B$ or not). A problem A is NP-complete under $\leq_T^p$ reduction if $A \in$ NP and for all $B \in$ NP, $B \leq_T^p A$. In the analogous way, we define NP-completeness under $\leq_m^p$ reduction. A problem is NP-complete if it is NP-complete under $\leq_m^p$ reduction.

The theory of NP-completeness starts with Cook's theorem from 1971 stating that SAT is NP-complete. Since then hundreds of problems have been shown to be NP-complete. Such problems arise from a large variety of domains and many of them are quite important for real-world applications.

The complexity classes E and EXP are important because they capture, typically, a natural category of "non-smart" algorithms that simply operate an exhaustive search to find the solution to a problem. EXP is also important because it is the smallest deterministic time class known to contain NP.

This chapter undertakes a quantitative analysis of the classes P, NP, E, and EXP, and of the relations between them. We briefly overview the sections in this chapter.

Section 3.2 attempts to clarify what is currently known about the hardness of NP-complete problems. There is a common perception that the only way to tackle an NP-complete problem is to consider the exponentially large list of all possible candidate solutions and to patiently try all of them till we hit a good one. This leads to the standard exponential time algorithm for NP problems. In fact, better search strategies have been found for many important NP-complete problems, with significantly faster running time (although still exponential). We present such a method in Section 3.2 for the 3-SAT problem. More precisely we describe a probabilistic algorithm for 3-SAT that runs in time $p(n)(3/4)^n$, for some polynomial p, which is much better than the standard $\Omega(2^n)$ algorithm.

It is widely conjectured that P is properly included in NP. If this conjecture stands, what can we say quantitatively about NP $-$ P? It is known that P is properly included in E. How "large" is E $-$ P? These are the type of questions that we explore in Section 3.3 and in Section 3.4. In Section 3.3, we use the apparatus of topology, more precisely, the effective Baire classification schemas and the superset topology (see Section 1.2). The main results are that if P $\neq$ NP, then NP $-$ P is effectively of the second Baire category (thus, topology-wise, it is not small), while the class of NP-complete problems under Cook reducibility is effectively of the first Baire category (thus, it is small). It follows that if P $\neq$ NP, then there are (from a topological point of view) many problems that are neither in P nor NP-complete.

In Section 3.4, we turn to measure-theoretical tools. In the context of analyzing classes such as P, NP, and E, it is meaningful to consider the variant of effective measure theory given by polynomial-time computable martingales (see Section 1.2). It can be seen that, in this framework, E does not have measure zero, while P does have measure zero. In fact, we present results that show that many classes that generalize P in quite various ways (such as, to give just one example, the class of P-selective sets) also have measure zero. These results show that, quantitatively speaking and using the yardsticks of effective measure theory, E $-$ P is quite large.

What about the measure-theoretical quantitative analysis of NP? Alas, NP remains evasive from this angle too: It is not known whether the effective measure of NP is zero (which would make NP similar to P) or not (which would make NP similar to E). Researchers working in this area have conjectured that the effective measure of NP is not zero (again when the effectivity is based on polynomial-time martingales). This conjecture implies P $\neq$ NP (because the measure of P is zero). More interestingly, as a result in Section 3.4 shows, the conjecture also implies that there exist problems that are NP-complete with respect to Cook reducibility, but not NP-complete with respect to Karp reducibility (a separation which is not known to follow from the hypothesis P $\neq$ NP).

Section 3.5 is dedicated to the quantitative analysis of the relation between relativized P and relativized NP. For any set, P^A (called P relativized with oracle A) denotes the class of sets computable by deterministic polynomial-time oracle machines working with oracle set A, and NP^A (called NP relativized with oracle A) denotes the class of sets computable by nondeterministic polynomial-time oracle machines working with oracle set A. It is known that there are oracle set A and B, such that $\mathrm{P}^A \neq \mathrm{NP}^A$ and $\mathrm{P}^B = \mathrm{NP}^B$. The main result in Section 3.5 shows that if the oracle set A is taken at random, then NP^A differs from P^A in a very strong way: There is a set $T(A)$ in NP^A that cannot be even poorly approximated by any P^A algorithm, in the sense that any P^A algorithm is correct on only a fraction of $(1/2 \pm \epsilon)$ of the input strings of length at most n, for all sufficiently large n.

The definitions of the classes DTIME$[f(n)]$ and NTIME$[f(n)]$, based on which the canonical complexity classes P, NP, E, and EXP are built, are done in the framework of the worst-case complexity analysis of problems. Of course, it is quite interesting to analyze how difficult a problem is on average, with respect to some relevant distributions on input strings at a given length. Section 3.6 introduces elements of average case complexity theory and presents the analogues of P, NP, and NP-completeness in the framework of this theory.

3.2 Upper bound for 3-Satisfiability

IN BRIEF: A probabilistic algorithm for 3-SAT is presented that runs in time $p(n)(\frac{3}{4})^n$, for some polynomial p, and which is correct with probability at least $1 - e^{-n}$.

The satisfiability (SAT) problem is in many ways the canonical representative of the class NP. To list just a few of the reasons, we recall that SAT has a simple formulation, it has been the first problem discovered to be NP-complete, and any nondeterministic polynomial-time computation can be transparently encoded as an instance of SAT (cf. the proof of Cook's Theorem). In particular, many practical NP-complete problems can be reduced quite directly to the SAT problem. It is thus important to study the complexity of this problem. No interesting general lower bound is known for the time needed to solve SAT (there are some good lower bounds for specific approaches to solving SAT). What about upper bounds? A brute-force algorithm runs in $O(n2^n)$ time on an instance ϕ with n variables by trying all possible truth assignments. A polynomial-time algorithm for SAT, of course, does not seem possible (it would imply P = NP). Remaining in the realm of exponential time algorithms, it is of interest to find algorithms for k-SAT [2] that work in time $O(c^n)$ for exponents $c > 1$ that are small. In this section we will present a probabilistic algorithm for 3-SAT that works in time $p(n)(\frac{3}{4})^n$, for some polynomial p, and which gives the correct answer with probability at least $1-e^{-n}$.

We recall that $\wedge$ denotes the boolean operation *and*, $\vee$ denotes the boolean operation *or*, T denotes the boolean value *true*, and F denotes the boolean value *false*. A boolean formula ϕ with variables $x_1, \ldots, x_n$, is in the *3-Conjunctive Normal Form* (briefly, 3-CNF) if it has the form $\phi = C_1 \wedge C_2 \wedge \ldots \wedge C_m$, and each C_i (called a *clause*) has the form $C_i = y_{i_1} \vee y_{i_2} \vee y_{i_3}$, where each y_{i_j} is either a variable $x_k \in \{x_1, \ldots, x_n\}$ or the negation $\bar{x}_k$ of a variable. For example, the following formula ϕ is in 3-CNF:

$$\phi = (x_1 \vee \bar{x}_2 \vee x_4) \wedge (\bar{x}_1 \vee x_3 \vee \bar{x}_4) \wedge (x_2 \vee \bar{x}_3 \vee x_4).$$

The formula ϕ is satisfiable because the assignment $x_1 = T$, $x_2 = T$, $x_3 = F$, and $x_4 = F$ makes the formula true (there are some other satisfying assignments as well). The 3-SAT problem is defined as follows.

Problem 3.2.1 3-SAT Problem:

Input: A formula ϕ in 3-CNF.

Question: Is ϕ satisfiable?

[2]The k-SAT problem is a variant of SAT, in which each clause is restricted to have at most k literals. For $k \geq 3$, k-SAT is still NP-complete

We will first sketch a very simple probabilistic algorithm that, for some polynomial p, runs in time $p(n)(\frac{3}{2})^n$ (which is already much better than $O(n2^n)$) and is correct with high probability. Let ϕ be a formula in 3-CNF. We will assume that ϕ is satisfiable, because this turns out to be the interesting case (in case ϕ is not satisfiable, the algorithm will always be correct). Let a^* be a truth assignment that satisfies ϕ. We first pick a a truth assignment uniformly at random. If a satisfies ϕ, we are done. If a does not satisfy ϕ, there must be a clause C that is not satisfied by a, but that is satisfied by a^*. Thus, if we pick uniformly at random a variable in C and we flip its truth value in a, we obtain a new a, which, with probability at least 1/3, is closer by one bit to a^*. Suppose that the initial a and a^* differ in j bits, for some $0 \leq j \leq n$. It follows that, if we repeat the above procedure n times, we obtain with probability at least $(1/3)^j$ an assignment that satisfies ϕ (this assignment may be a^*, but we may also hit another satisfying assignment before getting to a^*). There are 2^n possible initial assignments, and $\binom{n}{j}$ of these assignments differ from a^* in j bits. Therefore the probability of success (i.e., of finding a satisfying assignment) is

$$\frac{1}{2^n}\sum_{j=0}^{n}\binom{n}{j}\left(\frac{1}{3}\right)^j = \frac{1}{2^n}\left(1+\frac{1}{3}\right)^n = \left(\frac{2}{3}\right)^n.$$

Consequently, the probability of failure is at most $1-(\frac{2}{3})^n$. Therefore, if we repeat the whole procedure $n \cdot (\frac{3}{2})^n$ times, the error probability is at most

$$\left(1-\left(\frac{2}{3}\right)^n\right)^{n(\frac{3}{2})^n} \leq e^{-n}.$$

We will slightly modify the above algorithm and refine the analysis. We will obtain a probabilistic algorithm that runs in time $p(n)\cdot(\frac{4}{3})^n$, for some polynomial p.

Theorem 3.2.2 *There is a probabilistic algorithm for* 3-SAT *that on an instance ϕ, with n variables, runs in $O(n^{7/2}\left(\frac{4}{3}\right)^n)$ and has error probability at most e^{-n}.*

Proof. The input for the algorithm is a boolean formula ϕ in 3-CNF with n variables. The algorithm consists in repeating a polynomial number of times the following routine, which we call the *basic iteration.*

BASIC ITERATION
Step 1. Pick uniformly at random an initial assignment $a_0 \in \{0,1\}^n$
(as usual, we identify 0 with F and 1 with T).
Step 2. Repeat for $i = 0, \ldots, 3n-1$:
If a_i satisfies ϕ, stop the whole run of the algorithm and accept.
Let C be a clause not satisfied by ϕ.
Pick one of its three literals uniformly at random and flip its value in a_i.
Call the new assignment a_{i+1}.

As before, we consider that ϕ is satisfiable, because this is the interesting case. We focus on one fixed basic iteration. Let us call "Success" the event that the fixed basic iteration finds a satisfying assignment for ϕ. Let a^* be a fixed satisfying assignment for ϕ. Then

$$\text{Prob}(\text{“Success”}) \geq \text{Prob}(a_0 \text{ turns into } a^* \text{ in the basic iteration}). \tag{3.1}$$

We want to find a lower bound for the right hand side of the above inequality.

The *Hamming distance* between two strings a_1 and a_2 of equal lengths, denoted $\text{dist}(a_1, a_2)$, is, by definition, the number of positions in which the two strings differ. At an arbitrary iteration i of *Step 2*, assuming a_i is not a satisfying assignment,

$$\text{dist}(a_{i+1}, a^*) = \text{dist}(a_i, a^*) \pm 1.$$

We call iteration i "good", if

$$\text{dist}(a_{i+1}, a^*) = \text{dist}(a_i, a^*) - 1,$$

and we call it "bad", if

$$\text{dist}(a_{i+1}, a^*) = \text{dist}(a_i, a^*) + 1.$$

As before, we have that the probability that iteration i is "good" is at least $1/3$, because there must be at least one variable in C whose truth value is different in a_i from a^*. The probability that iteration i is "bad" is, of course, at most $2/3$. If we assume that the probability of a "good" iteration is $1/3$ and that the probability of a "bad" iteration is $2/3$, then we decrease the probability of turning a_0 into a^*. Therefore, since we are interested in finding a lower bound for the right hand side of the inequality (3.1), we make this assumption. Let us suppose that $\text{dist}(a_0, a^*) = j$, for some $0 \leq j \leq n$. We make one more simplification: In order to find a lower bound, it is enough if we consider only those transitions from a_0 to a^* that have i "bad" iterations and $j + i$ "good" iterations, for $0 \leq i \leq j$ (we must have j more "good" iterations than "bad" iterations to reach a^*). Note that the total number of iterations is $j + 2i \leq 3j \leq 3n$. This is why we do $3n$ iterations in *Step 2*. The probability of a transition from a_0 to a^* with $j + i$ "good" iterations and i "bad" iterations, in which the positions of "good" and "bad" iterations are fixed, is, under our assumption, $(\frac{1}{3})^{j+i} \cdot (\frac{2}{3})^i$. We need to determine the number of transitions of this type. This can be done using a particular form of the Ballot Theorem (see [Fel68]).

Claim 3.2.3 *The number of transitions from a_0 to a^* with $j+i$ "good" iterations and i "bad" iterations is* $\frac{j}{j+2i}\binom{j+2i}{i}$.

Proof. We consider the grid of pairs (x, y), with x and y integer numbers. The points of the grid form an oriented graph in which each node (x, y) has two outgoing edges to the nodes $(x + 1, y + 1)$ and $(x + 1, y - 1)$. The first type of edge is called an "up" edge, and an edge of the second type is called a "down" edge. Edges in this graph correspond to iterations in Step 2. In a pair (x, y), we view x as being the iteration number, and y as being the Hamming distance between the current assignment (at iteration x) and a^*. The assignment a_0 corresponds to the node $(0, j)$ and a^* corresponds to the node $(j + 2i, 0)$. Observe that just before arriving at $(j+2i, 0)$, we must be at $(j+2i-1, 1)$. Thus the number of transitions from a_0 to a^* with $j+i$ "good" iterations and i "bad" iterations is equal to the number of paths from node $(0, j)$ to $(j + 2i - 1, 1)$ that do not touch the x-axis (this is true because if at some iteration earlier than $j + 2i$ we touch the x-axis, it means that we have reached a^* before the $j + 2i$ stipulated iterations).

We will determine A, the total number of paths from $(0, j)$ to $(j + 2i - 1, 1)$, and B, the number of paths from $(0, j)$ to $(j + 2i - 1, 1)$ which touch and maybe even cross the x-axis. Next we will calculate $A - B$, which will give us the desired result.

For A, observe that each path from $(0, j)$ to $(j+2i-1, 1)$ has $j+2i-1$ edges, and that the difference between "down" edges and "up" edges must be $j - 1$. Therefore, there are $j+i-1$ "down" edges and i "up" edges. Since the "up" edges can occur anywhere in the path, it follows that

$$A = \binom{j + 2i - 1}{i}.$$

For B, observe that there is a 1-to-1 correspondence between the set of paths from $(0, j)$ to $(j+2i-1, 1)$ that touch or cross the x-axis and the (total) number of paths from $(0, -j)$ to the same $(j+2i-1, 1)$. Indeed, notice that $(0, -j)$ is the symmetric of $(0, j)$ with respect to the x-axis. Consider a path from $(0, j)$ to $(j + 2i - 1, 1)$ that touches the x-axis and let us consider the initial segment of this path until (and including) the first touch of the x-axis. If we reflect this segment on the x-axis and leave the rest unmodified, we obtain, in a 1-to-1 manner, a path from $(0, -j)$ to $(j + 2i - 1, 1)$. Therefore, to determine B, we need to count how many paths are there from $(0, -j)$ to $(j + 2i - 1, 1)$. Such a path has $j + 2i - 1$ edges, and the difference between "up" edges and "down" edges must be $j + 1$. Consequently, there are $j + i$ "up" edges, and $i - 1$ "down" edges. Since the "down" edges can occur anywhere in the path, it follows that

$$B = \binom{j + 2i - 1}{i - 1}.$$

Thus,

$$A - B = \binom{j + 2i - 1}{i} - \binom{j + 2i - 1}{i - 1} = \frac{j}{j + 2i}\binom{j + 2i}{i}.$$

This concludes the proof of Fact 3.2.3. ∎

Let p_j be the probability that there is a transition from a_0 to a^* with $j + i$ "good" iterations and i "bad" iterations, with $0 \leq i \leq j$, under our assumption that a "good" iteration happens with probability $1/3$, and a "bad" iteration happens with probability 2/3. According to our discussion above,

$$\begin{aligned} p_j &= \sum_{i=0}^{j} \frac{j}{j+2i} \binom{j+2i}{i} \left(\frac{1}{3}\right)^{j+i} \left(\frac{2}{3}\right)^{i} \\ &\geq \frac{1}{3} \sum_{i=0}^{j} \binom{j+2i}{i} \left(\frac{1}{3}\right)^{j+i} \left(\frac{2}{3}\right)^{i} \\ &> \frac{1}{3} \binom{3j}{j} \left(\frac{1}{3}\right)^{2j} \left(\frac{2}{3}\right)^{j}. \end{aligned}$$

The last inequality has been obtained by retaining only the last term in the sum. It is known (see for example [Bol85]) that for $0 < \lambda < 1$ and for any m,

$$\binom{m}{\lambda m} \geq \frac{1}{2J(\lambda, m)} 2^{h(\lambda)m} = \frac{1}{2J(\lambda, m)} \left(\frac{1}{\lambda}\right)^{\lambda m} \cdot \left(\frac{1}{1-\lambda}\right)^{(1-\lambda)m},$$

where $h(\lambda) \stackrel{\text{def}}{=} -\lambda \log_2 \lambda - (1-\lambda)\log_2(1-\lambda)$ is the binary entropy function, and $J(\lambda, m) = \sqrt{2\lambda(1-\lambda)\pi m}$. Therefore,

$$\begin{aligned} \binom{3j}{j} &\geq \frac{1}{2J(1/3, 3j)} 2^{h(1/3)\cdot(3j)} \\ &\geq \frac{1}{2J(1/3, 3n)} \cdot 3^j \left(\frac{3}{2}\right)^{2j}. \end{aligned}$$

Thus,

$$\begin{aligned} p_j &> \frac{1}{3} \cdot \frac{1}{2J(1/3, 3n)} \cdot 3^j \cdot \left(\frac{3}{2}\right)^{2j} \cdot \left(\frac{1}{3}\right)^{2j} \cdot \left(\frac{2}{3}\right)^{j} \\ &= \frac{1}{6J(1/3, 3n)} \cdot \frac{1}{2^j}. \end{aligned}$$

The probability that the initial assignment a_0 has $\text{dist}(a_0, a^*) = j$ is $\binom{n}{j}/(2^n)$. Therefore, the probability of the event "Success" (i.e., of the fact that in the fixed basic iteration we find a satisfying assignment) is at least

$$\begin{aligned} \sum_{j=0}^{n} \frac{\binom{n}{j}}{2^n} \cdot \frac{1}{6J(1/3, 3n)} \cdot \frac{1}{2^j} &= \frac{1}{6J(1/3, 3n)} \cdot \left[\frac{1}{2}\left(1 + \frac{1}{2}\right)^n\right] \\ &= \frac{1}{6J(1/3, 3n)} \cdot \left(\frac{3}{4}\right)^n. \end{aligned}$$

The probability of failure is at most $1 - \frac{1}{6J(1/3,3n)} \cdot \left(\frac{3}{4}\right)^n$. Consequently, if we make $6nJ(1/3, 3n) \cdot \left(\frac{4}{3}\right)^n$ basic iterations, the probability that we fail to find a satisfying assignment is at most

$$\left(1 - \frac{1}{6J(1/3,3n)} \cdot \left(\frac{3}{4}\right)^n\right)^{6nJ(1/3,3n)\cdot\left(\frac{4}{3}\right)^n} \leq e^{-n},$$

for sufficiently large n. ∎

3.3 NP vs. P—the topological view

IN BRIEF: Assume P $\neq$ NP. Then NP $-$ P is topologically not small. On the other hand, the class of NP-complete problems is topologically small.

The question whether P is equal or not to NP is the most outstanding open question in computational complexity. Solving this problem is beyond reach at this moment. However, there is strong evidence that P is properly included in NP (for example, just think about the hundreds of natural NP-complete problems for which no polynomial-time algorithm is known). It is thus reasonable to assume that P $\neq$ NP and to develop a theory based on this hypothesis.

In this section we undertake a topological analysis of some important subclasses of NP (assuming that P $\neq$ NP). As pointed out in Section 1.2.1, in principle, two topologies are relevant for such an analysis: The Cantor topology and the superset topology. The Cantor topology is more natural but, unfortunately, it is not adequate in this case. If we attempt to use Definition 2.1.2, we see easily that NP itself is "small" (more exactly, NP is effectively of first category). Consequently, the Cantor topology approach classifies as small all the classes inside NP and, thus, is not capable of differentiating the relative sizes of such classes. Therefore, we consider the superset topology. This approach will allow us to compare the size of some interesting classes inside NP such as, to name just two, NP $-$ P and the class of NP-complete problems. We show in this section that NP $-$ P, if not empty, is not small with respect to effective superset topology. More precisely, if not empty, NP $-$ P is of the second Baire category.

Superset topology has been introduced in Section 1.2.1. We recall that we consider the binary alphabet $\Sigma = \{0, 1\}$. Σ^* denotes the set of finite binary strings, and Σ^∞ denotes the set of infinite binary sequences. For $i \geq 1$, s_i denotes the i-th string in Σ^* in lexicographical order. The length of $v \in \Sigma^*$ is the number of symbols from Σ^* in v and is denoted by $|v|$. For $v \in \Sigma^*$ or in Σ^∞, for any natural number $i \geq 1$ (and $i \leq |v|$, in case $v \in \Sigma^*$), $v(i)$ denotes the i-th symbol in v. Thus, $v = v(1)v(2)\ldots v(|v|)$, for $v \in \Sigma^*$. The base $\mathcal{B}^S = (U_v^S)_{v\in\Sigma^*}$ of open sets in the superset topology is defined by

$$U_v^S = \{w \in \Sigma^\infty \mid \forall i\,((1 \leq i \leq |v| \text{ and } v(i) = 1) \Rightarrow (v(i) = w(i) = 1))\}.$$

Definition 3.3.1 (Baire classification with respect to the superset topology)
(1) *A class* $A \subseteq \Sigma^\infty$ *is* effectively nowhere dense with respect to the superset topology, *if there is a computable function* $f: \Sigma^* \to \Sigma^*$ *such that for every* v *in* Σ^*,

(i) v *is a prefix of* $f(v)$, *and*

(ii) $A \cap U^S_{f(v)} = \emptyset$.

(2) *A class* $A \subseteq \Sigma^\infty$ *is* effectively of first Baire category with respect to the superset topology, *if there is a countable decomposition* $A = \bigcup_{i \in \mathbb{N}} A_i$ *and a computable function* $f: \mathbb{N} \times \Sigma^* \to \Sigma^*$ *such that, for every* $i \in \mathbb{N}$ *and for every* $v \in \Sigma^*$,

(i) v *is a prefix of* $f(i, v)$, *and*

(ii) $A_i \cap U^S_{f(i,v)} = \emptyset$.

(3) *A class* $A \subseteq \Sigma^\infty$ *is* effectively of second Baire category with respect to the superset topology *if it is not effectively of first Baire category with respect to the superset topology.*

Sometimes, when the context is clear and for the sake of simplicity, we will just say first or second Baire category, or even first or second category. For the motivations behind these definitions, the reader is invited to go back to Section 1.2.1.

We present first a technical lemma that will be at the core of all proofs showing that various classes are of the second Baire category. The lemma utilizes some classical concepts from computable function theory, which we introduce next. We fix an enumeration $\{M_i\}_{i \in \mathbb{N}}$ of all standard Turing machines having as input one binary string. Let $(W_i)_{i \in \mathbb{N}}$ be the enumeration of the class of computably enumerable sets of binary strings defined by stipulating that, for each i, W_i is the domain of M_i, i.e., W_i is equal to the set of input strings on which M_i halts. Let $W_{i,s}$ be the set of strings enumerated in W_i within the first s steps of a dove-tailing simulation of M_i (see [Soa87]). We also fix an enumeration $\{\mathrm{P}_i\}_{i \in \mathbb{N}}$ of all deterministic polynomial-time machines, and an enumeration $\{\mathrm{NP}_i\}_{i \in \mathbb{N}}$ of all nondeterministic polynomial-time machines. For each i, $L(\mathrm{P}_i)$ ($L(\mathrm{NP}_i)$) denotes the language accepted by P_i (respectively, NP_i). If $x \in \Sigma^*$, $\mathrm{P}_i(x)$ ($\mathrm{NP}_i(x)$) denotes the output of the machine P_i (NP_i) on input x. We say that a set is co-finite if its complement is finite. Let $\mathrm{TOT} = \{x \mid W_x = \Sigma^*\}$ and $\text{Co-FIN} = \{x \mid \text{the complement of } W_x \text{ is finite }\}$. For a class A of computably enumerable sets, we say that TOT is m-reducible to $(A, \text{Co-FIN})$, notation

$$\mathrm{TOT} \leq_m (A, \text{Co-FIN}),$$

if there is a computable function $s: \mathbb{N} \to \mathbb{N}$ such that $i \in \mathrm{TOT}$ implies $W_{s(i)} \in A$ and $i \notin \mathrm{TOT}$ implies $s(i) \in \text{Co-FIN}$. It is well-known that TOT is complete for the Π_2 level of the arithmetic hierarchy (see [Soa87]).

The key observation, which is the content of the next lemma, is that if A is a class of the first Baire category, then (a) A is included in a set D which is in the Σ_2 level of the arithmetical hierarchy, and (b) A is included in the complement of Co-FIN. It follows that if A is a class that has the property that

TOT $\leq_m$ $(A, \text{Co-FIN})$, then A is necessarily of the second Baire category, because otherwise TOT would be reducible to a Σ_2 set (this is not possible because TOT is Π_2-complete).

Lemma 3.3.2 *Let A be a class of computably enumerable sets of strings with* TOT $\leq_m$ $(A, \text{Co-FIN})$*. Then A is effectively of the second Baire category with respect to the superset topology.*

Proof. Suppose A is of the first category. This means that there is a countable decomposition $A = \bigcup_{i \in \mathbb{N}} A_i$ and a computable function $f\colon \mathbb{N} \times \Sigma^* \to \Sigma^*$ such that, for every integer i and every string $w \in \Sigma^*$, w is a proper prefix of $f(i, w)$ and $U^S_{f(i,w)} \cap A_i = \emptyset$. Let

$$D = \{j \in \mathbb{N} \mid (\exists i)(\forall n)(\forall s)[\exists k \in \mathbb{N}, n < k \leq |f(i, 0^n)|, s_k \notin W_{j,s}]\}.$$

Clearly, Co-FIN is included in the complement of D and D is in the Σ_2 level of the arithmetic hierarchy. Observe that $W_j \in A$ implies $j \in D$. Indeed, if we assume that $W_j \in A$, then there exists some i such that $W_j \in A_i$. Also note that if, for some n, it holds true that, for all k with $n < k \leq |f(i, 0^n)|$, $s_k \in W_j$, then we can conclude that

$$W_j \in U^S_{f(i,0^n)},$$

which contradicts the fact that

$$A_i \cap U^s_{f(i,0^n)} = \emptyset,$$

for all integers i and n. It follows that

- $i \in \text{TOT}$ implies $W_{s(i)} \in A$, which implies $s(i) \in D$, and
- $i \notin \text{TOT}$ implies $s(i) \in \text{Co-FIN} \subseteq \bar{D}$.

($\bar{D}$ denotes the complement of D.) Consequently, TOT $\leq_m D$, which is a contradiction because TOT is Π_2-complete and D is in Σ_2. ∎

We show now that if $\text{NP} - \text{P} \neq \emptyset$, then $\text{NP} - \text{P}$ is of the second category. By Lemma 3.3.2, all we have to do is to show that TOT $\leq_m (\text{NP} - \text{P}, \text{Co-FIN})$.

Lemma 3.3.3 *If* $\text{NP} - \text{P} \neq \emptyset$*, then* TOT $\leq_m (\text{NP} - \text{P}, \text{Co-FIN})$.

Proof. Let $\{M_i\}_{i \in \mathbb{N}}$ be the fixed enumeration of all deterministic Turing machines such that W_i is the domain of M_i. For every Turing machine M_i we define a nondeterministic polynomial-time Turing machine $\text{NP}_{s(i)}$ such that:

- If $i \in \text{TOT}$, then $\text{NP}_{s(i)} \in \text{NP} - \text{P}$, and
- If $i \notin \text{TOT}$, then $\text{NP}_{s(i)}$ accepts all the inputs in Σ^* except for a finite set.

We first overview informally the construction. The computation of $\mathrm{NP}_{s(i)}$ is performed in stages, starting with *Stage* 0. The machine $\mathrm{NP}_{s(i)}$ has two kind of objectives depending on whether the current stage is even or odd. Namely, if the current stage is $2e$, for some integer e, $\mathrm{NP}_{s(i)}$ tries to find, in polynomial time, whether M_i accepts s_e. If and when this happens, we pass to the next stage, i.e., to *Stage* $2e+1$. In case $\mathrm{NP}_{s(i)}$ does not succeed in determining whether M_i accepts s_e, the current input is accepted. In this way, if $i \notin \mathrm{TOT}$, then clearly $\mathrm{NP}_{s(i)}$ remains stuck in an even stage and accepts a co-finite set. On the other hand, if the current stage is odd, say equal to $2e+1$ for some integer e, then $\mathrm{NP}_{s(i)}$ looks for a string y such that $\mathrm{NP}_{s(i)}(y) \neq \mathrm{P}_e(y)$. If no such y is found, the current input x is accepted if and only if $x \in SAT$. Consequently, if at *Stage* $2e+1$ only failures (to find an y as above) occur, then, starting from some string, $\mathrm{NP}_{s(i)}$ will be equal to SAT (formally, their characteristic functions will be equal; abusing notation, we often identify a set with its characteristic function). Hence, $\mathrm{NP}_{s(i)}$ will eventually find a string y such that $\mathrm{NP}_{s(i)}(y) \neq \mathrm{P}_e(y)$, because otherwise SAT would be, except for a finite number of inputs, equal to P_e and thus it would be in P, which contradicts our hypothesis that NP $\neq$ P. When this happens, stage is incremented to $2e+2$. We present next the complete construction.

Construction of $\mathrm{NP}_{s(i)}$

Initially, *Stage*$=0$. On input $x \in \Sigma^*$ of length n, the computation of $\mathrm{NP}_{s(i)}$ proceeds as follows:

(a) For n steps, $\mathrm{NP}_{s(i)}$ simulates deterministically the previous computations $\mathrm{NP}_{s(i)}(s_1), \mathrm{NP}_{s(i)}(s_2), \ldots$, and determines how many stages have been completed so far. The variable *Stage* contains the value of smallest uncompleted stage that $\mathrm{NP}_{s(i)}$ is able to find within the allowed n steps.

(b) *Case 1: Stage* $= 2e$. For n steps, $\mathrm{NP}_{s(i)}$ simulates M_i on input s_e. If M_i does not accept s_e in the allotted time, then x is accepted. Otherwise, *Stage* $= 2e+1$, and the input x is accepted.

Case 2: Stage $= 2e+1$. For n steps, $\mathrm{NP}_{s(i)}$ looks for a string $y \leq x$ such that $\mathrm{NP}_{s(i)}(y) \neq \mathrm{P}_e(y)$. During this search $\mathrm{NP}_{s(i)}$ is simulated deterministically on different inputs y. If such a y is found, then *Stage* $= 2e+2$ and x is accepted. Otherwise, x is accepted if and only if $x \in SAT$.

End of construction of $\mathrm{NP}_{s(i)}$

We need to show that (a) $\mathrm{NP}_{s(i)}$ is a polynomial-time nondeterministic machine, (b) if $i \in \mathrm{TOT}$ then $L(\mathrm{NP}_{s(i)})$ is not in P, and (c) if $i \notin \mathrm{TOT}$, then $\mathrm{NP}_{s(i)}$ accepts a co-finite set. These claims are proven in the following Claims.

Claim 3.3.4 $\mathrm{NP}_{s(i)}$ *is a polynomial-time nondeterministic algorithm.*

Proof. The only nondeterministic step in the computation of $\mathrm{NP}_{s(i)}$ on an input x occurs in (b), case 2. In this case, $\mathrm{NP}_{s(i)}$ has to determine if $x \in \mathrm{SAT}$, which can be done by a nondeterministic polynomial-time computation. All the other computations are performed in deterministic polynomial time (in fact, linear time). ∎

Theorem 3.3.11 INFORMAL STATEMENT: *If* NP *and* P *are different, then the class of problems that are neither in* P *nor* NP*-complete is not small from the point of view of superset topology.*

FORMAL STATEMENT: *Assume* NP $\neq$ P. *Then* (NP − P) − NPCOMP *is effectively of the second Baire category with respect to the superset topology.*

Proof. Suppose that (NP − P) − NPCOMP is of the first Baire category. It is easy to see that the union of two sets of first category is of first category as well. Since NP − P = ((NP − P) − NPCOMP) ∪ NPCOMP, we obtain that NP − P is of the first Baire category, which contradicts Theorem 3.3.9. ∎

Using the same tools, we can investigate classes of sets achieving a stronger form of separation between P and NP. Recall that an NP problem is a decision problem, i.e., a problem for which the solution on any input instance is *yes* or *no*. Given the strong evidence for the existence of NP problems admitting no polynomial time algorithm that gives the correct answer on all inputs, we may want to reconsider our demands and look for a polynomial-time algorithm for which only the *yes* answers are always correct, or perhaps only the *no* answers are always correct. The sets for which even these more modest objectives are not achievable are called P*-immune sets*, and, respectively, P*-simple sets*.

Definition 3.3.12 (a) *A set A is* P*-immune if A is infinite and has no infinite subset in* P.

(b) *A set A is* P*-simple if the complement of A is infinite and has no infinite subset in* P *(i.e., the complement of A is* P*-immune).*

Even if we assume that P $\neq$ NP, it is not known if there exists a set in NP that is P-immune set, or P-simple. However, we will show that if there exists one P-immune set in NP, then there exist many such sets, and that the similar assertion holds for P-simple sets.

Theorem 3.3.13 INFORMAL STATEMENT: *The class of* P*-immune sets in* NP *is either empty or not small from the point of view of superset topology.*

FORMAL STATEMENT: *If there exists a* P*-immune set in* NP, *then the class of* P*-immune sets in* NP *is effectively of the second Baire category with respect to the superset topology.*

Proof. Let $A = \{B \in \mathrm{NP} \mid B$ is P-immune$\}$. By hypothesis, A is not empty, so we fix a set H in A. We will construct, in an effective way, for each i a nondeterministic polynomial-time machine $\mathrm{NP}_{s(i)}$ such that if $i \in$ TOT then $L(\mathrm{NP}_{s(i)}) \in A$, and if $i \notin A$, then the language accepted by $\mathrm{NP}_{s(i)}$ is co-finite (i.e., $\mathrm{NP}_{s(i)}$ accepts all the strings except for a finite set). Then, by Lemma 3.3.2, the conclusion follows. Note that $\mathrm{NP}_{s(i)} \in A$ holds if for any deterministic polynomial-time machine P_k, either $L(\mathrm{P}_k)$ is finite, or $L(\mathrm{P}_k)$ is not an infinite subset of $L(\mathrm{NP}_{s(i)})$.

During the construction we use the variables *Stage*, *NextCandidate*, and a list, called *List*, which stores indices of deterministic polynomial-time machines that have been considered but not yet ruled out as accepting infinite subsets of

$L(\mathrm{NP}_{s(i)})$. *List* is viewed as a linear sequence of integers, so that we can speak about the first element of *List*, about the second element of *List*, and so forth. Insertions in *List* are made in the last positions, deletions can be made anywhere, but after a deletion the *List* is compactified so as not to contain any gap. We will take care not to increase the size of *List* too much, so that insertions and deletions can be realized in linear time. The variable *NextCandidate* keeps the index j of the next deterministic polynomial-time machine P_j that we attempt to introduce in *List*. *Stage* is a variable that records the current stage in the construction of $\mathrm{NP}_{s(i)}$, similarly to the construction in the proof of Lemma 3.3.3. If *Stage* has an even value, *Stage* $= 2e$, then $\mathrm{NP}_{s(i)}$ tries to find if M_i accepts s_e. When this happens, *Stage* is incremented to $2e+1$. If M_i does not accept s_e, then *Stage* remains perpetually at the value $2e$, causing $\mathrm{NP}_{s(i)}$ to accept all further inputs. If *Stage* has an odd value, then $\mathrm{NP}_{s(i)}$ tries to find for each k in *List* a string z such that $z \in L(\mathrm{P}_k) \cap \overline{L(\mathrm{NP}_{s(i)})}$ ($\overline{L(\mathrm{NP}_{s(i)})}$ denotes the complement of $L(\mathrm{NP}_{s(i)})$). In this attempt, $\mathrm{NP}_{s(i)}$ dedicates to each element of *List* a time that is proportional to its position in *List*, without exceeding n steps for the whole operation, where n is the length of the current input of $\mathrm{NP}_{s(i)}$. Specifically, if k is in position j, $n/(2^j)$ steps are spent in the effort of finding the desired string z. If no such z is found, then $\mathrm{NP}_{s(i)}$ accepts or rejects the current input x depending on whether $x \in H$, or, respectively $x \notin H$. However, as $\mathrm{NP}_{s(i)}$ processes increasingly longer inputs, and assuming that $L(\mathrm{P}_k)$ is infinite, there will be eventually enough time to discover a z such that $z \in L(\mathrm{P}_k) \cap \overline{L(\mathrm{NP}_{s(i)})}$. The existence of such a z follows from the P-immunity of H and from the fact that repeated failures make $L(\mathrm{NP}_{s(i)})$ be equal to H, modulo a finite set (i.e., their symmetric difference is a finite set).

Construction of $\mathrm{NP}_{s(i)}$

Initially, *Stage* $= 0$, *NextCandidate* $= 0$, *List* $= \emptyset$. On input $x \in \Sigma^*$ of length n, $\mathrm{NP}_{s(i)}$ runs as follows:

(a) For n steps, $\mathrm{NP}_{s(i)}$ simulates deterministically the previous computations (i.e., it computes $\mathrm{NP}_{s(i)}(s_1), \mathrm{NP}_{s(i)}(s_2), \ldots$, for as many inputs the time bound allows) and determines the current values of *Stage*, *NextCandidate*, and the content of *List*. It will become clear from the construction that all these values are the same on all the nondeterministic branches of the computations that are simulated, so there is no ambiguity in determining the above values of the three variables.

(b) *Case 1: Stage* $=2e$. The machine M_i on input s_e is simulated for n steps. If M_i accepts s_e in the allotted time, x is accepted, *Stage* is increased to $2e+1$, and $\mathrm{NP}_{s(i)}$ stops. Otherwise, x is also accepted, but *Stage* remains equal to $2e$. In both cases, $\mathrm{NP}_{s(i)}$ stops.

Case 2: Stage $= 2e+1$. Let m be the number of elements in *List*. If $m > n$, then x is rejected and $\mathrm{NP}_{s(i)}$ stops. Otherwise, *NextCandidate* is incremented by 1.

For $j \in \{1, \ldots, m+1\}$, let *List*[j] denote the value of the j-th element in *List*. Then for every $j \in \{1, \ldots, m+1\}$, for $n/(2^j)$ steps, $\mathrm{NP}_{s(i)}$ looks for a string z such that $z \in L(\mathrm{P}_{List[j]})$ and $z \notin L(\mathrm{NP}_{s(i)})$. In doing this, $\mathrm{NP}_{s(i)}$ is simulated in a deterministic way.

Case 2.1: The search succeeds for some j. Then for all such j's, *List*$[j]$ is deleted from *List*, x is accepted, and *Stage* $:= 2e+2$.

Case 2.2: The search fails for all j. Then x is accepted if and only if $x \in H$.

End of construction of $\mathrm{NP}_{s(i)}$.

The proof follows by the next series of facts.

Claim 3.3.14 $\mathrm{NP}_{s(i)}$ *is a polynomial-time nondeterministic machine.*

Proof. The only nondeterministic step in the computation of $\mathrm{NP}_{s(i)}$ occurs in (b), *Case 2.2*. This step is realized by simulating the nondeterministic polynomial-time machine that accepts H. All the other operations are performed in a deterministic way in polynomial time (in fact, linear time). Observe that the size of *List* is not allowed to increase too much, so that the insertions and the deletions can be done in linear time in the size of the current input. ∎

Claim 3.3.15 *Suppose that at a certain moment in the construction of* $\mathrm{NP}_{s(i)}$, *Stage* $= 2e$ *and* M_i *accepts* s_e. *Then there exists a later moment when* Stage *is increased to* $2e+1$.

Proof. This follows from the fact that on a long enough input string x, $\mathrm{NP}_{s(i)}$ has enough time to simulate the accepting computation of M_i on input s_e. ∎

Claim 3.3.16 *Suppose that at a certain moment in the construction of* $\mathrm{NP}_{s(i)}$, *Stage* $= 2e+1$. *Then there exists a later moment when* Stage *is increased to* $2e+2$.

Proof. Let us suppose the contrary. Clearly, while the construction is at *Stage* $2e+1$, there exists a moment when some index k such that $L(\mathrm{P}_k)$ is infinite is inserted in *List*. Since H is P-immune, $L(\mathrm{P}_k) \cap \overline{H} \neq \emptyset$. Moreover, $L(\mathrm{P}_k) \cap \overline{H}$ is an infinite set. This is true because if $L(\mathrm{P}_k) \cap \overline{H} = B$ and B is a finite set, then $L(\mathrm{P}_k) - B \subseteq H$, but $L(\mathrm{P}_k) - B$ is an infinite set in P. This would contradict the P-immunity of H. By our assumption, it follows that there exists a string x_0 such that for every string $x \geq x_0$, $\mathrm{NP}_{s(i)}$ accepts x if and only if $x \in H$. It follows that $L(\mathrm{NP}_{s(i)})$ is itself P-immune. Therefore, on a long enough input, $\mathrm{NP}_{s(i)}$ has enough time to discover a string z such that $z \in L(\mathrm{P}_k) \cap \overline{L(\mathrm{NP}_{s(i)})}$. This implies the incrementation of *Stage* to the value $2e+2$, which contradicts our assumption. ∎

Claim 3.3.17 *If* $i \notin \mathrm{TOT}$, *then* $\mathrm{NP}_{s(i)}$ *accepts a co-finite language.*

Proof. Let e be minimal with the property that M_i does not accept s_e. By Claim 3.3.15 and Claim 3.3.16, it follows that *Stage* reaches the value $2e$. From this moment on, $\mathrm{NP}_{s(i)}$ accepts every input. ∎

Claim 3.3.18 *If* $i \in \mathrm{TOT}$ *and* $L(\mathrm{P}_k)$ *is infinite, then* $L(\mathrm{P}_k) \cap \overline{L(\mathrm{NP}_{s(i)})} \neq \emptyset$.

Proof. Since $i \in \mathrm{TOT}$, it follows from Claim 3.3.15 and Claim 3.3.16 that there exists a moment when k is inserted in *List*. Then, by the argument we used in the proof of Claim 3.3.16, we get that $L(\mathrm{P}_k) \cap \overline{L(\mathrm{NP}_{s(i)})} \neq \emptyset$. ∎

Claim 3.3.19 $L(\mathrm{NP}_{s(i)})$ *is infinite.*

Proof. If $i \notin$ TOT, the conclusion follows from Claim 3.3.17. If $i \in$ TOT, then Claim 3.3.15 and Claim 3.3.16 together imply that the value of *Stage* passes through all positive integers. Any increase of *Stage* from an even value is done in (b) *Case 1* and implies also the acceptance of the current input string x. ∎

From Claims 3.3.17, 3.3.18, and 3.3.19, it follows that TOT $\leq_m (A, \text{Co-FIN})$, where A is the class of P-immune sets in NP. By Lemma 3.3.2, we conclude that NP $-$ P is of second Baire category. ∎

The analogous result for the case of sets in NP that are P-simple is derived similarly.

Theorem 3.3.20 INFORMAL STATEMENT: *The class of* P*-simple sets in* NP *is either empty or not small from the point of view of superset topology.*

FORMAL STATEMENT: *If there exists a* P*-simple set in* NP*, then the class of* P*-simple sets in* NP *is effectively of the second Baire category with respect to the superset topology.*

Proof. The proof relies again on Lemma 3.3.2. We fix H, a P-simple set in NP. For every integer i, we define a nondeterministic polynomial-time machine $\mathrm{NP}_{s(i)}$ such that (a) $i \in$ TOT implies H is included in $L(\mathrm{NP}_{s(i)})$, and (b) $i \notin$ TOT implies $\mathrm{NP}_{s(i)}$ accepts a co-finite set. Note that (a) implies that $L(\mathrm{NP}_{s(i)})$ is P-simple, because if B is any infinite set in P, then $B \cap \overline{H} \neq \emptyset$, and, therefore, $B \cap \overline{L(\mathrm{NP}_{s(i)})} \neq \emptyset$.

Construction of $\mathrm{NP}_{s(i)}$

On an input x of length n, $\mathrm{NP}_{s(i)}$ does the following:

(a) For n steps, $\mathrm{NP}_{s(i)}$ simulates deterministically the previous computations $\mathrm{NP}_{s(i)}(s_1)$, $\mathrm{NP}_{s(i)}(s_2), \ldots$, for as many inputs the time bound allows. It may happen that on some input s_k the simulation of $\mathrm{NP}_{s(i)}(s_k)$ finds different values for the variable *Stage* on the originally nondeterministic branches of the computation of $\mathrm{NP}_{s(i)}(s_k)$. If this is the case, the least such value is selected for *Stage*.

(b) *Case 1: Stage* $= 2e$. Then *Stage* is increased to $2e+1$ and the nondeterministic polynomial-time machine N that accepts H is started on input x. If a computation of $N(x)$ that accepts x is discovered, then $\mathrm{NP}_{s(i)}$ accepts x and *Stage* is reset to the value $2e$. Otherwise, $\mathrm{NP}_{s(i)}$ rejects x. After these operations, $\mathrm{NP}_{s(i)}$ stops.

Case 2: Stage $= 2e+1$. For n steps, $\mathrm{NP}_{s(i)}$ simulates M_i on input s_e. If M_i accepts s_e in the allotted time, then x is accepted, *Stage* is increased to $2e+2$, and $\mathrm{NP}_{s(i)}$ stops. Otherwise, $\mathrm{NP}_{s(i)}$ just accepts x (observe that x is accepted anyway) and stops.

End of construction of $\mathrm{NP}_{s(i)}$

Suppose that $i \in$ TOT. Then the value of *Stage* passes through all positive integers k. This is proved by induction on k. Suppose $k = 2e$. There exists a moment when the input x is such that $x \notin H$ (because $\overline{H}$ is infinite). At this moment, *Stage* becomes $2e+1$ and is never decreased below this value later. In case $k = 2e+1$, since $i \in$ TOT, we conclude that for a sufficiently long input x, $\mathrm{NP}_{s(i)}$ has enough time to discover that M_i accepts s_e and, consequently, to increase *Stage* to the value $2e+2$. Observe that any permanent increase of *Stage* in (b) *Case 1* implies that $\mathrm{NP}_{s(i)}$ rejects the current input. Hence $\overline{L(\mathrm{NP}_{s(i)})}$ is infinite. On the other hand, H is included in $L(\mathrm{NP}_{s(i)})$, because $\mathrm{NP}_{s(i)}$ rejects an input x only in case $x \notin H$ (see (b) *Case 1*). Clearly, the computation described for $\mathrm{NP}_{s(i)}$ can be carried on in nondeterministic polynomial-time. We conclude that $L(\mathrm{NP}_{s(i)})$ is a set in NP that is P-simple.

In case $i \notin$ TOT, it is easy to see that *Stage* stabilizes itself to the value $2e+1$, where s_e is the minimal string that M does not accept. From that moment on, $\mathrm{NP}_{s(i)}$ accepts all further input strings. Hence, in this case, $L(\mathrm{NP}_{s(i)})$ is co-finite. ∎

At the end of this section, we note that classes which are effectively of the second Baire category with respect to the superset topology exhibit the same kind of logical independence whice we have seen in Theorem 2.3.11. The proofs of the following results are almost identical to the ones we have seen in Proposition 2.3.9 and Proposition 2.3.10, and, therefore, we will just state the results and sketch very briefly the proofs.

Proposition 3.3.21 *Consider $\mathcal{P}$ a property of computable predicates such that if f is a computable predicate having property $\mathcal{P}$ then $f(x) = 0$ for infinitely many x. Suppose that there is a sound deductive system $\mathcal{T}$ such that, for each predicate f having the property $\mathcal{P}$, there is a Turing machine M that calculates f for which the sentence "The function computed by M has property $\mathcal{P}$" is a theorem of $\mathcal{T}$. Then the set of predicates having property $\mathcal{P}$ is effectively of Baire first category with respect to the superset topology.*

Proof. (sketch) Similar to the proof of Proposition 2.3.10. ∎

Theorem 3.3.22 *Let $\mathcal{T}$ be any sound deductive system and let $\mathcal{C}$ be a class of languages which is effectively of the second Baire category with respect to the superset topology. Suppose that any language A in $\mathcal{C}$ is (a) computable, and (b) the complement of A is infinite. Then there is a language $A \in \mathcal{C}$ such that, for each machine M computing A, the sentence "The function computed by M belongs to $\mathcal{C}$" is not a theorem of $\mathcal{T}$. Moreover, the set of such languages A is effectively of the second Baire category with respect to the superset topology.*

Proof. For any language A in $\mathcal{C}$ consider the assertion

For some machine M computing A, the sentence "The function computed by M belongs to $\mathcal{C}$" is a theorem of $\mathcal{T}$.

For any language A in $\mathcal{C}$, the above assertion is either true or false. By Proposition 3.3.21, the languages for which the assertion is true form a class which is effectively of first Baire category with respect to the superset topology. Since $\mathcal{C}$ is of the second Baire category, it follows that the language for which the assertion is false is of the second Baire category (recall that the union of two classes of first Baire category is of the first Baire category as well). ∎

Taking $C = \mathrm{NP} - \mathrm{P}$, we obtain the following result.

Corollary 3.3.23 *Let $\mathcal{T}$ be any sound deduction system and assume* $\mathrm{P} \neq \mathrm{NP}$. *Then there is a set A in* $\mathrm{NP} - \mathrm{P}$ *such that, for any Turing machine M calculating the characteristic function of A, the sentence* "The function computed by M belongs to $\mathrm{NP} - \mathrm{P}$" *is not a theorem of $\mathcal{T}$.*

Similar results hold for all the other classes that have been shown to be effectively of second Baire category.

3.4 P, NP, E—the measure-theoretical view

IN BRIEF: A type of resource-bounded measure (PF measure) is considered, in which martingales are polynomial-time computable. It is shown that $\mathrm{E} = \bigcup_{c>0} \mathrm{DTIME}[2^{cn}]$ does not have PF-measure zero. On the other hand, classes of sets for which some very weak membership property is decidable in deterministic polynomial time have PF-measure zero. The PF-measure of NP is not known, but it is shown that if it is not zero, then NP many-one completeness and NP Turing completeness are different.

We now turn to the measure-theoretical approach (the reader may find useful to review Section 1.2.2 and also Section 1.2.1 for the different notations regarding binary strings). This time, the computational requierements will be stronger: We will ask our constructions not only to be doable effectively (i.e., performed via computable functions), but to actually be doable in polynomial time. More precisely, in this section we will consider martingales that run in polynomial time, and, consequently, we will use Definition 1.2.11 with $\Gamma = \mathrm{PF}$, where PF is the class of functions computable in polynomial time.[4] We will see that this approach is useful for investigating the size of different classes of languages inside the class $\mathrm{E} = \bigcup_{c>0} \mathrm{DTIME}[2^{cn}]$.

First we need to show that E itself does not have PF-measure zero (otherwise, every class inside E would have PF-measure zero).

Theorem 3.4.1 E *does not have* PF*-measure zero.*

Proof. We show that for every martingale $d \in \mathrm{PF}$, there is a language $A \in \mathrm{E}$ such that d does not succeed on A. This implies that there is no martingale that can succeed on all languages in E, and, thus, E does not have PF-measure zero.

---

[4]Taking Γ to be the class of computable functions, as we did in the previous measure-theoretical (and, mutatis mutandi, topological) analysis, is not adequate, because it results in all classes of interest having measure zero.

Let d be a martingale in PF and let c be a positive constant so that d is computed by a machine whose running time on any input of length n is bounded by $n^c + c$. We define the function $a\colon \Sigma^* \to \Sigma^*$ by

$$a(x) = \begin{cases} x0, & \text{if } d(x0) \leq d(x), \\ x1, & \text{otherwise.} \end{cases}$$

We denote, for all $i \geq 1$, $a^i(\lambda) = \underbrace{a(a(\ldots a(\lambda)\ldots))}_{i \text{ times}}$, where λ is the empty word. Thus, $a^1(\lambda) = a(\lambda), a^2(\lambda) = a(a(\lambda))$, and so forth. Note that, for each i, the length of $a^i(\lambda)$ is i. Let A be the set defined as follows: For all $i \geq 1$,

$$s_i \in A \Leftrightarrow \text{ the last bit of } a^i(\lambda) \text{ is } 1.$$

We first show that $A \in \mathrm{E}$. Recall that $|s_i| = \lfloor \log i \rfloor$, for all $i \geq 1$. To check if s_i is in A or not, we need to calculate $a^i(\lambda)$, which involves i calculations of the martingale d on strings of length at most $i - 1$. These calculations take time bounded by

$$\begin{aligned} i \cdot ((i-1)^c + c) \leq i^{c+1} \text{ (for } i \text{ sufficiently large)} &= 2^{(c+1)\log i} \\ &\leq 2^{(|s_i|+1)(c+1)} < 2^{(c+2)|s_i|} \text{ (for } |s_i| \text{ sufficiently large).} \end{aligned}$$

Therefore $A \in \mathrm{DTIME}[2^{(c+2)n}] \subseteq \mathrm{E}$.

Next we show that d does not succeed on A. Note that, by the definition of the function a, $d(A(s_1)) \leq d(\lambda)$, and

$$d(A(s_1)A(s_2)\ldots A(s_n)) \leq d(A(s_1)A(s_2)\ldots A(s_{n-1})),$$

for all $n \geq 2$. Thus, for all $n \geq 1$, $d(A(s_1)A(s_2)\ldots A(s_n)) \leq d(\lambda)$, which implies that d does not succeed on A. ∎

A basic result of complexity theory (also, perhaps, the best known one, even among people with otherwise feeble acquaintance of complexity) is that $\mathrm{P} \subsetneq \mathrm{E}$. We want to investigate the amplitude of this separation. We will demonstrate that the PF-measure of P is zero, which, together with the fact that E does not have measure zero, means that the class of problems solvable in polynomial time represents just a very small part of E. Moreover, we will consider classes $\mathcal{C}$ obtained through various relaxations of P, and we will show that these classes also have PF-measure zero.

We first establish a lemma which is useful in showing that different classes have PF-measure zero. It provides conditions under which a countable union of PF-measure zero sets have PF-measure zero.

Lemma 3.4.2 *Let $C = \bigcup_{i \in \mathbb{N}} C_i$, with $C_i \subseteq \Sigma^\infty$, for all $i \in \mathbb{N}$. Let $d\colon \mathbb{N} \times \Sigma^* \to \Sigma^*$ be a martingale system such that: (a) For all $i \in \mathbb{N}$, d_i succeeds on C_i, and (b) there is a constant $c > 0$ and a Turing machine M such that, for all $i \in \mathbb{N}$ and for all inputs x of length $n \geq i$, M computes $d_i(x)$ in time bounded by $n^c(\log n)^{ci}$, Then C has PF-measure zero.*

Proof. For each $i \in \mathbb{N}$, let $\widetilde{C}_{2^{2^i}} = C_i$ and $\widetilde{d}_{2^{2^i}} = d_i$. For $j \in \mathbb{N}$ which is not of the form 2^{2^l}, we define $\widetilde{C}_j = \emptyset$, and we let $\widetilde{d}_j$ be the constant martingale that assigns 1 to all inputs. Note that $C = \bigcup \widetilde{C}_i$ and that, for all $i \geq 2$, $\widetilde{d}_i$ can be calculated in time $n^c (\log n)^{c \log\log i}$, for all inputs of length $n \geq \log\log i$. Also, it is immediate to check that the family of functions $(\widetilde{d}_i)_{i\in\mathbb{N}}$ forms a martingale system, and that each $\widetilde{d}_i$ succeeds on $\widetilde{C}_i$. Observe that, for $i \geq 2$, $(\log n)^{\log\log i} \leq n$, for all $n \geq i$. Therefore, for all $i \in \mathbb{N}$, $\widetilde{d}_i$ can be calculated in time bounded by n^{2c}, for all inputs of length $n \geq i$. Next we proceed like in the proof of Proposition 1.2.12. For each $i \in \mathbb{N}$, let $\delta_i(x)$ be defined such that, for each $x \neq \lambda$, $\widetilde{d}_i(x) = \delta_i(x) \cdot \widetilde{d}_i(\text{pred}(x))$, where $\text{pred}(x)$ is the prefix of x of length $|x| - 1$. Also let

$$\hat{d}_i(x) = \begin{cases} 2^{-i} & \text{if } |x| < i, \\ \delta_i(x) \cdot \hat{d}_i(\text{pred}(x)) & \text{if } |x| \geq i. \end{cases}$$

One can check that, for all i, (a) $\hat{d}_i$ is a martingale, (b) $\hat{d}_i$ succeeds on $\widetilde{C}_i$, and (c) $\hat{d}_i(x)$ can be computed in time bounded by n^{2c+1}, for all inputs x of length $n \geq i$. Let $\hat{d}(x)$ be defined by $\hat{d}(x) = \sum_{i=0}^{\infty} \hat{d}_i(x)$. Then

$$\begin{aligned} \hat{d}(x) &= \sum_{i=0}^{\infty} \hat{d}_i(x) \\ &= \sum_{i=0}^{|x|} \hat{d}_i(x) + \sum_{i=|x|+1}^{\infty} 2^{-i} \\ &= \sum_{i=0}^{|x|} \hat{d}_i(x) + 2^{-|x|}. \end{aligned}$$

Then $\hat{d}$ is a martingale and, from the above expression, it can be seen that $\hat{d}$ is in PF. Also, for each $i \in \mathbb{N}$, for x with $|x| \geq i$, $d(x) > \hat{d}_i(x)$. Since $\hat{d}_i$ succeeds on $\widetilde{C}_i$, it follows that $\hat{d}$ succeeds on $\widetilde{C}_i$ as well. Since this holds for all $i \in \mathbb{N}$, we conclude that $\hat{d}$ succeeds on C, and, thus, C has PF-measure zero. ∎

Theorem 3.4.3 P *has* PF*-measure zero.*

Proof. Let $(M_i)_{i\in\mathbb{N}}$ be an effective enumeration[5] of polynomial-time machine Turing machines accepting languages such that, for all i, the running time of M_i is bounded by n^i for all inputs of length $n \geq i$. Let A_i be the language accepted by M_i. Then $\text{P} = \bigcup_{i\in\mathbb{N}}\{A_i\}$. For each i, we define a martingale d_i that succeeds on $\{A_i\}$ as follows: $d_i(\lambda) = 1$, and, for every $x \in \Sigma^* - \{\lambda\}$,

$$d_i(x0) = \begin{cases} 2d_i(x) & \text{if } s_{|x|+1} \notin A_i, \\ 0 & \text{if } s_{|x|+1} \in A_i, \end{cases}$$

[5]This means that there is one Turing machine such that, for all $i \in \mathbb{N}$ and for all $x \in \Sigma^*$, $M(i, x) = M_i(x)$.

and

$$d_i(x1) = \begin{cases} 0 & \text{if } s_{|x|+1} \in A_i, \\ 2d_i(x) & \text{if } s_{|x|+1} \notin A_i. \end{cases}$$

It follows that, for every n, $d_i(A_i(s_1)A_i(s_2)\ldots A_i(s_n)) = 2^n$, and, thus, d_i succeeds on $\{A_i\}$. The value of d_i on input x of length n can be calculated in time bounded by $|s_{|x|}|^i \leq (\log n)^i$. By Lemma 3.4.2, it follows that P has PF-measure zero. ∎

The next result is a significant strenghtening of Theorem 3.4.3. P is the class of languages A for which membership in the language (i.e., answering the question "Is x in A?") is solvable in polynomial time. Researchers have considered generalizations of P obtained by relaxing the membership question. We will study to what extent Theorem 3.4.3 extends to these generalizations of P.

We list below the most important such generalizations of P with references to the papers where they have been introduced (some of these concepts have generated extensive literature, but it is beyond our scope to review it here).

A set $A \subseteq \Sigma^*$ is P-*selective* [Sel82] if there exists $f \in$ PF such that, for all pair (x_1, x_2), $f(x_1, x_2) \in \{x_1, x_2\}$ and $(x_1 \in A) \vee (x_2 \in A) \Rightarrow f(x_1, x_2) \in A$.

A set $A \subseteq \Sigma^*$ is P-*multiselective* [HJRW97] if there exists $f \in$ PF and a natural number constant $q \geq 1$ such that, for all q-tuple $(x_1, \ldots, x_q)$, $f(x_1, \ldots, x_q) \in \{x_1, \ldots, x_q\}$ and $(x_1 \in A) \vee \ldots \vee (x_q \in A) \Rightarrow f(x_1, \ldots, x_q) \in A$.

A set $A \subseteq \Sigma^*$ is *cheatable* [Bei87] if there exists $f \in$ PF and a natural number constant $q \geq 1$ such that, for all q-tuple $(x_1, \ldots, x_q)$, $f(x_1, \ldots, x_q)$ outputs a set $D \subseteq \Sigma^q$ of size q that contains $A(x_1) \ldots A(x_q)$.

A set $A \subseteq \Sigma^*$ is *easily countable* [HN93] if there exists $f \in$ PF and a natural number constant $q \geq 1$ such that, for all q-tuple $(x_1, \ldots, x_q)$, $f(x_1, \ldots, x_q) \in \{0, \ldots, q\}$ and $f(x_1, \ldots, x_q)$ is not equal to the cardinality of $A \cap \{x_1, \ldots, x_q\}$.

A set $A \subseteq \Sigma^*$ is *easily approximable* [KS91, BKS94] if there exists $f \in$ PF and a natural number constant $q \geq 1$ such that, for all q-tuple $(x_1, \ldots, x_q)$, $f(x_1, \ldots, x_q)$ outputs a q-bit vector $(y_1, \ldots, y_q)$ for which at least half the numbers y_i are equal to $A(x_i)$.

A set A is *near testable* [GHJY91] if there is $f \in$ PF such that, for each $\ell \in \mathbb{N}$, $f(s_\ell)$ computes the truth value of the predicate $(s_\ell \in A) \oplus (s_{\ell+1} \in A)$, where $\oplus$ represents the "exclusive or."

A set A is *nearly near testable* [HH91] if there is $f \in$ PF such that, for each $l \in \mathbb{N}$, $f(s_\ell)$ outputs the truth value of one of the following two predicates: (a) $s_\ell \in A$ or (b) $(s_\ell \in A) \oplus (s_{\ell+1} \in A)$.

A set A is *locally self-reducible* [BS95] if there is a constant $q \geq 1$ and a polynomial-time deterministic oracle machine M that recognizes A and, for all natural number $i \geq 1$, M on input s_i queries only elements of the set $\{s_{i-1}, s_{i-2}, \ldots, s_{i-q}\}$.

A set A is P-*approximable* [BKS94, Ogi94] if there exists some constant q such that, for all q-tuple $(x_1, \ldots, x_q)$, one can exclude in polynomial time one possibility of how the characteristic function of A is defined on $x_1, \ldots, x_q$.

The properties used in the above definitions are called collectively *polynomial-time weak membership properties*. We consider another type of sets, defined in the same spirit.

Definition 3.4.4 *(*P-quasi-approximable sets*) A set A is* P-quasi-approximable *if there exists a constant q and a polynomial-time algorithm M which takes as inputs q-tuples of strings and outputs either a q-long binary string or "I don't know" (denoted '?') and which satisfies the following property: For infinitely many q-tuples $(x_1, \ldots, x_q)$, with x_{i+1} being the lexicographical successor of x_i for $i = 1, \ldots, q-1$, M outputs a q-long binary string, and whenever this happens the q-long binary ouput string is different from $A(x_1)\ldots A(x_q)$.*

All the above types of sets are obtained by stating that some property of the characteristic function can be decided in polynomial time. It is easy to check that P-quasi-approximability is at least as weak as any of the other properties (e.g., if a set is P-selective then it is P-quasi-approximable, etc.), and, thus, the class of P-quasi-approximable sets includes the class of sets having any of the other polynomial-time weak membership properties. The next theorem shows that the class of sets that are P-isomorphic to some P-quasi-approximable set has PF - measure zero.

Definition 3.4.5 *(*P-isomorphism*) Two sets $A, B \subseteq \Sigma^*$ are* P-isomorphic *if there is a bijection $h\colon \Sigma^* \to \Sigma^*$ such that (a) both h and its inverse h^{-1} are in* PF, *and (b) for all $x \in \Sigma^*$, $x \in A$ if and only if $h(x) \in B$.*

Theorem 3.4.6 *The closure under* P*-isomorphism of the class of* P*-quasi-approximable sets has* PF*-measure zero.*

Proof. Let $(h_i, h_j)_{i,j \in \mathbb{N}}$ be an enumeration of all pairs of polynomial-time functions $h_i, h_j : \Sigma^* \to \Sigma^*$, and, for all natural numbers $q \geq 1$, let $(f_i^q)_{i \in \mathbb{N}}$ be an enumeration of polynomial-time computable functions $f_i^q\colon (\{0,1\}^*)^q \to \{0,1\}^q \cup \{?\}$. We will assume without loss of generality that, for all i, h_i and f_i are computable in time bounded by $n^{i^{(1/3)}}$, for all inputs of length $n \geq i$. The closure under P-isomorphism of the class of P-quasi-approximable sets is equal to the union of the classes $(A_t)_{t \in \mathbb{N}}$ defined as follows:

Let us consider a bijection $\langle \cdot,\cdot,\cdot,\cdot \rangle\colon \mathbb{N}^4 \to \mathbb{N}$. For $t = \langle i,j,m,q \rangle$, in case h_i is a bijection and h_j is the inverse of h_i, we let A_t be the class of sets A that are P-isomorphic via h_i to some set that is P-quasi-approximable via q and f_m^q; otherwise, we let A_t be the empty set.

By Lemma 3.4.2, it is enough if there is a martingale system $(d_t)_{t \in \mathbb{N}}$ such that each class A_t has PF-measure zero via d_t (i.e., d_t succeeds on A_t) and d_t runs in time $O(n(\log n)^t)$. It is thus sufficient to design such a martingale system $(d_t)_{t \in \mathbb{N}}$. We fix i, j, m, q and $t = \langle i,j,m,q \rangle$, and write more simply h, h^{-1}, f, d, A instead of $h_i, h_j, f_m^q, d_t, A_t$. We can take the bijection such that $t \geq \max(i,j,m,q)$. In what follows, we assume that h_i is a bijection and h_j is its inverse, otherwise it is clear that d succeeds on A. The martingale d will take advantage of the fact that from

Case 2.2. Both $(s_n, \hat{d}(s_n), x_n)$ or $(s_n, \hat{d}(s_n), 1 - x_n)$ are not in LIST(x'). Then $d(x) = d(x')$, LIST(x) = LIST(x') and x is marked *active*.

The following Claims show that d achieves the purported goals.

Claim 3.4.7 *$d(x)$ can be computed in time $O(n \cdot (\log n)^t)$, where $n = |x|$.*

Proof. The computation of $d(x)$ involves an autonomous part and the computation of $d(y)$ for all strict prefixes y of x. Since there are $|x|$ such prefixes, we only have to show that the autonomous part can be computed in $O(\log n)$. If Case 1 is entered, we have to compute $h(s_{|x|})$, check the TEST, and, if Case 1.2 occurs, find $z_1, \ldots, z_q$, insert q triplets in LIST(x) and do some easy computations. One can check that these operations take time $O((\log n)^t)$. The operations required by Case 2 take $O(t)$ time, since there are a constant number (namely $q \leq t$) of elements in LIST(x'). ∎

Claim 3.4.8 *$d(\cdot)$ is a martingale.*

Proof. Let $x = x_1 x_2 \ldots x_n$, $x' = x_1 x_2 \ldots x_{n-1}$ and $x'' = x'(1 - x_n)$, where $x_i \in \{0, 1\}$, for all $i \in \{1, \ldots, n\}$. We show that

$$d(x') = \frac{d(x) + d(x'')}{2}, \tag{3.3}$$

for all $x \in \{0, 1\}^*$. We focus on the computation of $d(x)$. If x' is marked *inactive*, then $d(x'')$ will also find x' to be *inactive* and the TEST evaluates the same in the computation of $d(x)$ and $d(x'')$. Now, relation 3.3 can be easily checked.

Suppose next that x' is marked *active* and let

$$\text{LIST}(x') = ((s_{i_1}, \hat{d}(s_{i_1}), b_1), \ldots, (s_{i_q}, \hat{d}(s_{i_q}), b_q)).$$

It is clear that the same case among Case 2.1 and Case 2.2 applies to both $d(x)$ and $d(x'')$. If Case 2.2 applies to both $d(x)$ and $d(x'')$, relation 3.3 is checked immediately. Suppose that Case 2.1 applies to both $d(x)$ and $d(x'')$, with $(s_n, \hat{d}(s_n), x_n)$ in LIST(x') (the other situation, $(s_n, \hat{d}(s_n), 1 - x_n)$ in LIST(x'), is symmetric). It follows that there is some p such that $n = i_p$ and $x_n = b_p$.

We claim that $p \neq 1$ (i.e., $(s_n, \hat{d}(n), b_n)$ is not the first triplet in LIST(x')). For the sake of obtaining a contradiction, assume that $p = 1$. This means that $i_1 = n$, which implies that x and x'', both being strings of length n, are triggering the current redistribution task. This contradicts the fact that x' is marked inactive. Therefore, $p \neq 1$. It is clear that for prefixes $x(1 : r)$ of x, with $i_{p-1} < r < i_p$ (if there are any), $d(x(1 : r))$ is computed according to Case 2.2, and, thus, $d(x(1 : r)) = d(x(1 : i_{p-1}))$. Now, either $x' = x_{i_{p-1}}$ or x' is a $x(1 : r)$ as above. In both cases

$$d(x') = d(x'(1 : i_{p-1})) = (1/(2^q - 1)) \cdot (\sum_{h=p-1}^{q-1} 2^h) \cdot d(x(1 : (i_1 - 1))).$$

Since

$$d(x) = (1/(2^q - 1)) \cdot (\sum_{h=p}^{q-1} 2^h) \cdot d(x(1:(i_1 - 1)))$$

and

$$d(x'') = (2^q/(2^q - 1)) \cdot d(x(1:(i_1 - 1))),$$

relation 3.3 is verified. ∎

Claim 3.4.9 *d succeeds on A.*

Proof. We inductively define the infinite sequence of integers $(\ell_i)_{i \in \mathbb{N}}$ as follows. Let $\ell_0 = 1$ and ℓ_{i+1} = "the smallest value $\ell > \ell_i$ that is selected as $\ell(x)$ in Case 1.2 in the computation of d on some $x \in \{0,1\}^*$." By the properties of f, it is clear that ℓ_i is defined for all i. For a value ℓ in the above sequence, let $m_\ell = \min\{\mathrm{pos}(h^{-1}(s_\ell)), \ldots, \mathrm{pos}(h^{-1}(s_{\ell+q-1}))\}$ and $M_\ell = \max\{\mathrm{pos}(h^{-1}(s_\ell)), \ldots, \mathrm{pos}(h^{-1}(s_{\ell+q-1}))\}$. Since for all sets B in A

$$f(s_\ell, \ldots, s_{\ell+q-1}) \neq B(h^{-1}(s_\ell)) \ldots B(h^{-1}(s_{\ell+q-1})),$$

it follows that $d(B(1:M_\ell)) = (1 + 1/(2^q - 1))d(B(1:m_\ell - 1))$. For any n, let T_n be such that $(1 + 1/(2^q - 1))^{T_n} \geq n$, we conclude that for all sets B in A, $d(B(1:M_{\ell_{T_n}})) = (1 + 1/(2^q - 1))^{T_n} d(\lambda) \geq n$, because $d(B(1:m_{\ell_{i-1}} - 1)) = d(B(1:M_{\ell_i}))$. ∎

Claims 3.4.7, 3.4.8, and 3.4.9 show that the requirements of Lemma 3.4.2 are satisfied for the clas of P-quasi-approximable set,and, therefore, this class has PF-measure zero. ∎

Since the classes with weak membership properties mentioned in this section, as well as the class of sets that are not P-bi-immune[6], are all included in the class of P-quasi-approximable sets, Theorem 3.4.6 has the following immediate corollary.

Corollary 3.4.10 *The following classes have* PF*-measure zero:*

(1) *the class of* P*-selective sets,*

(2) *the class of* P*-multiselective sets,*

(3) *the class of cheatable sets,*

(4) *the class of easily countable sets,*

(5) *the class of easily approximable sets,*

(6) *the class of near-testable sets,*

(7) *the class of nearly near-testable sets,*

(8) *the class of locally self-reducible sets,*

(9) *the class of* P*-approximable sets,*

(10) *the class of sets that are not* P*-bi-immune.*

---

[6] A set A is P-bi-immune if neither A nor its complement contain an infinite subset in P.

The class of sets that are in E and that are P-quasi-approximable also have PF-measure zero, simply because it is included in the class of P-quasi-approximable sets. Keeping in mind that E does not have PF-measue zero, it means that "most" (in the sense of PF-measure) sets in E are not P-quasi-approximable. More precisely the PF-measure of the class of sets which are in E and which are not P-quasi-approximable is not zero. In follows that "most" (again, in the sense of PF-measure) sets in E do not have any of the polynomial-time weak membership properties.

We show next that Theorem 3.4.6 does not extend to the closure of the class of P-quasi-approximable sets under many-one polynomial-time equivalences.

Definition 3.4.11 *Two sets* $A, B \subseteq \Sigma^*$ *are many-one polynomial-time equivalent if there are two functions* $f, g\colon \Sigma^* \to \Sigma^*$ *such that (a)* f *and* g *are computable in polynomial-time, (b)* $x \in A \Leftrightarrow f(x) \in B$*, and (c)* $x \in B \Leftrightarrow g(x) \in A$*.*

Theorem 3.4.12 *The class of sets that are many-one equivalent to some* P*-quasi-approximable set does not have* PF*-measure zero.*

Proof. Let

$$\mathcal{A} = \{A \in \mathrm{E} \mid A \text{ has infinitely many strings of the form } 0^n \text{ for some } n \in \mathbb{N}\}.$$

Observe that if A is a set in $\mathrm{E} - \mathcal{A}$, then A is not P-bi-immune, because there is an n_0 such that the set $D = \{0^n \mid n \geq n_0\}$ is infinite, in P, and included in the complement of A. Hence, by Corollary 3.4.10 (10), the class $\mathrm{E} - \mathcal{A}$ has PF-measure zero. Since the union of two PF-measure zero sets has PF-measure zero (this follows from Lemma 3.4.2) and since E does not have PF-measure zero, it follows that $\mathcal{A}$ does not have PF-measure zero.

We will show that any set A in $\mathcal{A}$ is polynomial-time equivalent to some set B which is not P-bi-immune (thus, B is P-quasi-approximable). Let $\{S_1 < S_2 < \ldots\}$ be the lexicographical ordering of $\Sigma^* - \{0^{2^n} \mid n \in \mathbb{N}\}$. For each $A \in \mathcal{A}$, we define B by: (1) $S_i \in B \Leftrightarrow s_i \in A$, for all $i \in \mathbb{N}$, and (2) $0^{2^n} \in B \Leftrightarrow 0^n \in A$, for all $n \in \mathbb{N}$. We take the function $f\colon \Sigma^* \to \Sigma^*$ to be defined by $f(s_i) = S_i$, for all $i \in \mathbb{N}$. We take the function $g\colon \Sigma^* \to \Sigma^*$ to be defined by $g(S_i) = s_i$, for all $i \in \mathbb{N}$, and, for all $n \in \mathbb{N}$, $g(0^{2^n}) = 0^n$. Clearly, f is computable in polynomial time, and, for all $x \in \Sigma^*$, $x \in A \Leftrightarrow f(x) \in B$. Also, g is computable in polynomial time, and, for all $x \in \Sigma^*$, $x \in B \Leftrightarrow g(x) \in A$. Consequently, A is one-one polynomial-time equivalent to B. Also B contains the set $\{0^{2^n} \mid 0^n \in A\}$, which is infinite and in P. Thus, B is not P-bi-immune. ∎

Thus the PF-measurability shows a big difference between exponential-time computation and polynomial-time computation, even if the latter is used to decide very weak properties of sets. The next natural question is: What is the PF-measure of NP? The problem is open and probably beyond the current state of affairs in complexity theory. Indeed, if NP has PF-measure zero then $\mathrm{NP} \subsetneq \bigcup_{k \in \mathbb{N}} \mathrm{DTIME}[2^{n^k}]$, and, if it does not have PF-measure zero, then $\mathrm{P} \subsetneq \mathrm{NP}$.

Given our belief in the deep separation between polynomial-time computation and nondeterministic polynomial-time computation, the hypothesis that NP does not have PF-measure zero looks plausible. It is interesting to see some consequences of this hypothesis which are not known to be implied by the weaker hypothesis $\mathrm{P} \subsetneq \mathrm{NP}$. The next theorem presents a very important such consequence: It shows that $\leq_m^p$ NP-completeness and $\leq_T^p$ NP-completeness are different.

Theorem 3.4.13 *Assume that* NP *does not have* PF*-measure zero. Then there is a set which is* NP*-complete under* $\leq_T^p$ *reduction and it is not* NP*-complete under* $\leq_m^p$ *reduction.*

Proof. We introduce first some notation. For any set $A \subset \Sigma^*$, we denote $A_0 = \{x \mid 0x \in A\}$, and $A_1 = \{x \mid 1x \in A\}$. SAT is the set of boolean formulas encoded in some standard way as binary strings. As it is well-known, SAT is $\leq_m^p$-complete for NP. If B and C are two sets in Σ^*, the tagged union of B and C is a set denoted $B \oplus C$ defined by

$$B \oplus C = \{0x \mid x \in B\} \cup \{1x \mid x \in C\}.$$

Claim 3.4.14 *Let* $A \in \mathrm{NP}$. *The set* $A_0 \oplus ((A_0 \cap \mathrm{SAT}) \oplus (A_0 \cup \mathrm{SAT}))$ *is* NP*-complete under* $\leq_T^p$ *reduction.*

Proof. Since A is in NP, A_0 is in NP as well. NP is closed under $\oplus$, $\cap$, and $\cup$; whence A is in NP. To show $\leq_T^p$-completeness, it is enough to demonstrate that $\mathrm{SAT} \leq_T^p A_0 \oplus ((A_0 \cap \mathrm{SAT}) \oplus (A_0 \cup \mathrm{SAT}))$ (because SAT is NP-complete). Observe that, for any $x \in \Sigma^*$,

$$x \in \mathrm{SAT} \iff (x \in A_0 \text{ and } x \in A_0 \cap \mathrm{SAT}) \text{ or } (x \notin A_0 \text{ and } x \in A_0 \cup \mathrm{SAT}).$$

Therefore,

$$x \in \mathrm{SAT} \iff (0x \in A \text{ and } 10x \in A) \text{ or } (0x \notin A \text{ and } 11x \in A).$$

Thus, with two easy-computable queries to A (of which the second depends on the answer to the first one), we can determine if $x \in \mathrm{SAT}$ or not. ∎

Claim 3.4.15 *The class* $\mathcal{A} = \{A \mid A_1 \leq_m^p A_0 \oplus ((A_0 \cap \mathrm{SAT}) \oplus (A_0 \cup \mathrm{SAT}))\}$ *has* PF*-measure zero.*

This will end the proof because from the assumption that NP does not have PF-measure zero, it follows that there is a set $A \in \mathrm{NP}$ such that $A \notin \mathcal{A}$. Consequently, $A_0 \oplus ((A_0 \cap \mathrm{SAT}) \oplus (A_0 \cup \mathrm{SAT}))$ is not $\leq_m^p$ complete for NP because A_1 is not $\leq_m^p$ reducible to it and A_1, as one can easily see, is in NP.

Proof of Claim 3.4.15. The proof is based on the fact that if $A_1 \leq_m^p A_0 \oplus ((A_0 \cap \mathrm{SAT}) \oplus (A_0 \cup \mathrm{SAT}))$, then there is some dependency between A_0 and A_1

$$h_{i,11}(x) = \begin{cases} y, & \text{if } x \in B_{11} \text{ and } h_i(x) = 11y \\ x_0, & \text{otherwise} \end{cases}$$

We note that $A_1 \cap B_0 \leq_m^p A_0$ via $h_{i,0}$ because

$$\begin{aligned} x \in A_1 \cap B_0 &\leftrightarrow (h_i(x) \in A_0 \oplus ((A_0 \cap \text{SAT}) \oplus (A_0 \cup \text{SAT})) \\ &\qquad \text{and } h_i(x) \text{ starts with } 0 \\ &\leftrightarrow h_{i,0}(x) \in A_0. \end{aligned}$$

Similarly, $A_1 \cap B_{10} \leq_m^p A_0 \cap \text{SAT}$ via $h_{i,10}$, and $A_1 \cap B_{11} \leq_m^p A_0 \cup \text{SAT}$ via $h_{i,11}$. For any $z \in \{0, 10, 11\}$ and for any x, $|h_{i,z}(x)| \geq |h_i(x)| - 2$. Since $B_0 \cup B_{10} \cup B_{11} = \Sigma^*$, by our observation above, there exists $z \in \{0, 10, 11\}$ such that for an infinity of $z \in B_z$, $|h_{i,z}(x)| \geq |x| - 2$. Depending on z, we have three cases.

Case 2.1. (z = 0) There are an infinity of $x \in B_0$ such that $|h_{i,0}(x)| \geq |x| - 2$. Since $A_1 \cap B_0 \leq_m^p A_0$ via $h_{i,0}$, for any $x \in B_0$, $x \in A_1 \iff h_{i,0}(x) \in A_0$, and thus there are an infinity of x such that $A(1x) = A(0h_{i,0}(x))$. Therefore, for such an x,

$$(A(1x), A(0h_{i,0}(x))) \neq (1,0)$$

The function f_i is defined by the following algorithm. On input y of length n, the algorithm looks for an x that is lexicographically larger than s_n such that $h_i(x)$ starts with 0 (i.e., $x \in B_0$) and $|h_{i,0}(x)| \geq |s_n|$. If no such x is found within $n^2(\log n)^i$ steps, the output is '?'. Otherwise the output is $(1x, 1), (0h_{i,0}(x), 0))$.

Note that, since we are in Case 2.1, for infinitely many y an x as above exists among the strings of length at most $|s_n| + 2 = (\lfloor \log n \rfloor + 2)$. Since checking an x takes time $O(|x|^i) = O((\log n)^i)$, the function will output a pair $((1x, 1), (0h_{i,0}(x), 0))$ infinitely many times. It follows that f_i satisfies conditions (1)–(4).

Case 2.2. There are an infinity of $x \in B_{10}$ such that $|h_{i,10}(x)| \geq |x| - 2$. Since $A_1 \cap B_{10} \leq_m^p A_0 \cap \text{SAT}$ via $h_{i,10}$, for no $x \in B_{10}$ is it possible that $x \in A_1$ and $h_{i,10}(x) \notin A_0$. Therefore, for such an x,

$$(A(1x), A(0h_{i,0}(x))) \neq (1,0).$$

Similarly to Case 2.1, the function f_i is defined by the following algorithm. On input y of length n, the algorithm looks for an x that is lexicographically larger than s_n such that $h_i(x)$ starts with 10 (i.e., $x \in B_{10}$) and $|h_{i,10}(x)| \geq |s_n|$. If no such x is found within $n^2(\log n)^i$ steps, the output is '?'. Otherwise the output is $(1x, 1), (0h_{i,0}(x), 0))$. As in Case 2.1, the function f_i satisfies (1)–(4).

Case 2.3. There are an infinity of $x \in B_{11}$ such that $|h_{i,11}(x)| \geq |x| - 2$. Since $A_1 \cap B_{11} \leq_m^p A_0 \cup \text{SAT}$ via $h_{i,11}$, for no $x \in B_{11}$ is it possible that $x \notin A_1$ and $h_{i,11}(x) \in A_0$. Therefore, for such an x,

$$(A(1x), A(0h_{i,0}(x))) \neq (0,1).$$

Similarly to Case 2.1, the function f_i is defined by the following algorithm. On input y of length n, the algorithm looks for an x that is lexicographically larger than s_n such that $h_i(x)$ starts with 11 (i.e., $x \in B_{11}$) and $|h_{i,11}(x)| \geq |s_n|$. If no such x is found within $n^2(\log n)^i$ steps, the output is '?'. Otherwise the output is $(1x, 0), (0h_{i,0}(x), 1))$. As in Case 2.1, the function f_i satisfies (1)–(4).

This ends the proof of Claim 3.4.16 and of Theorem 3.4.13. ∎

3.5 Strong relativized separation of P and NP

IN BRIEF: For almost all oracle sets A, there is a set L in NP^A with the following property: Any deterministic polynomial-time machine with access to A, that attempts to determine if a string x is in L, is correct on only half the inputs x of length at most n, for all sufficiently large n.

In some precise sense, it is *provable* that nondeterministic polynomial-time computation can do tasks that deterministic polynomial-time computation can not.

The catch is that we allow both type of computations to have access to an additional set, called the *oracle set*, which is viewed as a data base that can be queried. We say that the computations are done relative to an oracle set (for details, see Section 3.1). Thus, questions of the type "Is x in the oracle set?" receive an immediate answer. The oracle set may contain a lot of information that is available for free, and, consequently, computations relative to an oracle set can be much more powerful than computations that are done "from scratch." For example, any computably enumerable set can be solved in deterministic polynomial time relative to the oracle set that encodes the HALTING PROBLEM. Nevertheless, in spite of the distortion introduced by oracles, the question of what can be done relative to various oracle sets is a viable topic worthy of scientific investigation.

For a set A, we denote by P^A the class of languages that can be solved in deterministic polynomial time relative to A. We denote by NP^A the class of languages that can be solved in nondeterministic polynomial time relative to A. There are sets A such that $\mathrm{P}^A = \mathrm{NP}^A$. For example, if A is a set that is $\leq_T^p$-complete for PSPACE, then $\mathrm{PSPACE} \subseteq \mathrm{P}^A \subseteq \mathrm{NP}^A \subseteq \mathrm{PSPACE}^A = \mathrm{PSPACE}$, and, thus, $\mathrm{P}^A = \mathrm{NP}^A$. There also exist sets B such that $\mathrm{P}^B \subsetneqq \mathrm{NP}^B$ (this is the result that we have referred to in the first paragraph). In this section we will prove a result that strengthens quantitatively the relativized separation of P from NP in two quite different directions. Given the (apparently) conflicting views resulting from different oracle sets, it is natural to ponder which of the relations $\mathrm{P}^A = \mathrm{NP}^A$ and $\mathrm{P}^A \subsetneqq \mathrm{NP}^A$ happens for "most" oracles, i.e., what happens when A is chosen at random. We will see that the answer is that for "most" sets A, $\mathrm{P}^A \subsetneqq \mathrm{NP}^A$. This is the first direction of the generalization which regards the size of the set of oracles relative to which we have the separation of P and NP. The second direction, refers to a quantitative aspect of the separation itself. We will show that for "most" oracles A, there is a language L in NP^A such that no deterministic polynomial-time algorithm can answer correctly the question "Is x in L?" but for, roughly speaking, half of the inputs x. Since the answer is either YES or NO, it can not be worse than this. Such a separation is called a separation with *balanced immunity.*

Definition 3.5.1 (P-balanced immunity) *Let* $A \subseteq \Sigma^*$. *We say* A *is* P-balanced immune *if both* A *and its complement are infinite and each infinite set* $B \in$ P *satisfies the property that* $\lim_{n\to\infty} \frac{\|(B\cap A)^{\leq n}\|}{\|B^{\leq n}\|}$ *is defined and equals* 1/2. *We say a class* C *is* P *balanced immune if there is a set* $A \in C$ *that is* P *balanced immune.*

To define what we mean by "most oracle sets," we utilize the apparatus of measure theory introduced in Section 1.2.2. We recall that a set $A \subseteq \Sigma^*$ is identified with the infinite binary sequence $A(s_1)A(s_2)\ldots A(s_n)\ldots \in \Sigma^\infty$. Such a sequence is also identified with a real number in the interval $[0,1]$ by associating to the above infinite binary sequence, the real number having the binary representation $0.A(s_1)A(s_2)\ldots A(s_n)\ldots$. Via this representation, Σ^∞ can be viewed as the interval $[0,1]$. Therefore, a class of sets of binary strings represents a subset of $[0,1]$. Hence, such classes can be measured using the Lebesgue measure on $[0,1]$. This approach is very natural because the Lebesgue measure of the entire interval $[0,1]$ is one, and, thus, the measure is a probability measure on Σ^∞, which we denote Prob. Also, recall from Section 1.2.2, that the Lebesgue measure can be constructed starting with the basic intervals $(B_x)_{x\in\Sigma^*}$, where for $x = x_1x_2\ldots x_n$, $x_i \in \{0,1\}$, $B_x = [0.x_1x_2\ldots x_n, 0.x_1x_2\ldots x_n11\ldots]$. A natural way to define a random set A is to flip a fair coin infinitely main times and to use the i-th flip as the value of $A(s_i)$ (by considering, say, that head represents 0 and tail represents 1). Intuitively, the probability that the random set A belongs to B_x is $2^{-|x|}$. Since the length of the interval B_x is $2^{-|x|}$, we see that the Lebesgue measure on [0,1] corresponds to the above method method of building random sets of strings. Our previous informal statements asserting that some property holds for "most" oracles, means, formally, that the set of oracle sets for which that property is true has Lebesgue measure one. Or, in other words, if we build a set A by flipping a fair coin for each $x \in \Sigma^*$ to decide whether to put x in A or not, with probability one we obtain a set for which the property is true. The following terminology is very common. If $P()$ is a property that depends on an oracle set A and if the set $\{A \mid P(A) \text{ holds true}\}$ has measure one, we say that property P holds relative to a random oracle.

Theorem 3.5.2 INFORMAL STATEMENT: *For almost all oracle sets* A, *there is a set in* NP^A *that splits into half all infinite sets in* P^A *at all sufficiently large lengths.*

FORMAL STATEMENT: NP *is* P *balanced immune relative to a random oracle.*[7]

Proof. For each oracle set A, we build a language $T(A)$ and we show that (a) for all A, $T(A) \in \mathrm{NP}^A$, and (b) for a set of oracle sets A having measure one, $T(A)$ is P^A-balanced immune. In the construction, we split the characteristic sequence

[7]The notion of a P balanced immune can be relativized in the obvious way, i.e., by letting the set B in the Definition 3.5.1 be in relativized P.

of the oracle set into disjoint blocks that we attach to each x. Namely, for any x in Σ^*, let

$$\text{Block}(x) = \{y \mid (\exists u \in \Sigma^*)\,[y = xu \text{ and } |y| = 9|x| \text{ and } y \text{ is among the first } \lfloor (\ln 2)2^{8|x|} \rfloor \text{strings of length } 9|x| \text{ of the form } y = xu\,]\}.$$

For $y = xu$ in Block(x), we define $\xi^A(y) = A(xu1)A(xu10)\ldots A(xu10^{|u|-1})$. The language $T(A)$ is defined by

$$T(A) = \{x \mid (\exists y \in \text{Block}(x))[\xi^A(y) = \text{EIGHT}(x)]\},$$

where EIGHT(x) denotes the string obtained by concatenating x with itself eight times. Clearly $T(A)$ is in NP^A, for all oracles A, and thus objective (a) is realized.

We fix a deterministic polynomial-time oracle machine M and we let $L(M^A)$ be the language accpted by M with oracle set A. We look at the set of oracles A relative to which either $L(M^A)$ is finite or $\lim_{n\to\infty} \frac{\|(L(M^A)\cap T(A))^{\leq n}\|}{\|L(M^A)^{\leq n}\|}$ exists and is equal to $1/2$. We will show that this set has measure one. The intersection of all these sets taken over all deterministic polynomial-time machines has measure one as well because it is a countable intersection of measure one sets. Hence, for any oracle set A in this intersection, NP^A is P^A-balanced immune.

One important technical difficulty is that it is possible that, infinitely many times, M^A queries on some input v strings that may cause some string $w > v$ to be in $T(A)$ and this affects the independence of some random variables that will be considered later. We will first show that this can happen only for a set of oracle sets that has measure zero.

A string $y = xu$ that is in Block(x), for some x, is said to be *examined* by $M^A(w)$ if during the computation of $M^A(w)$ the oracle is queried about any string of the form $xu10^k$ for some $k < |u|$. Define

$$\text{EXAM}(A, w) = \{y \mid y \text{ examined by } M^A(w) \text{ and not examined by } M^A(v) \text{ for } v < w\},$$

and

$$\text{EVIDENCE}(A) = \bigcup_w \{y \mid y \in \text{Block}(x) \text{ and } \xi^A(y) = \text{EIGHT}(x) \text{ and } (\exists w \leq x)[y \in \text{EXAM}(A, w)]\}.$$

Let $\mathcal{A}_1 = \{A \mid \text{EVIDENCE}(A) \text{ is finite }\}$.

Claim 3.5.3 $\text{Prob}(\mathcal{A}_1) = 1$.

Proof. Since M^A can make only a polynomial number of queries, it follows that, for all w sufficiently long, EXAM(A, w) contains fewer than $2^{|w|}$ elements. The probability that a fixed $y \in \text{EXAM}(A, w)$ satisfies $\xi^A(y) = \text{EIGHT}(x)$, for some $x \geq w$, is $2^{-7|x|} \leq 2^{-7|w|}$. Let $E(w)$ be the event that there is y in EXAM(A, w)

The evaluation of the generic term in the third sum is

$$\begin{aligned}
&E[Y_i^2 Y_j Y_k \mid \mathcal{A}_1] \\
&\quad = \text{Prob}(x_j(A) \in T(A) \ \text{ and } x_k(A) \in T(A) \mid \mathcal{A}_1) \\
&\qquad + \text{Prob}(x_j(A) \notin T(A) \text{ and } x_k(A) \notin T(A) \mid \mathcal{A}_1) \\
&\qquad - \text{Prob}(x_j(A) \in T(A) \text{ and } x_k(A) \notin T(A) \mid \mathcal{A}_1) \\
&\qquad - \text{Prob}(x_j(A) \notin T(A) \text{ and } x_k(A) \in T(A) \mid \mathcal{A}_1) \\
&\quad = \text{Prob}(x_j(A) \in T(A) \mid \mathcal{A}_1)\text{Prob}(x_k(A) \in T(A) \mid \mathcal{A}_1 \text{ and } x_j(A) \in T(A)) \\
&\qquad + \text{Prob}(x_j(A) \notin T(A) \mid \mathcal{A}_1)\text{Prob}(x_k(A) \notin T(A) \mid \mathcal{A}_1 \text{ and } x_j(A) \notin T(A)) \\
&\qquad - \text{Prob}(x_j(A) \in T(A) \mid \mathcal{A}_1)\text{Prob}(x_k(A) \notin T(A) \mid \mathcal{A}_1 \text{ and } x_j(A) \in T(A)) \\
&\qquad - \text{Prob}(x_j(A) \notin T(A) \mid \mathcal{A}_1)\text{Prob}(x_k(A) \in T(A) \mid \mathcal{A}_1 \text{ and } x_j(A) \notin T(A)).
\end{aligned}$$

By Claim 3.5.4,

$$\begin{aligned}
E[Y_i^2 Y_j Y_k] &\leq 2 \cdot \left(\frac{1}{2} + \frac{16}{j^4}\right)\left(\frac{1}{2} + \frac{16}{k^4}\right) - 2 \cdot \left(\frac{1}{2} - \frac{16}{j^4}\right)\left(\frac{1}{2} - \frac{16}{k^4}\right) \\
&= 32\left(\frac{1}{j^4} + \frac{1}{k^4}\right) \leq 64 \cdot \frac{1}{m^4}.
\end{aligned}$$

Thus,

$$\sum_{km+1 \leq i,j,k \leq (k+1)m,\, j<k} E[Y_i^2 Y_j Y_k] \leq 64m\binom{m}{2} \cdot \frac{1}{m^4} \leq 32\frac{1}{m}.$$

Next we consider the generic term in the fourth sum and in a similar way we obtain

$$\begin{aligned}
E[Y_i Y_j Y_k Y_\ell] &\leq 8\left(\frac{1}{2} + \frac{16}{m^4}\right)^4 - 8\left(\frac{1}{2} - \frac{16}{m^4}\right)^4 \\
&= 8 \cdot \frac{16}{m^4} + 32 \cdot \frac{16^3}{m^{12}} \leq (2^7 + 2^{17}) \cdot \frac{1}{m^4}.
\end{aligned}$$

Thus,

$$\sum_{km+1<i<j<k<\ell\leq(k+1)m} E[Y_i Y_j Y_k Y_\ell] \leq m^4 \cdot (2^7 + 2^{17}) \cdot \frac{1}{m^4} = (2^7 + 2^{17}).$$

By substituting these evaluations in the inequality (3.4), we obtain

$$\begin{aligned}
\text{Prob}\left(\left|\frac{S_{k,m}(A)}{m}\right| > \epsilon \mid \mathcal{A}_1\right) &\leq \frac{1}{\epsilon^4 m^4}\left(m + 6\frac{m(m-1)}{2} + 12\frac{32}{m} + 4!(2^7 + 2^{17})\right) \\
&\leq \frac{c}{\epsilon^4 \cdot m^2},
\end{aligned}$$

for some constant c. ∎

Since the series $\sum_{m=1}^{\infty} \frac{c}{\epsilon^4 \cdot m^2}$ is convergent, using the Borel-Cantelli Lemma, we infer that, for every $\epsilon > 0$ and every $k \geq 1$,

$$\text{Prob}\left(\left|\frac{S_{k,m}(A)}{m}\right| < \epsilon \text{ for almost every } m \mid \mathcal{A}_1\right) = 1.$$

Since $\text{Prob}(\mathcal{A}_1) = 1$, it follows that, for every $\epsilon > 0$ and every $k \geq 1$,

$$\text{Prob}\left(\left|\frac{S_{k,m}(A)}{m}\right| < \epsilon \text{ for almost every } m\right) = 1. \tag{3.5}$$

Let $\mathcal{A}_{\epsilon,k}$ be the measure-one set of oracle set for which the event in the above probability expression holds. We denote

$$\text{IN}^A(m,n) = \|\{x_{m+1}(A), \ldots, x_n(A)\} \cap T(A)\|,$$

and

$$\text{OUT}^A(m,n) = \|\{x_{m+1}(A), \ldots, x_n(A)\} \cap \overline{T(A)}\|,$$

where $\overline{T(A)}$ is the complement of $T(A)$. Let us fix an arbitrary $\epsilon > 0$ and $k \geq 1$. Relation (3.5) implies that for each A in $\mathcal{A}_{\epsilon,k}$, there is m_0 such that, for all $m \geq m_0$,

$$|\text{IN}^A(km,(k+1)m) - \text{OUT}^A(km,(k+1)m)| \leq \epsilon \cdot m.$$

The set $\mathcal{A}_\epsilon = \bigcap_{k \geq 1} \mathcal{A}_{\epsilon,k}$ has measure one. Since $\text{OUT}^A(km,(k+1)m) + \text{IN}^A(km,(k+1)m) = m$, for any $\epsilon > 0$, for every oracle set $A \in \mathcal{A}_\epsilon$, for any $k \geq 2$, and for m sufficiently large,

$$\begin{aligned}
m(1-\epsilon) \leq\ & 2 \cdot \text{IN}^A(m,2m) && \leq m(1+\epsilon) \\
m(1-\epsilon) \leq\ & 2 \cdot \text{IN}^A(2m,3m) && \leq m(1+\epsilon) \\
& \vdots \\
m(1-\epsilon) \leq\ & 2 \cdot \text{IN}^A((k-1)m,km) && \leq m(1+\epsilon).
\end{aligned}$$

Summing up these inequalities, we obtain

$$(k-1)m(1-\epsilon) \leq 2 \cdot \text{IN}^A(m,km) \leq (k-1)m(1+\epsilon),$$

which implies that for all $\epsilon > 0$, for all $A \in \mathcal{A}_\epsilon$, for all k, for sufficiently large m,

$$\frac{1}{2} - \frac{\epsilon}{2} \leq \frac{\text{IN}^A(m,km)}{(k-1)m} \leq \frac{1}{2} + \frac{\epsilon}{2}. \tag{3.6}$$

We also have

$$\frac{\text{IN}^A(m,km)}{km} \leq \frac{\text{IN}^A(0,km)}{km} \leq \frac{m + \text{IN}^A(m,km)}{km},$$

which, combined with relation (3.6), yields

$$\frac{1}{2} - \frac{\epsilon}{2} - \frac{1}{k} \leq \frac{\mathrm{IN}^A(0, km)}{km} \leq \frac{1}{2} + \frac{\epsilon}{2} + \frac{1}{k}.$$

For $km \leq n < (k+1)m$,

$$\frac{\mathrm{IN}^A(0, km)}{n} \leq \frac{\mathrm{IN}^A(0, n)}{n} \leq \frac{\mathrm{IN}^A(0, km) + m}{n},$$

which implies

$$\frac{1}{2} - \frac{\epsilon}{2} - \frac{1}{k} \leq \frac{\mathrm{IN}^A(0, n)}{n} \leq \frac{1}{2} + \frac{\epsilon}{2} + \frac{2}{k}.$$

The above relation holds for all $\epsilon > 0$, for all $A \in \mathcal{A}_\epsilon$, for all k, and for all n sufficiently large. The set $\mathcal{A} = \bigcap \mathcal{A}_\epsilon$ has measure one. Hence, for all $A \in \mathcal{A}$ and thus with probability one of A, $\lim_{n \to \infty} \frac{\mathrm{IN}^A(0,n)}{n}$ exists and it is equal to 1/2. As noted, this concludes the proof of Theorem 3.5.2. ∎

3.6 Average-case complexity

IN BRIEF: A theory of average-case complexity is developed and the average-case analogues of the classes P and NP are defined. It is shown that there are NP-complete problems that are easy on average. A natural example of a problem that is complete for the average-case analogue of NP is exhibited.

An NP complete problem is considered to be a hard problem. However, NP-completeness only implies that there are some input instances on which the problem is unfeasible (of course assuming that P $\neq$ NP). It is possible that these instances are few, rare, and perhaps irrelevant in the sense that a casual user may never be interested to solve these instances. In many applications it is more meaningful to know that a problem is hard or easy "on average."

To tackle the issue of average complexity, we must first introduce a class of relevant probability distributions for the input instances. As usual, instances are encoded as strings over the binary alphabet $\Sigma = \{0, 1\}$. We consider the lexicographical ordering over Σ^* and for $x, y \in \Sigma^*$, $x < y$ means that x precedes y in this order. We denote the predecessor of x by $x - 1$ for any non-empty string x. A distribution on Σ^* can be given by either a distribution function or by a density function.

Definition 3.6.1 (Distribution function) *A distribution function is a function* $\mu\colon \Sigma^* \to [0, 1]$ *such that*

(a) *is non-decreasing, i.e., for all* $x, y \in \Sigma^*$, $x < y$ *implies* $\mu(x) \leq \mu(y)$, *and*

(b) *converges to 1, i.e.,* $\lim_{x \to \infty} \mu(x) = 1$.

Definition 3.6.2 (Density function) *A density function is a function* $\mu' : \Sigma^* \to [0,1]$ *such that* $\sum_{x \in \Sigma^*} \mu'(x) = 1$.

For any distribution function μ there is an associated density function μ' defined by

$$\mu'(x) = \begin{cases} \mu(\lambda) & \text{, if } x = \lambda \text{ (the empty string)} \\ \mu(x) - \mu(x-1) & \text{, if } x \neq \lambda. \end{cases}$$

Also, for any density function μ' there is an associated distribution function μ defined by

$$\mu(x) = \sum_{y \leq x} \mu'(y).$$

Therefore a pair (μ, μ'), with μ and μ' associated to each other, represents a unique object called a *distribution*.

Definition 3.6.3 (Distribution) *A distribution* μ^* *is a pair* (μ, μ'), *where* μ *is a distribution function,* μ' *is a density function, and* μ *and* μ' *are associated to each other as above.*

For technical convenience, we will assume that if μ is a distribution function, $\mu(\lambda) = 0$. Also, we will allow density functions for which $\sum_{x \in \Sigma^*} \mu'(x)$ is equal to a constant $c > 0$ that may be different from 1 (or distribution functions with the limit in Definition 3.6.1 (b) equal to an arbitrary constant $c > 0$), because they are easy to modify to satisfy the formal definition. For example, let us define a distribution on Σ^* based on the following random experiment: (a) First we pick a natural number n at random with some probability p_n, and (b) next we pick uniformly at random a string of length n. Thus, the probability that a given string x is chosen is $p_x = p_n \cdot \frac{1}{2^n}$, where $n = |x|$. In principle to obtain a density function we need

$$1 = \sum_{x \in \Sigma^*} p_x = \sum_{n \geq 1} p_n \cdot \sum_{x \in \Sigma^n} \frac{1}{2^n} = \sum_{n \geq 1} p_n.$$

If we take $p_n = \frac{1}{n^2}$, we have that $\sum_{n \geq 1} p_n = c \neq 1$ (actually $c = \frac{\pi^2}{6}$). The probabilities can be normalized by defining $p'_n = \frac{1}{c} \cdot \frac{1}{n^2}$. However, we will consider p_n acceptable as it is. The distribution defined by $\mu'(x) = \frac{1}{|x|^2} \frac{1}{2^{|x|}}$ is called the *standard uniform distribution* of Σ^*.

Definition 3.6.4 (Distributional problem) *A distributional problem is a pair* (A, μ^*), *where* A *is a language (equivalently, a decision problem), and* μ^* *is a distribution.*

It is clear that some restrictions on distributions must be imposed; otherwise it is always possible to have the worst-case complexity be the same as the average-case complexity. It seems reasonable to require that the density function is polynomial-time computable. Such distributions are said to be P-samplable.

Definition 3.6.5 (P-samplable distribution) *A distribution is* P*-samplable if there is a polynomial-time algorithm M that calculates the associated density function μ'. This means that, for all $x \in \Sigma^*$, $\mu'(x)$ has a finite binary expansion and $M(x)$ outputs this expansion in time that is polynomial in $|x|$.*

Unfortunately, this definition does not allow the development of a useful theory of average-case complexity. Ben-David et. al. [BDCGL92] have shown that for every standard NP-complete problem it is possible to build a P-samplable distribution relative to which the problem is hard on average.

Most commonly used distributions satisfy a stronger property: Their distribution function is computable in polynomial-time.

Definition 3.6.6 (P-computable distribution) *A distribution is* P*-computable if there is a polynomial-time algorithm M that calculates the associated distribution function μ. This means that, for all $x \in \Sigma^*$, $\mu(x)$ has a finite binary expansion and $M(x)$ outputs this expansion in time that is polynomial in $|x|$.*

Clearly, a distribution that is P-computable is also P-samplable (because $\mu'(x) = \mu(x) - \mu(x-1)$). The converse is probably not true.

Proposition 3.6.7 *If* P $\neq$ NP, *then there is a distribution that is* P*-samplable but not* P*-computable.*

Proof. We consider triples (ϕ, a, b), where ϕ is a boolean formula in CNF, a is a truth assignment for the variables of ϕ, and $b \in \{0,1\}$. We encode such triples via a 1-to-1 mapping as binary strings, and we denote by $\langle \phi, a, b\rangle$ the encoding of (ϕ, a, b). It can be easily arranged that both encoding and decoding can be done in polynomial time and that for all formulas ϕ and for all assignments a for ϕ, $\langle \phi, a, 1\rangle$ is lexicographically between $\langle \phi, (0, \ldots, 0), 1\rangle$ and $\langle \phi, (1, \ldots, 1), 1\rangle$. Let $|\phi|$ denote the length of some fixed natural encoding of the formula ϕ, let $|a|$ be the number of variables to which a assigns truth values, and let $t(\phi, a)$ be the truth value of ϕ under the assignment a. Let us consider the function

$$\mu'(x) = \begin{cases} \frac{1}{2^{2|\phi|}2^{|a|}} & \text{, if } x = \langle \phi, a, b\rangle \text{ and } t(\phi, a) = b \\ 0 & \text{, otherwise.} \end{cases}$$

Clearly, the encoding $\langle \phi, a, b\rangle$ can be taken such that μ' is computable in polynomial time. We have

$$\begin{aligned} \sum_{x\in\Sigma^*} \mu'(x) &= \sum_{n\geq 1} \sum_{\{\phi \mid |\phi|=n\}} \sum_{\{a,b \mid t(\phi,a)=b\}} \frac{1}{2^{2|\phi|}2^{|a|}} \\ &= \sum_{n\geq 1} \frac{1}{2^{2n}} \sum_{\{\phi \mid |\phi|=n\}} 2^{|a|} \cdot \frac{1}{2^{|a|}} \\ &\leq \sum_{n\geq 1} \frac{1}{2^{2n}} \cdot 2^n = \sum_{n\geq 1} \frac{1}{2^n} = 1. \end{aligned}$$

Thus μ' is a density function and the distribution associated to μ' is P-samplable. Note that if μ is the distribution function associated to μ' then $\mu(\langle\phi,(1,\ldots,1),1\rangle) - \mu(\langle\phi,(0,\ldots,0),1\rangle) \neq 0$ if and only if there is a satisfying assignment for ϕ (recall that for any truth assignment a for a formula ϕ, $\langle\phi,a,1\rangle$ is lexicographically between $\langle\phi,(0,\ldots,0),1\rangle$ and $\langle\phi,(1,\ldots,1),1\rangle$). Thus if μ were computable in polynomial time, it would imply SAT $\in$ P, and thus P $=$ NP. ∎

We define next what it means for a (decision) problem to be feasible on average, i.e., to be solvable in polynomial time on average. At a first sight, we should simply require that the expected running time over all inputs of a given length is bounded by a fixed polynomial, i.e., require that the running time $t_A(x)$ of an algorithm for a distributional problem (A,μ^*) satisfies for some fixed k and c,

$$\sum_{x\in\Sigma^n} \mu'(x)\cdot t_A(x) < c\cdot n^k, \text{ for all } n\in\mathbb{N}. \tag{3.7}$$

Unfortunately, this attempt, though natural, has serious deficiencies that make it unsuitable for developing a theory of average-case complexity. To illustrate the problems with this definition, let us consider the function

$$f(x) = \begin{cases} 2^n, & \text{if } x = 0^n \\ 0, & \text{otherwise.} \end{cases}$$

The expected value for inputs x of length n under the uniform distribution is

$$\sum_{x\in\Sigma^n} \frac{1}{2^n} f(x) = 1$$

However the expected value of f^2 is

$$\sum_{x\in\Sigma^n} \frac{1}{2^n} f^2(x) = 2^n.$$

Thus the class of functions with a polynomially-bounded expected value is not closed under squaring, and, in general, under multiplication. A definition of average-case complexity based on Equation (3.7) would be dependent on the type of machine that we are considering, because converting from one model to another (for example from a Turing machine with two tapes to a Turing machine with one tape) usually implies a polynomial slow-down of the running time. Moreover, even for a fixed model of computation, there would be serious problems. For instance, if we compose two functions t_A and t_B, both satisfying the relation in Equation (3.7), the resulting function may not satisfy that relation. Composing two functions is an operation that is needed when, to give just one example, we reduce one problem to another.

Therefore, Levin [Lev86] has proposed another definition which avoids all these problems and which is now widely accepted as the right definition for average polynomial time.

Definition 3.6.8 *Let μ^* be a distribution. A function f is polynomial on μ^*-average if there is a constant $\epsilon > 0$ such that*

$$\sum_{x \neq \lambda} \mu'(x) \frac{(f(x))^{\epsilon}}{|x|} < \infty,$$

where μ' is the density function associated to μ.

Note that this definition states that, for some $\epsilon > 0$, $(f(x))^{\epsilon}$ is linear on average. Let us check that the class of functions that are polynomial on μ^*-average is closed under multiplication. Let us consider two functions f and g which, for some constants $\epsilon_1 > 0$ and $\epsilon_2 > 0$, satisfy

$$\sum_{x \neq \lambda} \mu'(x) \frac{(f(x))^{\epsilon_1}}{|x|} < \infty, \text{ and}$$

$$\sum_{x \neq \lambda} \mu'(x) \frac{(g(x))^{\epsilon_2}}{|x|} < \infty.$$

Consider $\epsilon = (1/2)\min(\epsilon_1, \epsilon_2)$. Then,

$$\sum_{x \neq \lambda} \mu'(x) \frac{(f(x) \cdot g(x))^{\epsilon}}{|x|} \leq \sum_{x \neq \lambda} \mu'(x) \frac{(f(x))^{2\epsilon} + (g(x))^{2\epsilon}}{|x|}$$

$$\leq \sum_{x \neq \lambda} \mu'(x) \frac{(f(x))^{\epsilon_1}}{|x|} + \sum_{x \neq \lambda} \mu'(x) \frac{(g(x))^{\epsilon_2}}{|x|} < \infty.$$

Thus, Definition 3.6.8 avoids the problem we have seen before (as well as other deficiencies of our first attempt) and it provides the basis for defining feasibility on average.

Definition 3.6.9 (AP) AP *is the class of distributional problems (A, μ^*) that can be solved by a deterministic algorithm having a running time polynomial on μ^*-average.*

We next show that there are NP-complete problems that are feasible in the average-sense with respect to quite natural distributions. We consider the following problem, which is one distributional version of the well-known NP-complete problem 3COLORABILITY.

Problem 3.6.10 D-3COL Problem:

Input: A graph G.

Question: Is there a 3-coloring of the graph G? In other words, can the nodes of G be colored with 3 colors such that no pair of adjacent vertices are colored with the same color?

Distribution: The density function μ' is defined as follows: A natural number n is picked randomly with probability $1/(n^2)$. Next a graph with vertices labeled $1, \ldots, n$ is picked by taking independently for every two nodes i and j an edge (i, j) with probability $1/2$.

Proposition 3.6.11 D-3COL *is in* AP.

Proof. The proof is based on the fact that most graphs have K_4 as a subgraph (K_4 is the complete graph with four vertices). Such a graph, obviously, is not 3-colorable. Therefore, our algorithm on input a graph G, first checks for the presence of K_4 as a subgraph of G. If K_4 is detected (and this happens most of the times), then the verdict comes immediately: The graph is not 3-colorable. If K_4 is not detected, then in a brute-force manner, we try all possible 3-colorings. This takes a long time, but because it is done only rarely, the average running time will be polynomial.

Let us do the calculations. The probability that four given vertices form a K_4 subgraph is $(1/2)^6 = 1/64$ (because there are six possible pairs of vertices). Suppose the number n of vertices has been fixed. We group the vertices into disjoint groups of four. The probability that no group is a K_4 is $\left(1 - \frac{1}{64}\right)^{n/4} = \left(\frac{63}{64}\right)^{n/4}$, and therefore

$$\text{Prob}(G \text{ has } n \text{ vertices and contains no } K_4) \leq \frac{1}{n^2} \cdot \left(\frac{63}{64}\right)^{n/4}.$$

Let H_n be the set of graphs with n vertices that do not contain K_4, i.e., the event in the equation above. If the input graph G with n vertices has a K_4 subgraph, the running time $t(G)$ of the algorithm is bounded by a polynomial $p(n)$, because we only need to check the $\binom{n}{4} < n^4$ subsets of four vertices to find the K_4 subgraph. If the input graph G with n vertices does not contain K_4, then in addition to the $p(n)$ steps above, the algorithm is going over all 3^n possible 3-colorings. Thus, in this case, the running time $t(G)$ is bounded by $3^n \cdot q(n)$, for some polynomial q, and this is less than 4^n for n sufficiently large.

We take k such that (a) $p(n)^{1/k}$ is less than the length of the encoding of a graph G with n vertices (this length is denoted by $|G|$) and (b) $4^{1/k} \cdot \left(\frac{63}{64}\right)^{1/4} \leq \alpha < 1$ (for some constant α). To finish the proof, it is sufficient to show that

$$\sum_G \frac{t(G)^{1/k}}{|G|} \cdot \mu'(G) < \infty.$$

We calculate a truncation of this series, discarding a finite number of initial terms corresponding to graphs for which $|G|$ is too small and does not satisfy the above inequalities. Clearly, since we are omitting a finite number of terms, it is sufficient to show the convergence of the truncated series.

$$\sum_G \frac{t(G)^{1/k}}{|G|} \cdot \mu'(G) = \sum_{G \text{ has } K_4} \frac{t(G)^{1/k}}{|G|} \cdot \mu'(G) + \sum_{G \text{ has no } K_4} \frac{t(G)^{1/k}}{|G|} \cdot \mu'(G).$$

For the first term observe that $t(G)^{1/k} < p(n)^{1/k} < |G|$, and therefore

$$\sum_{G \text{ has } K_4} \frac{t(G)^{1/k}}{|G|} \cdot \mu'(G) \leq \sum_{G \text{ has } K_4} \mu'(G) \leq \infty.$$

For the second term, we have

$$\sum_{G \text{ has no } K_4} \frac{t(G)^{1/k}}{|G|} \cdot \mu'(G) \leq \sum_n \sum_{G \in H_n} \frac{4^{n/k}}{|G|} \cdot \mu'(G)$$
$$\leq \sum_n 4^{n/k} \cdot \sum_{G \in H_n} \mu'(G)$$
$$\leq \sum_n 4^{n/k} \cdot \frac{1}{n^2} \cdot \left(\frac{63}{64}\right)^{n/4}$$
$$= \sum \frac{1}{n^2} \cdot \left(4^{1/k} \cdot \left(\frac{63}{64}\right)^{1/4}\right)^n \leq \sum_n \frac{1}{n^2} \cdot \alpha^n < 1.$$

This ends the proof of Proposition 3.6.11. ∎

Thus, there are problems that are hard (in our example, hard meaning NP-complete) in the worst-case and easy on average. There exist also problems that remain hard on average as well. As in the case of worst-case analysis, a notion of completeness is helpful to describe this phenomenon. We first define an analogue of NP for the average-case.

Definition 3.6.12 (DistNP) DistNP *is the class of distributional problems* (A, μ^*) *having the property that* A *is a decision problem in* NP *and* μ^* *is a* P*-computable distribution.*

We also need a notion of reducibility between distributional problems. The main requierements are (a) the transitivity of the reduction relation, and (b) the fact that if (A, μ^*) reduces to (B, ν^*) and (B, ν^*) is in AP, then (A, μ^*) is also in AP. To obtain these properties, in addition to the normal relation between the decision problems A and B, we also need to ensure that the reduction does not map many instances A ("many" according to μ^*) into few instances of B ("few" according to ν^*). Otherwise, it would be possible that most instances of A are mapped to a few hard instances of B, and, thus, even if (B, ν^*) is in AP, it would not follow that (A, μ^*) is in AP. The needed technical concept is that of domination between distributions.

Definition 3.6.13 (Domination) *Let* μ^* *and* ν^* *be two distributions and* μ' *and* ν' *be, respectively, their associated density functions. We say that* ν^* *dominates* μ^* *(or* μ^* *is dominated by* ν^**), and we write* $\mu^* \preceq \nu^*$*, if there is a polynomial* p *such that, for all* $x \in \Sigma^*$, $\mu'(x) \leq p(|x|) \cdot \nu'(x)$.

Definition 3.6.14 (Average-case reduction) *Let* (A, μ^*) *and* (B, ν^*) *be two distributional problems, and let* μ' *and* ν' *be the density functions of* μ^* *and respectively* ν^*. *We say that* (A, μ^*) *is polynomial-time reducible to* (B, ν^*) *(notation* $(A, \mu^*) \leq^p (B, \nu^*)$*) if there is a polynomial-time computable function* f *such that*

(1) *for all* x, $x \in A \Leftrightarrow f(x) \in B$, *and*

(2) *there is a distribution τ^* such that $\mu^* \leq \tau^*$, and, for all y in the range of f, $\nu'(y) = \sum_{x \in f^{-1}(y)} \tau'(x)$ (where τ' is the density function associated to τ^*).*

We show that this notion of reducibility has the desired properties.

Proposition 3.6.15 (1) *If $(A_1, \mu_1^*) \leq^p (A_2, \mu_2^*)$ and $(A_2, \mu_2^*) \leq^p (A_3, \mu_3^*)$, then $(A_1, \mu_1^*) \leq^p (A_3, \mu_3^*)$.*

(2) *If $(A, \mu^*) \leq^p (B, \nu^*)$ and $(B, \nu^*) \in$ AP, then $(A, \mu^*) \in$ AP.*

Proof. (1) Let f and g be two functions such that (A_1, μ_1^*) is reducible to (A_2, μ_2^*) via f and (A_2, μ_2^*) is reducible to (A_3, μ_3^*) via g. We show that $g \circ f$ reduces (A_1, μ_1^*) to (A_3, μ_3^*). Clearly, $x \in A_1 \Leftrightarrow f(x) \in A_2 \Leftrightarrow g(f(x)) \in A_3$. Since $(A_1, \mu_1^*) \leq^p (A_2, \mu_2^*)$ and $(A_2, \mu_2^*) \leq^p (A_3, \mu_3^*)$, we also know that there exist a polynomial p and two density functions ξ_1' and ξ_2' such that for all x, y, and z, (a) $|f(x)| \leq p(|x|)$, (b) $|g(x)| \leq p(|x|)$, (c) $\mu_1'(x) \leq p(|x|)\xi_1'(x)$, (d) $\mu_2'(y) \leq p(|y|)\xi_1'(y)$, (e) $\mu_2'(y) = \sum_{x \in f^{-1}(y)} \xi_1'(x)$ and (f) $\mu_3'(z) = \sum_{y \in g^{-1}(z)} \xi_2'(y)$.

For any z in the range of $g \circ f$ we have

$$\begin{aligned}
\mu_3'(z) &= \sum_{y \in g^{-1}(z) \cap \text{Range}(f)} \xi_2'(y) \\
&\geq \sum_{y \in g^{-1}(z) \cap \text{Range}(f)} \frac{1}{p(|y|)} \cdot \mu_2'(y) \\
&= \sum_{y \in g^{-1}(z) \cap \text{Range}(f)} \frac{1}{p(|y|)} \cdot \sum_{x \in f^{-1}(y)} \xi_1'(x) \\
&\geq \sum_{y \in g^{-1}(z) \cap \text{Range}(f)} \frac{1}{p(|y|)} \cdot \sum_{x \in f^{-1}(y)} \frac{1}{p(|x|)} \cdot \mu_1'(x) \\
&\geq \sum_{x \in (g \circ f)^{-1}(z)} \frac{1}{p(p(|x|))} \cdot \frac{1}{p(|x|)} \cdot \mu_1'(x).
\end{aligned}$$

Let

$$c(z) = \sum_{x \in (g \circ f)^{-1}(z)} \frac{1}{p(p(|x|))} \cdot \frac{1}{p(|x|)} \cdot \mu_1'(x).$$

The relation established above shows that $\mu'(g(f(x)) \geq c(g(f(x))$, for all x. Consider the distribution ξ_3^* having the associated density function

$$\xi_3'(x) = \frac{1}{p(p(|x|))p(|x|)} \cdot \frac{\mu_3'(g(f(x)))}{c(g(f(x)))} \cdot \mu_1'(x).$$

Then, clearly, for any z in the range of $g \circ f$,

$$\mu_3'(z) = \sum_{x \in (g \circ f)^{-1}(x)} \xi_3'(x),$$

and

$$\mu_1'(x) \leq p(p(|x|)) \cdot p(|x|) \cdot \xi_3'(x).$$

Therefore $(A_1, \mu_1^*) \leq^p (A_3, \mu_3^*)$, which ends the proof of (1).

(2) Let f be the polynomial-time reduction from (A, μ^*) to (B, ν^*). Because of the domination property, there exist a distribution density μ_1' and a polynomial p such that, for all x, $\mu'(x) \leq p(|x|) \cdot \mu_1'(x)$, and for all y in the range of f, $\nu'(y) = \sum_{x \in f^{-1}(y)} \mu_1'(x)$. Since (B, ν^*) is in AP, there is an algorithm M with running time t satisfying, for some $\epsilon > 0$,

$$\sum_y \frac{t^\epsilon(y)}{|y|} \cdot \nu'(y) < \infty.$$

Without loss of generality we can assume $\frac{t^\epsilon(y)}{|y|} < 1$. We know that $x \in A$ if and only if $f(x) \in B$ and, thus, the determination of whether $x \in A$ can be done by (a) calculating $f(x)$, and (b) running M on $f(x)$. The time to do (a) is polynomial for all x, so it is sufficient to show that the time to do (b), which is $t(f(x))$, is polynomial on ν^*-average.

There exists $k > 0$ such that $|f(x)| < |x|^k$ for all but finitely many x. Let $h(x) = \frac{t(f(x))}{p(|x|)^{k/\epsilon}}$. We show that $h(x)$ is polynomial on μ^*-average and from here it follows that $t(f(x)) = h(x) \cdot p(|x|)^{k/\epsilon}$ is polynomial on μ^*-average (because of the closure under multiplication of the class of functions polynomial under μ^*-average). We have

$$\begin{aligned}
\sum_x \frac{h(x)^{\epsilon/k}}{|x|} \cdot \mu'(x) &= \sum_x \frac{t(f(x))^{\epsilon/k} \cdot \frac{1}{p(|x|)}}{|x|} \cdot \mu'(x) \\
&\leq \sum_x \frac{(t(f(x)))^{\epsilon/k}}{|x|} \cdot \mu_1'(x) \\
&= \sum_{y \in \text{Range}(f)} \sum_{x \in f^{-1}(y)} \frac{(t(y))^{\epsilon/k}}{|x|} \cdot \mu_1'(x) \\
&\leq \sum_{y \in \text{Range}(f)} \frac{(t(y))^{\epsilon/k}}{|y|^{1/k}} \sum_{x \in f^{-1}(y)} \mu_1'(x) \\
&= \sum_{y \in \text{Range}(f)} \frac{(t(y))^{\epsilon/k}}{|y|^{1/k}} \cdot \nu'(y) \\
&= \sum_{y \in \text{Range}(f)} \left(\frac{(t(y))^\epsilon}{|y|} \right)^{1/k} \cdot \nu'(y) \\
&= \sum_y \nu'(y) < \infty.
\end{aligned}$$

This ends the proof of Proposition 3.6.15. ∎

Equipped with a reducibility, we can show that there are problems complete for DistNP.

Problem 3.6.16 Distributional Bounded Halting (D-BH) Problem:

Input: A triplet $(N, x, 1^k)$, where N is a nondeterministic machine N, x is an input string for N, and k is a natural number.

Question: Does N halt on input x within k steps?

Distribution: $\mu'_{\text{D-BH}}((N, x, 1^k)) = \frac{1}{|N|^2 \cdot 2^{|N|}} \cdot \frac{1}{|x|^2 \cdot 2^{|x|}} \cdot \frac{1}{k^2}$. (This corresponds to choosing N, x, and 1^k independently according to the standard uniform distribution.)

The *Bounded Halting Problem* (BH) (which is D-BH without the distribution) is easily shown to be NP-complete (in the standard sense). Indeed, let A be a problem in NP. Then there is a nondeterministic polynomial-time machine N_A which solves A and which runs in time $p(n)$, for some polynomial p. Then $x \in A$ if and only if $(N_A, x, 1^{p(|x|)}) \in$ BH. Showing completeness in the average case is more delicate because we have to consider all NP-complete problems A and, in addition, all P-computable distributions. It is possible that according to such a distribution a string x has density much greater than $2^{-|x|}$, while the triplet x is mapped to by the standard reduction seen above has $\mu'_{\text{D-BH}}$ density less than $2^{-|x|}$. This violates the domination rule for a reduction among distributional problems. The problem is overcome by first mapping strings with high density into short strings. More precisely, a string x is mapped into a string whose length is $\leq 1 + \log \frac{1}{\mu'(x)}$. This is achieved in the following lemma.

Lemma 3.6.17 *Let μ^* be a* P*-computable distribution and μ' its associated density function. There exists a function* code : $\Sigma^* \to \Sigma^*$ *such that*

(1) code *is 1-to-1,*

(2) code *is computable in polynomial time, and*

(3) *for every x,* $|\text{code}(x)| \leq 1 + \min\left\{|x|, \log \frac{1}{\mu'(x)}\right\}$.

Proof. There are two categories of strings: (a) strings x with $\mu'(x) \leq 2^{-|x|}$, and (b) strings x with $\mu'(x) > 2^{-|x|}$. We use two different encodings for the two categories. To keep the coding 1-to-1, the encoding of strings in category (a) starts with 0, and the encoding of strings in category (b) starts with 1.

For strings in category (a), $\text{code}(x) = 0x$. It is clear that conditions (1), (2), and (3) are verified.

Let μ be the distribution function associated to μ^*. For strings in category (b), $\text{code}(x)$ is of the form $1z$, where z is taken to be the binary expansion of a certain value in the interval $[\mu(x-1), \mu(x))$. This ensures the 1-to-1 property of the mapping (because the intervals $[\mu(x-1), \mu(x))$ are disjoint). The string z is the longest common prefix of the binary representation of $\mu(x)$ and $\mu(x-1)$. This ensures that $\text{code}(x)$ is computable in polynomial time. We still need to check property (3). Note that, since $\mu'(x) = \mu(x) - \mu(x-1)$ and $\mu'(x) > 2^{-|x|}$, we have $|z| < |x|$ and

$$\begin{aligned}\mu'(x) &= \mu(x) - \mu(x-1) \\ &\leq \left(\sum_{i=1}^{|z|} z(i)\cdot 2^{-i} + \sum_{i=|z|+1}^{\infty} 2^{-i}\right) - \left(\sum_{i=1}^{|z|} z(i)\cdot 2^{-i}\right) \\ &= \sum_{i=|z|+1}^{\infty} 2^{-i} = 2^{-|z|},\end{aligned}$$

because $\mu(x) \leq 0.z1\ldots1\ldots$ and $\mu(x-1) \geq 0.z$. Thus, $|z| \leq \log\frac{1}{\mu'(x)}$, and, therefore, $|\text{code}(x)| \leq 1 + \min\{|x|, \log\frac{1}{\mu'(x)}\}$. ∎

Theorem 3.6.18 D-BH *is complete for* DistNP.

Proof. Let (A, μ^*) be a distributional problem for DistNP and N_A be a nondeterministic polynomial-time machine that accepts A in time $p_A(|x|)$, where p_A is a polynomial. Consider $N_{A,\mu}$ the nondeterministic polynomial-time machine that on input y, guesses nondeterministically x such that $\text{code}(x) = y$, and then runs N_A on input x (if there is no such x, it will reject). Let $p(n) = n + p_{\text{code}}(n) + p_A(n)$, where $p_{\text{code}}(n)$ is the time required to calculate code(x), for a string x of length n. The reduction from (A, μ^*) to $(\text{D-BH}, \mu^*_{\text{D-BH}})$ is given by

$$f(x) = (N_{A,\mu}, \text{code}(x), 1^{p(|x|)}).$$

It can be checked immediately that $x \in A \Leftrightarrow f(x) \in \text{D-BH}$, and that f can be calculated in polynomial time. It remains to check the domination property. By Lemma 3.6.17, $\mu'(x) \leq 2\cdot 2^{-|\text{code}(x)|}$. Therefore,

$$\begin{aligned}\mu'_{\text{D-BH}}(N_{A,\mu}, \text{code}(x), 1^{p(|x|)}) &= \frac{1}{|N_{A,\mu}|^2\cdot 2^{|N_{A,\mu}|}}\cdot\frac{1}{|\text{code}(x)|^2\cdot 2^{|\text{code}(x)|}}\cdot\frac{1}{(p(|x|))^2} \\ &\leq \frac{c}{2}\cdot\frac{1}{|\text{code}(x)|^2}\cdot\frac{1}{(p(|x|))^2}\cdot\mu'(x),\end{aligned}$$

where $c = \frac{1}{|N_{A,\mu}|^2\cdot 2^{|N_{A,\mu}|}}$ (i.e., it does not depend on x). It follows that

$$\mu'(x) \leq \frac{2}{c}\cdot|\text{code}(x)|^2\cdot p(|x|)^2\cdot\mu'_{\text{D-BH}}(N_{a,\mu}, \text{code}(x), 1^{p(|x|)}).$$

Therefore the domination requirement is satisfied if we take

$$\mu'_1(x) = \mu'_{\text{D-BH}}(N_{a,\mu}, \text{code}(x), 1^{p(|x|)}).$$

(Note that since the coding is 1-to-1, x is the only element mapped into $(N_{a,\mu}, \text{code}(x), 1^{p(|x|)})$. ∎

D-BH is the generic complete problem for the class DistNP, in the sense that, being built from the Bounded Halting Problem, it simply encompasses all NP problems with all their inputs. Such problems are not very useful for showing the existence of other complete problems via reductions. More natural examples of problems that are complete for DistNP are known, but the list of such problems is currently far smaller than the list of NP-complete problems. We content to present (following the exposition in [Wan97a]) just one example of a more natural DistNP complete problem.

Problem 3.6.19 Distributional Post Correspondence Problem (D-PC):
Input: A nonempty set LIST $= ((\ell_1, r_i), \ldots, (\ell_m, r_m))$ of pairs of binary strings and a positive integer n written in the unary alphabet.
Question: Is there a sequence of at most n integers $i_1, \ldots, i_k$, $k \leq n$, such that $\ell_{i_1}\ell_{i_2}\ldots\ell_{i_k} = r_{i_1}r_{i_2}\ldots r_{i_k}$? (Such a sequence is called a solution of size k of the problem.)
Distribution:

$$\mu'_{\text{D-PC}}(((\ell_1, r_1), \ldots, (\ell_m, r_m)), 1^n) = \frac{1}{n^2}\frac{1}{m^2}\prod_{i=1}^{m}\frac{1}{|\ell_i|^2 \cdot |r_i|^2 \cdot 2^{|l_i|+|r_i|}},$$

(i.e., the uniform distribution).

Theorem 3.6.20 D-PC *is complete for* DistNP.

Proof. Let (A, μ^*) be a problem in DistNP. Thus, there is a nondeterministic polynomial-time Turing machine M_1 such that M_1 accepts A. We can assume without loss of generality that M_1 has only one accepting state and that, for all x, all the computation paths of M_1 on input x are bounded by some polynomial in $|x|$. We will be using the function *code* from Lemma 3.6.17 for the distribution function μ' associated to μ^*. Recall that for all x, we have $\mu'(x) \leq 2^{-|\text{code}(x)|}$. As in the proof of Theorem 3.6.18, from M_1 we build another nondeterministic Turing machine M as follows. M on input $1w$ guesses nondeterministically x such that $\text{code}(x) = w$. If the input does not start with 1 or if x is not found, M rejects immediately. Otherwise, M simulates M_1 on input x. Clearly, M also has exactly one accepting state, and there is a polynomial p such that, for all x, $x \in A$ if and only if $1\text{code}(x)$ is accepted by M in time at most $p(|x|)$. We can assume as well that, for all x, all the computation paths of M on input x are bounded by $p(|x|)$, and also that M has a single tape. Next we build the reduction function f. We fix x to be an input binary string for the problem A and we have to build an instance for the D-PC problem. For the machine M, let Q be the set of states, q_0 be the starting state, a be the (unique) accepting state, δ the transition function, Σ the alphabet. Let $z = 1\text{code}(x)$ and let $\Sigma_1 = Q \cup \Sigma \cup \{B, \triangle, \square, !\}$, where B is the blank symbol and $\triangle, \square$ and ! are new symbols. The reduction f will be 1-to-1 and, thus, we cannot have in the set LIST of the instance f(x) a string longer than $c|x|$, with $c > 1$, because otherwise the domination property cannot be satisfied ($\mu'_{\text{D-PC}}(f(x))$ would be less than polynomially smaller than $\mu'(x)$). To take care of this, all the strings in the D-PC instance that we are constructing will have length at most $|x| + O(\log(|x|))$. We need an additional encoding function, which depends on x, and which we describe now. We define a bijective function $d : \Sigma_1 \to S \subseteq \{0,1\}^L$, for some positive integer L, and we call the strings $d(s)$, with $s \in \Sigma_1$, *codewords.* The encoding d has the following properties:

(1) $L = O(\log(|x|))$,

(2) No codeword is a substring of x,

(3) The set of all proper prefixes of all codewords is disjoint from the set of all proper suffixes of all codewords,

(4) 1, 10, 000, 100 are not prefixes of any codeword.

Note that any string that starts with 1, and in particular $z = 1\text{code}(x)$, can be decomposed in a unique way as a concatenation of 1, 10, 000, and 100. The function d is built as follows. The codewords will belong to the regular set $R = 0100(00 + 11)^*11$. This ensures that conditions (3) and (4) hold true. The value L is taken to be the least even integer such that $2^{(L-6)/2} \geq |x| + \|\Sigma_1\|$. Therefore, $L = O(\log(|x|))$. Also, since the string x has at most $|x|$ substrings of length L, we can pick a set S of strings in R that does not have any substring of x and that can be put into a bijective correspondence, which is our d, with Σ_1 (note that R has $2^{(L-6)/2}$ strings of length L). The encoding d can be extended in the obvious way to Σ_1^*, i.e., for any $v \in \Sigma_1^*$, $d(v)$ is obtained by replacing each symbol in v with the corresponding codeword. We now build $f(x)$ as an instance for the D-PC problem. Thus, $f(x)$ consists of a set of pairs of words, LIST(x), and of a nonnegative value n written in unary. We will define n later, so let us focus for now on LIST(x). The set LIST(x) consists of six groups of pairs of binary strings.

Group 1: $(d(\triangle), d(\triangle)zd(\Box!q_0))$;

Group 2: $(u, d(u'))$, for all $u \in \{1, 10, 000, 100\}$, where u' is obtained from u by replacing each 0 with !0 and each 1 with !1;

Group 3: $(d(X!), d(!X))$, for each $X \in \Sigma \cup \{B, \Box\}$;

Group 4: For each $q \in Q - \{a\}$ and for each $X \in \Sigma \cup \{B\}$,

$$\begin{aligned} (d(q!X!), d(!Y!p)), &\text{ if } (p, Y, R) \in \delta(q, X);\\ (d(Z!q!X!), d(!p!Z!Y)), &\text{ if } (p, Y, L) \in \delta(q, X);\\ (d(q!\Box!), d(!Y!p!\Box)), &\text{ if } (p, Y, R) \in \delta(q, B);\\ (d(Z!q!\Box!), d(!p!Z!Y!\Box)), &\text{ if } (p, Y, L) \in \delta(q, B); \end{aligned}$$

Group 5: For all $X, Y \in \Sigma \cup \{B\}$,

$$(d(X!a!), d(!a)),\ (d(X!a!Y!), d(!a)),\ d(a!Y!), d(!a));$$

Group 6: $(d(a!\Box!\Box!), d(!\Box))$.

If $u \in \Sigma \cup \{B\}$, we denote by $\widetilde{d(u)}$ the string obtained by replacing in $d(u)$ each codeword $d(v)$ with $d(v!)$ and omitting the last $d(!)$. A partial solution of the problem is a pair of words (u, v) such that u is a prefix of v, and such that u and v are obtained from a sequence of (not necessarily distinct) pairs in LIST(x) by concatenating their left strings and, respectively (i.e., for v), their right strings. We will show that there is a sequence of partial solutions that describe in a natural way the computation of M on input z. The status of the machine M at a given time is described completely by the content of its tape at that moment, by the current state $q \in Q$, and by the position of the read/write head on the tape. All these elements taken together define the *configuration* of M at a given moment. If the content of the tape is $\alpha\beta$, with $\alpha, \beta \in \Sigma^*$, the read/write head is scanning the cell containing the rightmost symbol in α, and the current state is q, then the corresponding configuration can be represented by the string $\alpha q \beta$. For a configuration $C = \alpha q \beta$, we denote

$$\langle C \rangle = \widetilde{d(\alpha)} d(!q!) \widetilde{d(\beta)} d(!\Box).$$

The machine M starts in the initial configuration $C_0 = q_0 z$ and it moves succesively through a sequence of configurations. Let $\langle\text{START}\rangle$ be the string $d(\triangle z \Box !)$. Observe that if we try to build a solution for LIST(x), the only pair that can be used to start is the one from Group 1, $(d(\triangle), d(\triangle) z d(\Box ! q_0))$. Next, in order to build z in the left hand side of the solution, we can only use pairs from Group 2; we will append z to the left hand side, and in the right hand side we get $d(!)\widetilde{d(z)}$. Next, to place $d(\Box !)$ in the left hand side, we can only use the pair from Group 3 $(d(\Box !), d(!\Box))$. Concatenating these pairs, gives

$$(\langle\text{START}\rangle, \ \ \langle\text{START}\rangle\langle\text{C}_0\rangle),$$

and this is the only way to start building a solution. Observe that the numbers of pairs from LIST(x) used to build this partial solution is bounded by a polynomial in $|x|$.

Next, to continue building our solution, we need to append $\langle C_0 \rangle$ to the left hand side of our partial solution. If $q_0 = a$ (the unique accepting state), we can complete a solution by appending pairs from Group 5 and, at the end, the pair from Group 6. If $q_0 \neq a$, then we can only use a pair from Group 4 that corresponds to a legal move of M from configuration C_0. This legal move (if there is one) takes M into some configuration C_1. Next it can be checked that we can only use pairs from Group 3. This leads us to the partial solution

$$(\langle\text{START}\rangle\langle\text{C}_0\rangle, \ \ \langle\text{START}\rangle\langle\text{C}_0\rangle\langle\text{C}_1\rangle).$$

Observe that in the transition from the partial solution $(\langle\text{START}\rangle, \ \langle\text{START}\rangle\langle\text{C}_0\rangle)$ to the partial solution $(\langle\text{START}\rangle\langle\text{C}_0\rangle, \ \langle\text{START}\rangle\langle\text{C}_0\rangle\langle\text{C}_1\rangle)$, we have used a polynomial (in $|x|$) number of pairs from LIST(x).

In a similar way, it can be checked that, if we have built the partial solution

$$(\langle\text{START}\rangle\langle\text{C}_0\rangle \ldots \langle\text{C}_{k-1}\rangle, \ \ \langle\text{START}\rangle\langle\text{C}_0\rangle \ldots \langle\text{C}_{k-1}\rangle\langle\text{C}_k\rangle), \tag{3.8}$$

the only way to continue and place $\langle C_k \rangle$ in the left hand side is: (a) If C_k contains a, then we can complete a solution by using pairs from Groups 3 and 5 and at the end the pair of Group 6, and (b) if C_k does not contain a, then there must be a legal move taking the machine M from configuration C_k to configuration C_{k+1}, and, in this case, the only partial solution that we can obtain is

$$(\langle \mathrm{START} \rangle \langle \mathrm{C_0} \rangle \ldots \langle \mathrm{C_{k-1}} \rangle \langle \mathrm{C_k} \rangle, \ \langle \mathrm{START} \rangle \langle \mathrm{C_0} \rangle \ldots \langle \mathrm{C_k} \rangle \langle \mathrm{C_{k+1}} \rangle). \tag{3.9}$$

It can be again checked that, since C_k has size bounded by the fixed polynomial $p(|x|)$, to make the transition from the partial solution given in Equation (3.8) to the partial solution given in Equation (3.9), we have used a number of pairs from LIST(x) that is bounded by a fixed polynomial in $|x|$.

Therefore, the instance LIST(x) has a solution if and only if M on input $z = 1\mathrm{code}(x)$ has a computation path that goes through a sequence of consecutive configurations $C_0, C_1, \ldots, C_k$, and the last configuration, C_k, contains the unique accepting state a. The existence of such a computation path shows that M accepts x, and, as noted, this can only happen in at most $p(|x|)$ steps. Thus, if C_k has the accepting state a, then $k \leq p(|x|)$. Recall our estimation on the number of pairs from LIST(x) necessary to make the transition from one partial solution to the next one containing a new $\langle C \rangle$ in the left hand side. It follows that there is a polynomial q such that M accepts $z = 1\mathrm{code}(x)$ if and only if LIST(x) has a solution of size $g(|x|)$. So, our reduction is

$$f(x) = (\mathrm{LIST}(x), 1^{g(|x|)}).$$

It is easy to check that $f(x)$ is computable in polynomial time, and, using the above remarks, that $x \in A$ if and only if $f(x) \in$ D-PC.

It remains to check the domination property. Note that the reduction f is 1-to-1 (this follows from the pair in Group 1 and because code$(\cdot)$ is injective). Also note that the length of the string $d(\triangle)zd(\square!q_0)$ (which appears in Group 1) is bounded by $|\mathrm{code}(x)| + O(L)$, where $L = O(\log(|x|))$, and that the length of each of the other strings in LIST(x) is bounded by $O(L) = O(\log(|x|))$. Consequently,

$$|f(x)| = |\mathrm{code}(x)| + O(\log(|x|)) + g(|x|).$$

It follows that for some polynomial r,

$$\mu'_{DPC}(f(x)) \geq 2^{-|\mathrm{code}(x)|} \cdot \frac{1}{r(|x|)} \geq \mu'(x) \cdot \frac{1}{r(|x|)},$$

and, therefore, μ^*_{DPC} dominates μ^*. ∎

3.7 Comments and bibliographical notes

The probabilistic algorithm for 3-SAT in Section 3.2 is due to Schöning [Sch99]. It has been slightly improved several times, and, at the time of this writing, the most efficient probabilistic algorithm for 3-SAT runs in time $O(1.32793^n)$ and has been developed by Rolf [Rol03]. Non-trivial exact algorithms for several NP-complete problems have been found and the article of Woeginger [Woe03] is an informative survey of this area.

As mentioned in Section 2.7, the classification schemas induced by effective Baire category concepts have been introduced in computation and complexity theory by Mehlhorn [Meh73]. Mehlhorn's approach has been extended in several directions primarily by considering different types of open set extensions and by limiting the computing power of the extension functions (see the articles by Lutz [Lut90], Fenner [Fen95], Ambos-Spies [AS96], Ambos-Spies and Reimann [ASR96]). The idea of using the superset topology in the context of effective Baire classification of classes inside NP is due to Zimand [Zim93]. The results from Section 3.3 are from the same paper [Zim93]. The technique used to demonstrate Theorem 3.3.9 and several other related results is called *delayed diagonalization* and has been invented by Ladner [Lad75] to show that if $\mathrm{P} \neq \mathrm{NP}$, then there exist sets in NP that are neither NP-complete nor in P.

The main concepts of resource-bounded measure theory have been developed by Lutz [Lut90, Lut92]. He has brought to light some earlier studies of Schnorr [Sch73] and has shown the applicability of this theory in the exploration of some quantitative issues in computational complexity. It is now a mature area with its own ramifications, open problems, and all the other attributes of a vital theory. The survey papers of Lutz [Lut97] and Ambos-Spies and Mayordomo [ASM97] provide a good coverage of the core directions. Theorem 3.4.1 and Theorem 3.4.3 state simple and basic facts of resource-bounded measure theory. The class of P-quasi-approximable sets has been introduced by Zimand [Zim98]. It is a generalization of a large number of classes (see the list in Corollary 3.4.10) that capture in various ways the idea of a polynomial-time weak membership property. Theorem 3.4.6 and Theorem 3.4.12 have been shown by Zimand [Zim98]. The fact that the hypothesis "NP does not have PF measure zero" implies that NP-completeness under Cook reductions differs from NP-completeness under Karp reductions (i.e., Theorem 3.4.13) has been shown by Lutz and Mayordomo [LM96]. This result has been extended to other reductions by Ambos-Spies and Bentzien [ASB97].

Relativization is a basic notion in computability theory. It has first been used in complexity theory by Baker, Gill, and Solovay [BGS75]. Their article shows the existence of oracle sets A and B such that $\mathrm{P}^A = \mathrm{NP}^A$ and $\mathrm{P}^B \subsetneq \mathrm{NP}^B$. The study of complexity classes relativized with random oracles has been initiated by Bennett and Gill [BG81]. They have shown that relative to a random oracle A, NP^A is P^A bi-immune. Theorem 3.5.2 is a strenghtening of this result and has been obtained by Hemaspaandra and Zimand [HZ96]. The notion of P-balanced immunity has been introduced by Müller [Mü93]. Kautz and Miltersen [KM94] have shown that relative to a random oracle A, NP^A does not have effective measure zero with respect to P^A-computable martingales.

The theory of average-case complexity was initiated by Levin [Lev86]. Levin's paper is very concise and it does not elucidate the motivation behind some of the

tasks that can be done exponentially faster on almost every input on a quantum computer.[1]

It should be noted that building a quantum computer faces some formidable technological challenges. This is an aspect that is not within the scope of this book.

There is a major difference at a very basic level between the states of a quantum computer and those of a classical computer. A classical computer at a given moment is in exactly one configuration (assuming that the program is loaded in the memory, a configuration consists of the content of the memory at the given time). Considering, for simplicity, that all input data has been entered and stored in the initial configuration, the configuration of a classical machine at time t precisely dictates the configuration at time $t+1$. Therefore, the computation of a classical computer can be fully described by a sequence of configurations

$$C_0 \rightarrow C_1 \rightarrow \ldots \rightarrow C_n,$$

where each transition $C_t \rightarrow C_{t+1}$ is done according to the controlling program and complies to the laws of classical physics. Each configuration can be written (encoded) as a string of characters, where each character is 0 or 1. Such a character is called a *bit*. Thus, a bit is exactly either 0 or 1, and a configuration can be given by presenting (separately, one after the other) a sequence of bits. These are very basic observations about the nature of a classical computation that, usually, we take for granted without giving them much thought. However, some of these features are no longer true for quantum computation. According to quantum theory, small particles, which supposedly describe the state of a quantum computer, do not have at a given time a fixed position and velocity. Consequently, at a given moment, a quantum computer can be in a *superposition* of configurations. Each configuration contributes with a certain complex-numerical weight, called the *amplitude*, to the superposition. Intuitively, we should perceive a quantum computer as being at a given time simultaneously and to varying degrees in several configurations at once.

The computation of a quantum computer can be described as a sequence of superpositions

$$\phi_0 \rightarrow \phi_1 \rightarrow \ldots \rightarrow \phi_n,$$

where the transitions $\phi_t \rightarrow \phi_{t+1}$ must be done in a way that satisfies the laws of quantum mechanics.

Some transition steps can be *measurements* which make a special type of transitions that do not have a classical counterpart. In a measurement, a device from outside the quantum system is observing the quantum computer or just a part of it. The measurement of a superposition will capture only one configuration that contributes to the superposition. We also say that a configuration is *observed*. More precisely, a configuration C of a superposition ϕ is observed with a probability that

[1]The task that we show to exhibit an exponential speed-up on a quantum computer is a relativized one that has access to a black-box function. Nevertheless, it does prove the capacity of quantum computation to be exponentially superior to classical computation.

is equal to the square of the absolute value of the amplitude of C in ϕ. It follows from here that the sum of the squares of the absolute values of all the configurations in a superposition must be equal to one. In addition, subsequent to the measurement, the superposition of the machine collapses to the unique configuration that has been observed. As we will see later, it is also possible to measure just a part of a quantum computer, such as a register. In this case, after the measurement, that part (in our example, the register) will be in a unique subconfiguration and the rest will be in a superposition of subconfigurations compatible with the fixed part.

It is now the time to transpose these concepts into a mathematical formalism. Similarly to the way in which the state of a classical computer can be described by bits, the superposition in which a quantum machine is at a given time is described by quantum bits, or *qubits*.

A qubit is a unit vector in the two dimensional complex vector space $\mathbb{C}^2$ for which a particular basis, denoted by $\{|0\rangle, |1\rangle\}$ has been fixed. The vector space $\mathbb{C}$ must also be equipped with an inner product, denoted $(\cdot, \cdot)$, and the vector basis, $|0\rangle$ and $|1\rangle$, should be orthonormal, i.e., $(|0\rangle, |1\rangle) = 0$. Thus, a qubit is represented as

$$\alpha_0|0\rangle + \alpha_1|1\rangle,$$

where α_0 and α_1 are complex number such that

$$|\alpha_0|^2 + |\alpha_1|^2 = 1.$$

We are saying that the qubit is in the superposition of the states $|0\rangle$ and $|1\rangle$ with amplitudes α_0 and α_1, respectively. Physically, the orthonormal vectors $|0\rangle$ and $|1\rangle$ may correspond to the vertical polarization and respectively to the horizontal polarization of a photon, or, equally well, they may correspond to the spin-up and spin-down states of an electron.

The notation $|x\rangle$ is part of the bra/ket notation introduced by Dirac. A vector such as $|x\rangle$ is called a *ket* and should be viewed as a column vector. For example the orthonormal basis vectors $|0\rangle$ and $|1\rangle$ can be expressed, respectively, as

$$\begin{pmatrix} 1 \\ 0 \end{pmatrix} \text{ and } \begin{pmatrix} 0 \\ 1 \end{pmatrix}$$

Any complex linear combination of $|0\rangle$ and $|1\rangle$, $\alpha_0|0\rangle + \alpha_1|1\rangle$, (in other words, any qubit) can be written

$$\begin{pmatrix} \alpha_0 \\ \alpha_1 \end{pmatrix},$$

or $(\alpha_0, \alpha_1)^T$. For a ket $|x\rangle$, there is a matching *bra*, denoted $\langle x|$. The bra $\langle x|$ is, by definition, the conjugate transpose of $|x\rangle$. Thus, $\langle x|$ should be viewed as a row vector. A bra $\langle x|$ and a ket $|y\rangle$ can be combined in $\langle x||y\rangle$, usually written as $\langle x|y\rangle$, and this denotes the inner product of the two vectors. For instance since the basis

$\{|0\rangle, |1\rangle\}$ is orthonormal, we have

$$\begin{aligned}\langle 0|0\rangle &= 1\\ \langle 0|1\rangle &= 0\\ \langle 1|0\rangle &= 0\\ \langle 1|1\rangle &= 1.\end{aligned}$$

The notation $|x\rangle\langle y|$ is the outer product of $|x\rangle$ and $\langle y|$ and it denotes a linear function from $\mathbb{C}^2$ to $\mathbb{C}^2$. For example, $|0\rangle\langle 1|$ is the mapping that maps $|1\rangle$ to $|0\rangle = (1,0)^T$ and $|0\rangle$ to $(0,0)^T$. This is so because

$$|0\rangle\langle 1|(|1\rangle) = |0\rangle\langle 1|1\rangle = 1|0\rangle = \begin{pmatrix}1\\0\end{pmatrix}$$

and

$$|0\rangle\langle 1|(|0\rangle) = |0\rangle\langle 1|0\rangle = 0|0\rangle = \begin{pmatrix}0\\0\end{pmatrix}.$$

In matrix form, $|0\rangle\langle 1|$ can be written as

$$\begin{pmatrix}0 & 1\\ 0 & 0\end{pmatrix}.$$

In other words,

$$|0\rangle\langle 1| = \begin{pmatrix}1\\0\end{pmatrix} \otimes (0 \quad 1) = \begin{pmatrix}0 & 1\\ 0 & 0\end{pmatrix},$$

that is, we replace the 1 in the $(1,0)^T$ with $1 \cdot \langle 1| = (0,1)$, and the 0 from the same $(1,0)^T$ with $0 \cdot \langle 1| = (0,0)$.

This will become more clear after we introduce the tensor product $\otimes$. As we have mentioned earlier, if we measure a qubit $\alpha_0|0\rangle + \alpha_1|1\rangle$, we see either $|0\rangle$, and this can happen with probability $|\alpha_0|^2$, or we see $|1\rangle$, with probability $|\alpha_1|^2$ (this is why it is necessary that $|\alpha_0|^2 + |\alpha_1|^2 = 1$). In addition, the prior-to-measurement qubit is irrevocably lost and replaced with the observed "pure" qubit $|0\rangle$ or $|1\rangle$.

More precisely, the measurement axiom of quantum theory stipulates that any device measuring a 2-dimensional quantum system has an associated orthonormal basis with respect to which the measurement is done. The measurement of the system transforms its state into one of the measuring device's basis vectors. In our considerations, we will assume the fixed orthonormal basis $\{|0\rangle, |1\rangle\}$ for all quantum systems and measuring devices and thus we can adopt the simplified situation above.

Analogously to the classical case, qubits are used to describe the state of a quantum computer. In order to perform a non-trivial computation, the qubits must be transformed. Let us see what transformations can a quantum computer do on a single qubit. Considerations from quantum theory impose that any such transformation T is linear and preserves the norm, i.e., $||T(x)|| = ||x||$ for any vector x. The transformations with these properties are called *unitary transformations* and can also be described as follows. Any linear transformation on a finite dimensional complex vector space corresponds to a matrix M with complex coefficients (the i-th column of M consists of the coefficients of the vector into which the i-th vector of the base is mapped). Let M^* denote the conjugate transpose of the matrix M. A matrix M corresponds to a unitary transformation, we also say that the matrix M is *unitary*, if $M \cdot M^* = \mathrm{I}$, where I is the identity matrix.

To illustrate, we present a few useful single-qubit transformations. We indicate in each case how the basis vectors are transformed and (redundantly) present the corresponding matrix:

$$\mathrm{I}: \quad \begin{array}{l} |0\rangle \to |0\rangle \\ |1\rangle \to |1\rangle \end{array} \qquad \mathrm{I} = \begin{pmatrix} 1 & 0 \\ 0 & 1 \end{pmatrix}$$

$$\mathrm{NOT}: \quad \begin{array}{l} |0\rangle \to |1\rangle \\ |1\rangle \to |0\rangle \end{array} \qquad \mathrm{NOT} = \begin{pmatrix} 0 & 1 \\ 1 & 0 \end{pmatrix}$$

$$\mathrm{PhaseShift}: \quad \begin{array}{l} |0\rangle \to -|0\rangle \\ |1\rangle \to -|1\rangle \end{array} \qquad \mathrm{PhaseShift} = \begin{pmatrix} -1 & 0 \\ 0 & -1 \end{pmatrix}$$

$$\mathrm{H}: \quad \begin{array}{l} |0\rangle \to \frac{1}{\sqrt{2}}|0\rangle + \frac{1}{\sqrt{2}}|1\rangle \\ |1\rangle \to \frac{1}{\sqrt{2}}|0\rangle - \frac{1}{\sqrt{2}}|1\rangle \end{array} \qquad \mathrm{H} = \frac{1}{\sqrt{2}} \begin{pmatrix} 1 & 1 \\ 1 & -1 \end{pmatrix}. \tag{4.1}$$

It can be readily checked that all the above transformations are unitary. The transformation H is called the Hadamard transformation and it plays an important role in many quantum algorithms. When applied to the ket $|0\rangle$, the Hadamard transformation produces a superposition in which $|0\rangle$ and $|1\rangle$ have equal amplitude. Thus, if we make a measurement in the resulting superposition, we will observe $|0\rangle$ and $|1\rangle$ with equal probability. All the examples above do a transformation of a single qubit. Such transformations are called *quantum unary gates* by analogy with computational gates in a circuit that does classical computation. Note however that in classical computation there exists only one non-trivial unary gate, namely the (classical) NOT gate, whereas in quantum computation there are infinitely many quantum unary gates.

Unary quantum gates are not sufficient for building a universal quantum computer (that is a computer capable of executing all the possible quantum algorithms). Therefore we need to introduce systems of multiple qubits. In classical computation, since a bit is an element of the set $\{0, 1\}$, to obtain sequences of, say, n bits we make the cartesian product $\{0, 1\}^n$. A qubit is a vector in the vector space $\mathbb{C}^2$ and the analogue of the cartesian product for vector spaces is the tensor product. Without going into the details of the construction, we will just mention that if V_1 is an n-dimensional vector space and V_2 is a k-dimensional vector space, one can construct a new vector space denoted $V_1 \otimes V_2$ of dimension nk, called the *tensor product* of V_1 and V_2. For each pair of vectors $h \in V_1$, $k \in V_2$, there is an associated vector denoted $h \otimes k$ in $V_1 \otimes V_2$ (called the tensor product of h and k) and the following properties hold for all $h, h_1, h_2 \in V_1$, $k, k_1, k_2 \in V_2$, and $\lambda \in \mathbb{C}$:

$$\begin{aligned} (h_1 + h_2) \otimes k &= h_1 \otimes k + h_2 \otimes k, \\ h \otimes (k_1 + k_2) &= h \otimes k_1 + h \otimes k_2, \\ \lambda(h \otimes k) &= (\lambda h) \otimes k = h \otimes (\lambda k). \end{aligned}$$

It follows that if $\{e_1, e_2, \ldots, e_n\}$ and $\{f_1, f_2, \ldots, f_k\}$ are respectively basis in V_1 and V_2, then $\{e_i \otimes f_j \mid 1 \le i \le n, 1 \le j \le k\}$ is a basis of $V_1 \otimes V_2$. If V_1 and V_2 have both an inner product, $V_1 \otimes V_2$ inherits the inner product defined by

$$(h_1 \otimes k_1, h_2 \otimes k_2) = (h_1, h_2) \cdot (k_1, k_2).$$

It is also possible to construct the *tensor product of two linear applications*. If $A : V_1 \to V_1$ and $B : V_2 \to V_2$ are two linear applications, then the tensor product of A and B is a linear application $A \otimes B : V_1 \otimes V_2 \to V_1 \otimes V_2$, defined by

$$(A \otimes B)(x \otimes y) = (Ax) \otimes (By).$$

For matrices A, B, C, D, U, doing the corresponding mappings, the following hold:

$$(A \otimes B)(C \otimes D) = AC \otimes BD$$

and

$$\begin{pmatrix} A & B \\ C & D \end{pmatrix} \otimes U = \begin{pmatrix} A \otimes U & B \otimes U \\ C \otimes U & D \otimes U \end{pmatrix}$$

For instance if

$$A = \begin{pmatrix} 1 & 0 \\ 1 & 1 \end{pmatrix} \text{ and } B = \begin{pmatrix} 1 & 2 \\ 3 & 4 \end{pmatrix}$$

then

$$A \otimes B = \begin{pmatrix} 1 & 2 & 0 & 0 \\ 3 & 4 & 0 & 0 \\ 1 & 2 & 1 & 2 \\ 3 & 4 & 3 & 4 \end{pmatrix}$$

If A^* denotes the conjugate transpose of the matrix A, then $(A \otimes B)^* = A^* \otimes B^*$. The tensor product of a finite sequence of matrices is unitary if and only if each matrix in the sequence is unitary up to a constant and the product of the constants is 1. Formally, if

$$U = A_1 \otimes \ldots \otimes A_n,$$

then U is unitary if and only if

$$A_i^* A_i = k_i I, \qquad i = 1, \ldots, n$$

and

$$k_1 k_2 \dots k_n = 1.$$

Because of the distributive law, we can perform calculations such as the following

$$(\alpha_0|v_0\rangle + \alpha_1|v_1\rangle) \otimes (\beta_0|u_0\rangle + \beta_1|u_1\rangle) =$$
$$\alpha_0\beta_0|v_0\rangle \otimes |u_0\rangle + \alpha_1\beta_0|v_1\rangle \otimes |u_0\rangle + \alpha_0\beta_1|v_0\rangle \otimes |u_1\rangle + \alpha_1\beta_1|v_1\rangle \otimes |u_1\rangle.$$

The above constructions are easy to generalize to the n-fold tensor product of n vector spaces $V_1, V_2, \dots, V_n$.

To conclude, if we need systems of multiple qubits we have to consider tensor products of the form $\mathbb{C}^2 \otimes \mathbb{C}^2 \otimes \dots \otimes \mathbb{C}^2$. For example, if we want to represent a system of n qubits $q_1, q_2, \dots, q_n$, each of them equal to

$$\frac{1}{\sqrt{2}}(|0\rangle + |1\rangle),$$

then we will say that the the global system of n qubits is in the state $q_1 \otimes q_2 \otimes \dots \otimes q_n$, which is the vector

$$\frac{1}{\sqrt{2^n}}\left(|0\rangle \otimes \dots \otimes |0\rangle + |0\rangle \otimes \dots \otimes |1\rangle + \dots + |1\rangle \otimes \dots \otimes |1\rangle\right),$$

lying in the space $(\mathbb{C}^2)^{\otimes n}$. It is common to abbreviate $|0\rangle \otimes |0\rangle \otimes \dots \otimes |0\rangle$ by $|00\dots0\rangle$, and similar notations are used for the other combinations of the vector basis $|0\rangle$ and $|1\rangle$. Thus, in the short notation, the above global system is in the superposition

$$\frac{1}{\sqrt{2^n}}\left(|0\dots0\rangle + |0\dots1\rangle + \dots + |1\dots1\rangle\right).$$

It is important to remark still another difference between the states of a classical computer and the states of a quantum computer. To describe a classical state one can describe *separately* the bits that make that state. This is no longer necessarily true with systems of qubits. For instance, a 2-qubit system can be in the state $|00\rangle + |11\rangle$ and this system cannot be described by giving the two qubits separately. Formally, $|00\rangle + |11\rangle$ cannot be decomposed into the tensor products of two qubits. Indeed, suppose that there exist $a_1, a_2, b_1, b_2 \in \mathbb{C}$ such that

$$(a_1|0\rangle) + b_1|1\rangle) \otimes (a_2|0\rangle) + b_2|1\rangle) = |00\rangle + |11\rangle.$$

The right hand side is equal to

$$a_1a_2|00\rangle + a_1b_2|01\rangle + b_1a_2|10\rangle + b_1b_2|11\rangle.$$

To have this equal to $|00\rangle + |11\rangle$, we need $a_1b_2 = 0$ which implies that either $a_1a_2 = 0$ or $b_1b_2 = 0$. States like $|00\rangle + |11\rangle$ that cannot be decomposed into the tensor product of individual qubits are called *entangled states.* These states defy normal intuition but, on the other hand, seem to play an essential role in the way quantum computation transcends some inherent performance limitations of classical computation.

We can now describe more elaborate quantum computations that act on more qubits. As before, a system of n qubits can be modified by applying a unitary transformation. For n qubits, such a transformation is given by a 2^n-by-2^n matrix (because there are 2^n vector basis vectors for a system of n qubits). Sometimes, these transformations cannot be decomposed into a tensor product of smaller transformations.

The *controlled-NOT transformation*, C_{NOT} (more commonly called the controlled-NOT gate) is similar to an if statement. It acts on two qubits as follows: It leaves the first qubit unchanged and it flips the second qubit if and only if the first bit is $|1\rangle$. The vectors $|00\rangle, |01\rangle, |10\rangle$ and $|11\rangle$ form an orthonormal basis for $C \otimes C$ and the C_{NOT} gate transforms these basis vectors as follows:

$$C_{\text{NOT}}: \quad \begin{array}{l} |00\rangle \rightarrow |00\rangle \\ |01\rangle \rightarrow |01\rangle \\ |10\rangle \rightarrow |11\rangle \\ |11\rangle \rightarrow |10\rangle. \end{array}$$

If we associate to the basis $|00\rangle, |01\rangle, |10\rangle, |11\rangle$, in order, the following complex 4-tuples $(1,0,0,0)^T, (0,1,0,0)^T, (0,0,1,0)^T, (0,0,0,1)^T$, then, in matrix notation, C_{NOT} is

$$\begin{pmatrix} 1 & 0 & 0 & 0 \\ 0 & 1 & 0 & 0 \\ 0 & 0 & 0 & 1 \\ 0 & 0 & 1 & 0 \end{pmatrix}.$$

Observe that, as it should be the case, C_{NOT} is a unitary transformation, because $C^*_{\text{NOT}} = C_{\text{NOT}}$ and $C_{\text{NOT}} \cdot C_{\text{NOT}} = I$. Also, it can be shown that C_{NOT} cannot be decomposed into a tensor product of two single-qubit transformations.

The C_{NOT} gate is important because Barenco et. al [BBC$^+$95]) have shown that the set consisting of the C_{NOT} gate and all the 1-qubit quantum gates of the form

$$\begin{pmatrix} \cos\alpha & \sin\alpha \\ -\sin\alpha & \cos\alpha \end{pmatrix}, \qquad \begin{pmatrix} e^{i\alpha} & 0 \\ 0 & e^{i\alpha} \end{pmatrix}$$

with $\alpha \in [0, 2\pi)$ can simulate all the quantum transformations. In this respect, this set of gates is similar to the sets of classical gates {AND , NOT} or {OR, NOT}, which are sufficient for performing all classical transformations.

A natural question is whether classical computation can be done via quantum transformations. The answer is not obvious, because quantum transformations are unitary and thus reversible, but the computations done by an AND gate or an OR gate are not reversible. The problem is that these transformations are not injective: Given the output of an AND or of an OR gate, one cannot tell what the inputs were (the NOT gate is injective and, obviously, reversible). Fortunately, by using some extra inputs and outputs, any transformation can be transformed into a reversible transformation from which one can extract the original transformation. Indeed, if $x \to f(x)$ is the original transformation from $\{0,1\}^n$ to $\{0,1\}^m$, then $(x,y) \to (x, y \oplus f(x))$ is reversible, where $\oplus$ is here the bitwise modulo 2 addition. This is easy to see because if we apply the new transformation twice we get

$$(x,y) \to (x, y \oplus f(x)) \to (x, y \oplus f(x) \oplus f(x)) = (x,y).$$

Thus to any arbitrary classical transformation f with n binary inputs and m binary outputs, one can attach the transformation

$$U_f : |x,y\rangle \to |x, y \oplus f(x)\rangle, \tag{4.2}$$

which is unitary for any f. In the above expression we have abused the notation conveniently, because the x and the y in the kets are in fact the tensor products of the bits that make the (classical) binary strings x and y. Also the pair x, y in the ket is another notation for $x \otimes y$. It is sometimes useful to think that there are two registers, one containing the tensor product of the qubits of x, and the other one containing the tensor product of the qubits of y. To calculate $f(x)$, the transformation U_f is applied to $|x,0\rangle$ (that is x tensored with m zeros), because, from Equation (4.2), $U_f(|x,0\rangle) = |x, f(x)\rangle$. Using this technique, it can be shown that any classical calculation can be carried on a quantum computer and, moreover, the increase in time complexity is by at most a constant multiplicative factor. (The time complexity of a quantum computation is defined in Section 4.3.)

We next give a more formal treatment of the concept of a measurement. We have seen how a measurement (also called an observation) acts on one qubit. Let us consider now a two qubit system. The state of such a system can be expressed as

$$\alpha_0|00\rangle + \alpha_1|01\rangle + \alpha_2|10\rangle + \alpha_3|11\rangle, \tag{4.3}$$

with

$$|\alpha_0|^2 + |\alpha_1|^2 + |\alpha_2|^2 + |\alpha_3|^2 = 1.$$

Suppose we measure the first qubit of the system with respect to the standard basis $\{|0\rangle, |1\rangle\}$. We will rewrite the state in Equation 4.3 as a tensor product of the bit being measured (in our example, the first one) and a vector of length one. Let

$$\beta_0 = \sqrt{|\alpha_0|^2 + |\alpha_1|^2} \text{ and } \beta_1 = \sqrt{|\alpha_2|^2 + |\alpha_3|^2}.$$

Then,

$$\begin{aligned} &\alpha_0|00\rangle + \alpha_1|01\rangle + \alpha_2|10\rangle + \alpha_3|11\rangle \\ &\qquad = |0\rangle(\alpha_0|0\rangle + \alpha_1|1\rangle) + |1\rangle(\alpha_2|0\rangle + \alpha_3|1\rangle) \\ &\qquad = \beta_0|0\rangle\Big(\frac{\alpha_0}{\beta_0}|0\rangle + \frac{\alpha_1}{\beta_0}|1\rangle\Big) + \beta_1|1\rangle\Big(\frac{\alpha_2}{\beta_1}|0\rangle + \frac{\alpha_3}{\beta_1}|1\rangle\Big). \end{aligned}$$

When we measure the first bit, with probability $\beta_0^2 = |\alpha_0|^2 + |\alpha_1|^2$ we observe $|0\rangle$, and with probability $\beta_1^2 = |\alpha_1^2 + |\alpha_3|^2$ we observe $|1\rangle$. In the first case, the system collapses to

$$|0\rangle\left(\frac{\alpha_0}{\beta_0}|0\rangle + \frac{\alpha_1}{\beta_0}|1\rangle\right)$$

and, in the second case, the system collapses to

$$|1\rangle\left(\frac{\alpha_2}{\beta_1}|0\rangle + \frac{\alpha_3}{\beta_1}|1\rangle\right).$$

This example can be generalized to measuring k qubits in a system of n qubits. Other types of measurement are possible for a system of n qubits. The most general type is obtained via the concept of an *observable*. An observable is a partioning of the 2^n-dimensional space of n qubits into a number of orthogonal subspaces

$$H = S_1 \times S_2 \times \ldots \times S_m.$$

Each superposition in H can be written as a sum

$$\psi_1 + \psi_2 + \ldots + \psi_m,$$

with $\psi_i \in S_i, i = 1, \ldots, m$. The measurement randomly selects one subspace S_i with probability equal to the square of the amplitude of ψ_i and collapses the system to ψ_i, also scaling to give length one to the resulting new superposition.

4.2 Quantum finite automata

IN BRIEF: There are quantum finite automata having exponentially fewer states than the equivalent minimal deterministic finite automata.

We can now show that there are circumstances in which quantum computation can be much more efficient than classical computation. We consider here the simplest model of computation, namely that of a finite automaton. We recall that a (classical) *finite automaton* M is a device described by the tuple $(Q, \Sigma, \delta, q_0, Q_{\text{acc}})$ where Q is the finite set of states, Σ is the finite input alphabet, $\delta : Q \times \Sigma \rightarrow Q$ is the transition function, $q_0 \in Q$ is the starting state, and Q_{acc} is a subset of Q consisting of the accepting states. The finite automaton can be visualized as having a read-only one-way tape on which it is written the input $x = x_1 x_2 \ldots x_m$, $x_i \in \Sigma, i = 1, \ldots, m$. The execution of M on x begins in the start state $r_0 = q_0$ after which M moves successively in the states

$$r_1 = \delta(r_0, x_1), \quad r_2 = \delta(r_1, x_2), \ldots, r_m = \delta(r_{m-1}, x_m).$$

The input x is accepted if r_m, the last state in the chain of transitions, is in Q_{acc}; if $r_m \notin Q_{\text{acc}}$, x is rejected.

There is an alternative way to present the transition function which will be helpful when we move on to quantum finite automata. Let $Q = \{q_0, \ldots, q_{n-1}\}$. For each symbol $a \in \Sigma$, we consider the matrix V_a of size $n \times n$ given by

$$V_a(i,j) = \begin{cases} 1, & \text{if } \delta(q_i, a) = q_j \\ 0, & \text{otherwise.} \end{cases}$$

It is easy to see that the collection of matrices $(V_a)_{a\in\Sigma}$ is a complete description of the transition function δ.

A quantum finite automaton (QFA) is defined similarly, the major difference being that the transitions are described by unitary transformations. There are a few less important variations as well, needed for a neat presentation. Thus a QFA is a tuple $M = (Q, \Sigma, (V_a)_{a\in\Sigma}, q_0, Q_{\text{acc}}, Q_{\text{rej}})$, where $Q = \{q_0, \ldots, q_{n-1})$ is the finite set of states, Σ is the finite alphabet, $(V_a)_{a\in\Sigma}$ is a collection of unitary linear transformations that describe the transition function of M, q_0 is the starting state, and $Q_{\text{acc}} \subset Q$ and $Q_{\text{rej}} \subset Q$ are two disjoint set of states, called the accepting states and respectively the rejecting states. The states in $Q_{\text{non}} = Q - (Q_{\text{acc}} \cup Q_{\text{rej}})$ are called the non-halting states. The starting state q_0 is a non-halting state. We associate to the states $\{q_0, \ldots, q_{n-1}\}$ in a one-to-one manner a system of orthonormal vectors $\{|q_0\rangle, \ldots, |q_{n-1}\rangle\}$ chosen from a sufficiently large vector space over the complex field $\mathbb{C}$, endowed with an inner product. For concreteness, using the concepts introduced in the previous section, each state is encoded as a tensor product of $|0\rangle$s and $|1\rangle$s. At a given time, the QFA M will be in a superposition that is described by a vector in the span of the vectors $\{|q_0\rangle, \ldots, |q_{n-1}\rangle\}$ (i.e., a linear combination with complex coefficients of these vectors). Let H_n denote this vector space. Henceforth, we identify a state q_i with the corresponding vector $|q_i\rangle$. Each transformation $V_a, a \in \Sigma$, is a unitary transformation from H_n to H_n. Thus V_a can be described by an $n \times n$ unitary matrix with complex coefficients (the i-th column is the image via the transformation of $|q_{i-1}\rangle$). As in the case of a (classical) finite automaton, M will do a series of transitions dictated by the input string that is written on its tape and by the transition function described by the transformations $(V_a)_{a\in\Sigma}$. After each move, the QFA M makes a measurement to see if it is in a state in Q_{acc}, or in Q_{rej}, or in Q_{non}. More precisely, let

$$E_{\text{acc}} = \text{span}\{|q\rangle \mid q \in Q_{\text{acc}}\},$$

$$E_{\text{rej}} = \text{span}\{|q\rangle \mid q \in Q_{\text{rej}}\}, \quad \text{and}$$

$$E_{\text{non}} = \text{span}\{|q\rangle \mid q \in Q_{\text{non}}\}.$$

These three subspaces define an *observable* of H_n. The subspaces $E_{\text{acc}}, E_{\text{rej}}$, and E_{non} are orthogonal to each other and, thus, any vector $|v\rangle \in H_n$ can be written as the sum of its projections $|v_{\text{acc}}\rangle, |v_{\text{rej}}\rangle$, and $|v_{\text{non}}\rangle$ onto E_{acc}, E_{rej}, and E_{non}. By normalizing, we can assume that the norm of $|v\rangle$ is one and there are unique (up to a difference in phase) complex numbers $\alpha_{\text{acc}}, \alpha_{\text{rej}}$, and α_{non} such that

$$|v\rangle = \alpha_{\text{acc}}|v_{\text{acc}}\rangle + \alpha_{\text{rej}}|v_{\text{rej}}\rangle + \alpha_{\text{non}}|v_{\text{non}}\rangle$$

and

$$|\alpha_{\text{acc}}|^2 + |\alpha_{\text{rej}}|^2 + |\alpha_{\text{non}}|^2 = 1.$$

As explained in the previous section, the measurement in superposition $|v\rangle$ with respect to the observable $(E_{\text{acc}}, E_{\text{rej}}, E_{\text{non}})$ consists in the selection of one of the subspaces $E_{\text{acc}}, E_{\text{rej}}, E_{\text{non}}$ with probability, respectively, $|\alpha_{\text{acc}}|^2$, $|\alpha_{\text{rej}}|^2$, $|\alpha_{\text{non}}|^2$. After the measurement the machine collapses to the projection of $|v\rangle$ onto the subspace that has been selected.

Let us now explain the evolution of the QFA M on an input $x = x_1 \ldots x_m$. It is convenient to use a special symbol, $\$ \notin \Sigma$, to mark the end of the input string, and thus we assume that on the input tape there is the string $x_1 \ldots x_m\$$. The QFA M starts in the superposition $|q_0\rangle$. Then M reads one by one the symbols $x_1, \ldots, x_m$. Reading one symbol a implies the following actions:

(1) The transformation V_a is applied to the current superposition ϕ and thus the machine moves into the superposition $\phi' = V_a(\phi)$.

(2) The new superposition ϕ' is measured with respect to the observable

$$(E_{\text{acc}}, E_{\text{rej}}, E_{\text{non}}).$$

The machine moves into a new superposition ϕ'' which is the projection of ϕ' onto the subspace that has been selected by the measurement. If $\phi'' \in E_{\text{acc}}$, M stops and the input is accepted. If $\phi'' \in E_{\text{rej}}$, M stops and the input is rejected. If $\phi'' \in E_{\text{non}}$ then the next symbol is read if there is one. If there is no symbol left, M stops.

The input is accepted if the machine stops in a superposition in E_{acc}. Otherwise, the input is rejected. The strings that are accepted with probability bigger than 1/2 form the *language accepted by the QFA*.

Let us consider an example. We will take a one-letter alphabet $\Sigma = \{a\}$. The states are $Q = \{q_0, q_1, q_{\text{acc}}, q_{\text{rej}}\}$, with $Q_{\text{acc}} = \{q_{\text{acc}}\}$ and $Q_{\text{rej}} = \{q_{\text{rej}}\}$. The starting state is q_0. The unitary transformations V_a and $V_\$$ are given by

$$V_a = \begin{pmatrix} 0 & 0 & \frac{1}{\sqrt{2}} & -\frac{1}{\sqrt{2}} \\ \frac{1}{\sqrt{2}} & -\frac{1}{\sqrt{2}} & 0 & 0 \\ 0 & 0 & \frac{1}{\sqrt{2}} & \frac{1}{\sqrt{2}} \\ \frac{1}{\sqrt{2}} & \frac{1}{\sqrt{2}} & 0 & 0 \end{pmatrix} \quad \text{and } V_\$ = \begin{pmatrix} 1 & 0 & 0 & 0 \\ 0 & 0 & 1 & 0 \\ 0 & 1 & 0 & 0 \\ 0 & 0 & 0 & 1 \end{pmatrix},$$

where the rows and the columns are indexed in order by $|q_0\rangle, |q_1\rangle, |q_{\text{acc}}\rangle, |q_{\text{rej}}\rangle$. For instance, when reading an a in the superposition $|q_1\rangle$ we look at the second column (indexed by $|q_1\rangle$) of V_a and see that the machine moves to the superposition $-\frac{1}{\sqrt{2}}|q_1\rangle + \frac{1}{\sqrt{2}}|q_{\text{rej}}\rangle$. This move will be also denoted as

$$|q_1\rangle \xrightarrow{a} -\frac{1}{\sqrt{2}}|q_1\rangle + \frac{1}{\sqrt{2}}|q_{\text{rej}}\rangle.$$

This machine works as follows on the input aa.

Step 1. Recall first that, in fact, the string $aa\$$ is written on the tape of M. The machine starts in $|q_0\rangle$. By reading the first a the machine moves into the superposition $\frac{1}{\sqrt{2}}|q_1\rangle + \frac{1}{\sqrt{2}}|q_{\text{rej}}\rangle$. A measurement is next done and $|q_{\text{rej}}\rangle$ is observed with probability $(\frac{1}{\sqrt{2}})^2 = \frac{1}{2}$ and in this situation the machine stops with the reject verdict. With probability $(\frac{1}{\sqrt{2}})^2 = \frac{1}{2}$, $|q_1\rangle$ is observed, the machine collapses to $|q_1\rangle$ and the computation continues with Step 2.

Step 2. This step is reached with probability $\frac{1}{2}$. In superposition $|q_1\rangle$, the second a is read and, according to V_a, the machine moves into $-\frac{1}{\sqrt{2}}|q_1\rangle + \frac{1}{\sqrt{2}}|q_{\text{rej}}\rangle$. A measurement is done and with probability $(\frac{1}{\sqrt{2}})^2 = \frac{1}{2}$ (conditioned by the fact that we are in Step 2) $|q_{\text{rej}}\rangle$ is observed and the machine stops with the reject verdict. With conditional probability $(\frac{1}{\sqrt{2}})^2 = \frac{1}{2}$, $|q_1\rangle$ is observed, the machine collapses to $|q_1\rangle$ and the computation continues with Step 3.

Step 3 This step is reached with probability $\frac{1}{2} \cdot \frac{1}{2} = \frac{1}{4}$. The \$ is read and, according to $V_\$$ the machine moves into $|q_{\text{acc}}\rangle$. A measurement is done and with conditional probability 1, $|q_{\text{acc}}\rangle$ is observed and the machine stops with the accept verdict.

Thus the string aa is accepted with probability $\frac{1}{4}$ and rejected with probability $\frac{3}{4}$.

The natural question is: What are the quantum finite automata good for? It can be shown that any language accepted by a QFA is regular (thus accepted by a classical finite automaton) and, moreover, there are quite simple regular languages that cannot be accepted by QFAs. Thus classical finite automata should not be thrown away yet. However, as we will show next, there are situations in which QFAs can be much more efficient than their classical counterparts. When talking about finite automata, the most relevant computational resource is the number of states in the automaton. We will give an example of a family of languages that can be accepted quantically with a number of states that is exponentially smaller than the best (i.e., with the minimum number of states) classical finite automata. This family is formed by the languages (L_p), p prime number, where

$$L_p = \{\pounds a^i \mid i \text{ is divisible by } p\}$$

($\pounds$ is a special marking symbol). It is easy to see that any classical finite automaton that accepts L_p needs p states: There are p equivalence classes for the relation $\equiv_{L_p}$ given by $x \equiv_{L_p} y$ if and only if for every string z, $xz \in L_p \leftrightarrow yz \in L_p$, and as it is well known, the minimal classical finite automaton has one state for each equivalence class of $\equiv_{L_p}$. It can be shown that even probabilistic finite automata that accept L_p with probability $\frac{1}{2} + \epsilon$, for some fixed ϵ, need to have at least p states. In contrast, we show the following.

Theorem 4.2.1 *For any prime number p, for any $\epsilon > 0$, there is a QFA with $O(\log p)$ states that accepts L_p with error probability at most ϵ.*

Proof. Let p be a fixed prime number. We will first build a QFA M that accepts all the strings a^j with j a multiple of p with probability 1 and rejects the other strings of the form a^j with probability 1/8. In the second stage of the proof we will show how to amplify the probability of rejection to $1 - \epsilon$.

For the first stage of the proof we need $p - 1$ auxiliary automata M_k, $k \in \{1, \ldots, p - 1\}$. The automaton M_k has the states $Q = \{q_0, q_1, q_{\text{acc}}, q_{\text{rej}}\}$ with q_0 the starting state, and the transition function is given by the unitary transformations:

$$V_a = \begin{pmatrix} \cos(\frac{\pi k}{p}) & i\sin(\frac{\pi k}{p}) & 0 & 0 \\ i\sin(\frac{\pi k}{p}) & \cos(\frac{\pi k}{p}) & 0 & 0 \\ 0 & 0 & 1 & 0 \\ 0 & 0 & 0 & 1 \end{pmatrix} \text{ and } V_\$ = \begin{pmatrix} 0&0&1&0 \\ 0&0&0&1 \\ 1&0&0&0 \\ 0&1&0&0 \end{pmatrix}.$$

The columns and the rows of V_a and $V_\$$ are indexed in order by $|q_0\rangle$, $|q_1\rangle$, $|q_{\text{acc}}\rangle$, and $|q_{\text{rej}}\rangle$.

Claim 4.2.2 *For each integer $j \geq 0$, after reading a^j the automaton M_k is in the superposition* $\cos\left(\frac{\pi jk}{p}\right)|q_0\rangle + i\sin\left(\frac{\pi jk}{p}\right)|q_1\rangle$.

Proof. The proof is by induction on j. For $j = 0$ the relation is immediate. For the transition step $j \to j + 1$, notice that from the superposition

$$\cos\left(\frac{\pi jk}{p}\right)|q_0\rangle + i\sin\left(\frac{\pi jk}{p}\right)|q_1\rangle,$$

after reading an a, the machine moves to

$$\begin{aligned} &\cos\left(\frac{\pi jk}{p}\right)\left[\cos\left(\frac{\pi k}{p}\right)|q_0\rangle + i\sin\left(\frac{\pi k}{p}\right)|q_1\rangle\right] \\ &\quad + i\sin\left(\frac{\pi jk}{p}\right)\left[i\sin\left(\frac{\pi k}{p}\right)|q_0\rangle + \cos\left(\frac{\pi k}{p}\right)|q_1\rangle\right] \\ &\quad = \cos\left(\frac{\pi(j+1)k}{p}\right)|q_0\rangle + i\sin\left(\frac{\pi(j+1)k}{p}\right)|q_1\rangle. \quad \blacksquare \end{aligned}$$

Claim 4.2.3 *Let a^j be a string such that j is divisible by p. Then each QFA $M_k, k = 1, \ldots, p$, accepts a^j with probability 1.*

Proof. Fix $k \in \{1, \ldots, p - 1\}$. Reading a^j, the QFA M_k does the following transitions

$$|q_0\rangle \xrightarrow{a^j} \cos\left(\frac{\pi jk}{p}\right)|q_0\rangle + i\sin\left(\frac{\pi jk}{p}\right)|q_1\rangle.$$

The last superposition is $(\pm 1)|q_0\rangle$, because $\cos\left(\frac{\pi jk}{p}\right)$ is 1 or -1 and $\sin\left(\frac{\pi jk}{p}\right) = 0$. Therefore, when reading the end-marker \$ and making the transition dictated by $V_\$$, M_k moves into the superposition $(\pm 1)|q_{acc}\rangle$. Thus, M_k accepts a^j with probability 1. ∎

Claim 4.2.4 *Let a^j be a string such that j is prime with p. Then $\frac{p-1}{2}$ of the QFAs M_k reject a^j with probability at least 1/2.*

Proof. Reading a^j and then the end-marker \$, a QFA M_k does the following transitions

$$|q_0\rangle \xrightarrow{a^j} \cos\left(\frac{\pi jk}{p}\right)|q_0\rangle + i\sin\left(\frac{\pi jk}{p}\right)|q_1\rangle \xrightarrow{\$} \cos\left(\frac{\pi jk}{p}\right)|q_{acc}\rangle + i\sin\left(\frac{\pi jk}{p}\right)|q_{rej}\rangle.$$

Thus the probability of accepting is $\cos^2\left(\frac{\pi jk}{p}\right)$ and $\cos^2\left(\frac{\pi jk}{p}\right) \leq \frac{1}{2}$ if and only if

$$\frac{\pi jk}{p} \in \bigcup_{\ell}\left[\pi\ell + \frac{\pi}{4}, \pi\ell + \frac{3\pi}{4}\right].$$

Clearly,

$$\frac{\pi jk}{p} \in \bigcup_{\ell}\left[\pi\ell + \frac{\pi}{4}, \pi\ell + \frac{3\pi}{4}\right] \longleftrightarrow \frac{\pi((jk) \mod (p))}{p} \in \left[\frac{\pi}{4}, \frac{3\pi}{4}\right].$$

Since j is prime with p, the values $jk \mod (p)$, $k \in \{1, \ldots, p-1\}$, are just $1, 2, \ldots, p-1$ in a different order. It remains to count how many of the values

$$\frac{\pi}{p}, \frac{2\pi}{p}, \ldots, \frac{(p-1)\pi}{p}$$

are in the interval $\left[\frac{\pi}{4}, \frac{3\pi}{4}\right]$. If $p = 4m+1$, for some natural m, it can be checked that $\frac{j}{4m+1} \in \left[\frac{1}{4}, \frac{3}{4}\right]$ for $j \in \{m+1, \ldots, 3m\}$, and thus there are $2m = \frac{p-1}{2}$ values that are in the interval. If $p = 4m+3$ for some natural m, it can be checked that $\frac{j}{4m+3} \in \left[\frac{1}{4}, \frac{3}{4}\right]$ for $j \in \{m+1, \ldots, 3m+1\}$, and thus there are $2m+1 = \frac{p-1}{2}$ values that are in the interval as well. ∎

Thus, for a^j with j not a multiple of p, if we pick at random one of the QFAs M_k, with probability $\frac{1}{2}$ over the choice of the QFA, M_k will reject a^j with probability at least $\frac{1}{2}$. We need however a single automaton that rejects all the strings a^j with j prime with p with some significant (say, constant) probability. To obtain such an automaton we consider sequences of $r = \lceil 8 \ln p \rceil$ automata randomly chosen (with repetition) from the set of automata $M_k, k \in \{1, \ldots, p-1\}$. Such a sequence is said to be *good* for a string a^j if at least $\frac{1}{4}$ of all the machines in the sequence reject a^j with probability at least $\frac{1}{2}$.

Claim 4.2.5 *There is a sequence that is good for all strings a^j for which j is prime with p.*

Proof. Observe first that all the automata M_k behave in the same way on a^j and $a^{j'}$ with $j = j'$ mod (p) (because after reading a^p the machine is back to $|q_0\rangle$). So, we need to look only at $a, a^2, \ldots, a^{p-1}$ and find a sequence that is good for all these strings. Fix a^j with $j \in \{1, \ldots, p-1\}$. How many sequences are not good for a^j? Let us consider a random such sequence (i.e., each element is chosen randomly and independently of the other elements) $M_{k_1}, \ldots, M_{k_r}$. The probability that a fixed M_{k_i} is good for a^j is $\frac{1}{2}$. Using the Chernoff's bounds, the probability that a fraction of less than $\frac{1}{4} = \frac{1}{2} - \frac{1}{4}$ of the automata in the sequence is good for a^j is at most $e^{-2(\frac{1}{4})^2 \cdot 8 \ln p} = \frac{1}{p}$. The fraction of sequences that are not good for at least one a^j, $j \in \{1, \ldots, p-1\}$, is thus at most $\frac{p}{p-1} < 1$. So there is a sequence that is good for all a^j with j not a multiple of p. ∎

Let $M_{k_1}, \ldots, M_{k_r}$ be a good sequence for all a^j with j not a multiple of p (recall that $r = \lceil 8 \ln p \rceil$). We build an automaton M that "composes" these automata together. M consists of $M_{k_1}, \ldots, M_{k_r}$ plus an extra state which is the starting state. The accepting states will be the union of the accepting states of the automata in the sequence and the rejection states will be the union of the rejecting states of the automata in the sequence. Using a Hadamard transformation, upon reading the left end-marker $\pounds$, M passes from the starting state to a superposition of the starting states of $M_{k_1}, \ldots, M_{k_r}$ with equal amplitudes. After this first step, the transitions of each M_{k_i} are performed. If a word is in L_p, then it is accepted with probability 1, because each M_{k_i} accepts with probability 1. If $\pounds a^j \notin L_p$, at least a fraction of $\frac{1}{4}$ of the automata $M_{k_1}, \ldots, M_{k_r}$ will reject it with probability at least $\frac{1}{2}$. So such a string a^j is rejected with probability $\frac{1}{8}$.

We will next increase the probability of rejection from $\frac{1}{8}$ to $1 - \epsilon$. This is done by "iterating" each automaton M_k d times, for a well-chosen value of d which will be specified later. More precisely, for each $k \in \{1, \ldots, p-1\}$ we build an automaton M'_k having 2^d non-halting states labelled $q_{0\ldots0}, q_{0\ldots1}, \ldots, q_{1\ldots1}$ (each index is d bits long). If

$$D_k = \begin{pmatrix} \cos\left(\frac{\pi k}{p}\right) & i \sin\left(\frac{\pi k}{p}\right) \\ i \sin\left(\frac{\pi k}{p}\right) & \cos\left(\frac{\pi k}{p}\right) \end{pmatrix},$$

then the matrix describing the transition of M'_k on reading a is

$$\underbrace{D_k \otimes \ldots \otimes D_k}_{d \text{ times}}.$$

The QFA M'_k has also an accepting state $q_{\rm acc}$ and $2^d - 1$ rejecting states that correspond into a one-to-one manner to the states $q_{0\ldots1}, \ldots, q_{1\ldots1}$. The starting state is $q_{0\ldots0}$. When reading the end-mark \$ the automaton M'_k passes from $q_{0\ldots0}$ to the accepting state, and from any other state $q_{x_1\ldots x_d}$ to the corresponding (via the one-to-one mapping) rejecting state. Given the transition function, it can be seen that a state $q_{x_1\ldots x_d}$, where each $x_i \in \{0, 1\}$, can be viewed as the tensor product $q_{x_1} \otimes \ldots \otimes q_{x_d}$ of the states of the QFA M_k that we have considered before (recall that we identify q_i with $|q_i\rangle$). Then, analogously to Fact 4.2.2, it can be shown that, after reading a^j, M'_k is in the superposition

$$\underbrace{\left(\cos\left(\frac{\pi jk}{p}\right)|q_0\rangle + i\sin\left(\frac{\pi jk}{p}\right)|q_1\rangle\right)\otimes\ldots\otimes\left(\cos\left(\frac{\pi jk}{p}\right)|q_0\rangle + i\sin\left(\frac{\pi jk}{p}\right)|q_1\rangle\right)}_{d\text{ times}}.$$

The amplitude of $|q_0\rangle \otimes \ldots \otimes |q_0\rangle = q_{0\ldots0}$ in the above superposition is $\left(\cos\left(\frac{\pi jk}{p}\right)\right)^d$. If j is a multiple of p, this is still 1 or -1, and thus the strings a^j with j a multiple of p are accepted with probability 1.

Let $\gamma = \epsilon/3$. For any $\delta \in [0,1]$, it holds that

$$\left|\cos\left(\frac{\pi jk}{p}\right)\right| \leq 1-\delta$$

for

$$\frac{\pi jk}{p} \in \bigcup_{\ell} [\pi\ell + \arccos(1-\delta), \pi(\ell+1) - \arccos(1-\delta)].$$

As before we need to count how many elements in the sequence

$$\frac{\pi}{p}, \frac{2\pi}{p}, \ldots, \frac{(p-1)\pi}{p}$$

are in the interval

$$[\arccos(1-\delta), \pi - \arccos(1-\delta)].$$

There is $\delta \in [0,1]$, such that at least a fraction of $1-\gamma$ values from the above sequence verify the relation. We take d such that $(1-\delta)^d < \gamma$. So, at least a fraction of $1-\gamma$ of the machines M'_k accept a^j with probability at most γ, or, equivalently, reject a^j with probability at least $1-\gamma$. We call such a M'_k *good*. Next, we call a sequence of automata $M'_{k_1}, \ldots, M'_{k_m}$ *good* if at least $1-2\gamma$ of them are good. For a fixed a^j with j not divisible by p, by the Chernoff bounds, the probability that less than $1-2\gamma = (1-\gamma)-\gamma$ fraction of M'_k is good is at most $e^{-2\gamma^2 m} = \frac{1}{p}$, for $m = \lceil \frac{1}{\gamma^2}\ln p \rceil$. Thus, a fraction of at most $\frac{p-1}{p}$ sequences are not good for at least one a^j with $j \in \{1, \ldots, p-1\}$. Reasoning as before, there is a sequence $M'_{k_1}, \ldots, M'_{k_m}$, with $m = \lceil \frac{1}{\gamma^2}\ln p \rceil$, that is good for all a^j, j not divisible by p. We compound the automata in the good sequence as before. If a word is in L_p then it is accepted with probability 1. If $\pounds a^j \notin L_p$, at least a fraction of $(1-2\gamma)$ of the automata will reject it with probability at least $1-\gamma$. Thus the probability that such a string is rejected by the compound automata is $(1-2\gamma)(1-\gamma) > 1-3\gamma = 1-\epsilon$. Note that the compound automata has $1 + 2^{d+1}\cdot m = O(\log p)$ states. This finishes the proof. ∎

4.3 Polynomial-time quantum algorithms

IN BRIEF: There are tasks that can be done by quantum algorithms in polynomial time and that, with classical algorithms, require exponential time on almost every input.

In spite of Theorem 4.2.1, quantum finite automata are quite computationally-impaired devices: It has been shown that they cannot recognize even the class of regular languages.

Therefore to realize the potential of quantum computation, we will pass to full-powered quantum computers. From a previous discussion, we know that classical computation can be carried on a quantum computer. Moreover, by a careful analysis of the way in which a classical computation can be done in a reversible manner and thus implemented through quantum gates, it can be shown that a classical computation that takes $t(n)$ time can be run on a quantum computer in $O(t(n))$ time (see for example Berthiaume [Ber97]). On the other hand, we will show below that quantum computation can be time-wise exponentially more efficient than classical computation. In fact, in the spirit of this book, we will show a very strong separation between quantum and classical computation: There are problems solvable in polynomial time by a quantum machine that has access to a black-box function and that need exponentially many classical computation steps (with access to the black-box function) on almost every input.

We need to first describe the concept of a quantum computer. One appropriate model is that of a quantum Turing machine which, roughly said, can be built from a classical Turing machine similarly to the way in which we have defined a quantum finite automaton starting from a classical finite automaton.

Unfortunately, describing programs for quantum Turing machines is an extremely tedious enterprise. Moreover, the basic steps of the algorithms are hard to understand because the main driving ideas are cluttered in low-level technicalities. There is an alternative model, that of a *quantum register machine*, that allows for more natural descriptions and that tends to become the standard medium for presenting quantum algorithms.

A quantum register machine consists of a constant number of registers. Each register is capable of storing a number of qubits (this number depends on the length of the input). Initially, the first register contains the input encoded in binary and represented by the corresponding qubit combination (e.g., if the input is 101, then the first register is initialized with $|101\rangle$) and the rest of the registers are set to $|0\dots0\rangle$.

A *basic operation* on a quantum register machine is either a *simple unitary transformation* U or a measurement M.

A unitary transformation is said to be *simple* if it transforms one or two qubits and acts as the identity on the other qubits. Thus, a simple unitary transformation is of the form $V \otimes I$, where V is a unitary transformation of one or two qubits and

I is the identity transformation of the remaining qubits. In other words, a simple unitary transformation is given by a 1-qubit or a 2-qubit gate together with the specification of the qubit or of the pair of qubits to which it applies.

A measurement M can be applied to one or more registers. If the register (or the registers) that is being measured is in the superposition ψ, the effect of the observation is that the register (or the registers) collapses to one of the configuration that contributes to the superposition ψ with a probability equal to the square of the amplitude of the configuration. In other words, the registers just specify an observable on the space of all qubits of the machine.

A computation for a quantum register machine is a sequence of basic operations

$$C_1, \ldots, C_k,$$

and the time complexity of the computation is the number of basic operations in the sequence.

In principle, we should have considered as valid transformations only the unitary transformations that correspond to a universal set of quantum gates such as the set proven to be universal by Barenco et. al. [BBC$^+$95] consisting of the C_{NOT} gate and the rotations of a single qubit. However, for simplicity, we allow all unitary transformations of one and two qubits as basic operations. As presented above, the register quantum machine is a non-uniform model because it depends on the length of the input. One can define a uniform[2] version and Yao [Yao93] has shown that this modified model is equivalent with a quantum Turing machine.

In what follows, we consider only quantum computations that output either 1, meaning acceptance of the input, or 0, meaning rejection of the input. Due to the probabilistic nature of the measurement operation, these outcomes are probabilistic as well. In analogy with classical computation, we will say that a quantum computation is feasible if it can be done with a polynomial number of operations and if the result, 0 or 1, is achieved with a probability bounded away from 1/2 by a constant.

Definition 4.3.1 *A language $L \subseteq \Sigma^*$ is in the class* BQP *if there is a uniform quantum register machine M and a constant $\epsilon > 0$ such that on every input $x \in \Sigma^*$*

(1) *if $x \in L$, then M accepts x with probability at least $\frac{1}{2} + \epsilon$,*

(2) *if $x \notin L$, then M rejects x with probability at least $\frac{1}{2} + \epsilon$.*

We are interested in exploring the relation between (a) the class of problems that are feasible with classical computational means and (b) the class of problems that are feasible using quantum computation. The class in (a) is either the class P or,

[2]Uniform, in this context, means that there is a classic polynomial-time Turing machine that given an input x builds the initial setting of the registers of the quantum register machine with input $|x\rangle$ together with a table containing, in the order in which they are performed, the sequence of: (a) 1-qubit or 2-qubit quantum gates and the qubits to which they apply, and (b) the measurement operations

if we allow probabilistic computation and occasionaly wrong answers, BPP. The class in (b) is BQP.

According to our earlier discussion about executing classical computational operations on a quantum computer, it holds that $\mathrm{P} \subseteq \mathrm{BPP} \subseteq \mathrm{BQP}$. It can be shown that $\mathrm{BQP} \subseteq \mathrm{P}^{\#\mathrm{P}}$, where $\mathrm{P}^{\#\mathrm{P}}$ is the class of problems solvable in polynomial time by a (classical) Turing machine with access to an oracle in #P.[3] It is well-known that $\mathrm{P}^{\#\mathrm{P}}$ is contained in PSPACE. It follows that separating BQP and BPP will also separate P from PSPACE, a result which is beyond reach at this time.

Therefore we will content with comparing the relativized computational models, that is quantum and classical computation done with respect to an oracle. In this context, we will show a strong separation of the models: There is a task that can be done in polynomial time on a quantum computer, but that needs classical exponential time on almost every input.

In our discussion we use *function oracles.* Such an oracle is given by a function f that maps strings to strings. The oracle, when queried about an input x, provides in one computation step $f(x)$. Such an oracle is also called a *black-box* because the above mechanism is the only way by which the algorithm can obtain any information about the function. Similarly to Equation (4.2), the access to the oracle is modeled by allowing, for two special registers (the *oracle registers*), the transformation $|x, y\rangle \rightarrow |x, y \oplus f(x)\rangle$. Note that if y is the all zeros string, the transformation provides the value of $f(x)$ in the second register. This is a reversible transformation and will be counted as one computation step. It is noteworthy, that many of the known quantum algorithms, including the famous search algorithm of Grover [Gro96] (see Section 4.4), can be casted in the black-box model.

Task for which we prove the superiority of quantum computation

The computational problem that we consider is a variation of a problem first considered by Simon [Sim97]. The problem is defined in terms of a function oracle that satisfies a certain constraint. Namely, we consider function oracles A of the following form:

Definition 4.3.2 *A function oracle A* satisfies the Simon's promise *if A is a collection of functions $(f_{n,A})_{n\in\mathbb{N}^+}$ having, for each $n \geq 1$, the following properties:*

(i) $f_{n,A} : \{0,1\}^n \rightarrow \{0,1\}^{n-1}$,

(ii) *$f_{n,A}$ is 2-to-1,*

(iii) *there is a string $s_{n,A}$ in $\{0,1\}^n - \{0^n\}$ such that for all x of length n, $f_{n,A}(x \oplus s_{n,A}) = f_{n,A}(x)$.*

In the above relation, $\oplus$ denotes the bitwise exclusive-or.

---

[3]A function $f : \Sigma^* \rightarrow \mathbb{N}$ is in #P if there is a polynomial-time nondeterministic machine M such that, for all $x \in \Sigma^*$, $f(x)$ is the number of accepting computations of M on input x.

NOTATION. Let $\mathcal{A}$ be the class of functional oracles that satisfy Simon's promise. For $A \in \mathcal{A}$ and $x \in \Sigma^+$, $A(x)$ denotes $f_{|x|,A}(x)$.

Let $A \in \mathcal{A}$. To avoid some technical complications, the function oracle A is not defined at the empty word ϵ. Thus queries about the value of $A(\epsilon)$ are illegal. The parity of a binary string is 1 if the number of 1s in the string is odd, and it is 0 otherwise. We define the language L_A over $\{0, 1\}$ as follows:

$$L_A = \{w \in \{0,1\}^+ \mid \text{parity}(s_{4|w|,A}) = 1\} \tag{4.4}$$

i.e., $w \in L_A$ if and only if $w \neq \epsilon$ and there are an odd number of 1s in the unique string s other than $0^{4|w|}$ with the property that for all x of length $4|w|$,

$$A(x \oplus s) = A(x).$$

We show that (a) for all A in $\mathcal{A}$, the decision problem "Is x in L_A" can be solved efficiently via quantum computation, and (b) for most A in $\mathcal{A}$, any classical algorithms that solves the problem runs exponentially many steps on almost every input. We start with (a).

Theorem 4.3.3 *For any $A \in \mathcal{A}$, there is a polynomial-time quantum algorithm that accepts L_A with zero error probability.*

Proof. We will show that there is a polynomial-time quantum algorithm that has access to A and that on input w determines $s_{4|w|,A}$. Since L_A is just the set of those strings w for which the parity of $s_{4|w|,A}$ is odd, the conclusion follows immediately.

Let us fix an oracle $A \in \mathcal{A}$, a positive integer n, an input w of length n, and, for brevity, let us denote $s_{4n,A}$ by s and $f_{4n,A}$ by f. Strings of length $4n$ will be viewed as vectors in the vector space $(\mathbb{Z}_2)^{4n}$ over the field $\mathbb{Z}_2$. The algorithm will determine $4n - 1$ vectors in $(\mathbb{Z}_2)^{4n}$, $z_1, z_2, \ldots, z_{4n-1}$ such that

$$z_1 \cdot s = z_2 \cdot s = \ldots = z_{4n-1} \cdot s = 0,$$

where the operation "$\cdot$" is the inner product in $(\mathbb{Z}_2)^{4n}$. Moreover, we will choose the vectors $z_1, z_2, \ldots, z_{4n-1}$ so as to be linearly independent. Once we have $z_1, z_2, \ldots, z_{4n-1}$ with these properties, we solve (by classical means) the system of linear equations

$$\begin{aligned} z_1 \cdot x &= 0 \\ z_2 \cdot x &= 0 \\ &\vdots \\ z_{4n-1} \cdot x &= 0. \end{aligned}$$

Since the system has the solutions 0 and s, we will determine with zero error probability the vector s.

End of Procedure Description

The procedure is used to generate the vectors $z_1, \ldots, z_{4n-1}$ inductively as follows. Initially we start with some arbitrary t. We can assume that $t \neq s$, because otherwise we are done. Using the quantum procedure we determine a vector z_1 such that $z_1 \cdot s = 0$ and $z_1 \cdot t = 1$.

Suppose that we have found $\{z_1, \ldots, z_k\}$, a set of linearly independent vectors such that $z_1 \cdot s = z_2 \cdot s = \ldots = z_k \cdot s = 0$. We need to find z_{k+1} that is linearly independent with $\{z_1, \ldots, z_k\}$ and $z_{k+1} \cdot s = 0$. We first determine t such that $z_1 \cdot t = 0, z_2 \cdot t = 0, \ldots, z_k \cdot t = 0$, i.e., t is orthogonal to the subspace generated by $z_1, \ldots, z_k$.

If $t = s$, we are done. If not, we run the quantum procedure and we obtain a vector z. This vector z can be taken to be z_{k+1}. Indeed $z \cdot s = 0$, and z cannot be a linear combination $z_1, \ldots, z_k$, because in this case we would have $z \cdot t = 0$. Consequently, iterating this construction $4n - 1$ times, we obtain the desired sequence $z_1, \ldots, z_{4n-1}$ (or we find s). It remains to show how to do the sign flipping in Step 3 with 1-qubit quantum gates. We first apply the transformation H' to the fourth register, where H' is given by

$$H' : \begin{array}{l} |0\rangle \to \frac{1}{\sqrt{2}}|0\rangle - \frac{1}{\sqrt{2}}|1\rangle \\ |1\rangle \to \frac{1}{\sqrt{2}}|0\rangle + \frac{1}{\sqrt{2}}|1\rangle \end{array} \qquad H' = \frac{1}{\sqrt{2}} \begin{pmatrix} 1 & 1 \\ -1 & 1 \end{pmatrix}.$$

This does the transition

$$\frac{1}{\sqrt{2^{4n}}} \sum_{x \in \{0,1\}^{4n}} |t, x, \min(f(x), f(x \oplus t)), 0\rangle \to \frac{1}{\sqrt{2^{4n}}} \sum_{x \in \{0,1\}^{4n}} |t, x, \min(f(x), f(x \oplus t)), b\rangle, \quad (4.7)$$

with b being the qubit $b = \frac{1}{\sqrt{2}}|0\rangle - \frac{1}{\sqrt{2}}|1\rangle$. Let U_P be the gate that performs the classical computation $|t, x, b\rangle \to |t, x, b \oplus P(x,t)\rangle$. Let $X_0 = \{x \mid P(x,t) = 0\}$ and $X_1 = \{x \mid P(x,t) = 1\}$. We next apply the transformation U_P to registers 1, 2, and 4 in the above superposition and obtain

$$\begin{aligned} &U_P\Big(\frac{1}{\sqrt{2^{4n+1}}} \sum_{x \in \{0,1\}^{4n}} |t, x, \min(f(x), f(x \oplus t)), b\rangle\Big) \\ &= \frac{1}{\sqrt{2^{4n+1}}} U_P \Big(\sum_{x \in X_0} |t, x, \min(f(x), f(x \oplus t)), 0\rangle \\ &\quad + \sum_{x \in X_1} |t, x, \min(f(x), f(x \oplus t)), 0\rangle \\ &\quad - \sum_{x \in X_0} |t, x, \min(f(x), f(x \oplus t)), 1\rangle \end{aligned}$$

$$\left. - \sum_{x\in X_1} |t, x, \min(f(x), f(x\oplus t)), 1\rangle \right)$$

$$= \frac{1}{\sqrt{2^{4n+1}}} \left(\sum_{x\in X_0} |t, x, \min(f(x), f(x\oplus t)), 0\oplus 0\rangle \right.$$

$$+ \sum_{x\in X_1} |t, x, \min(f(x), f(x\oplus t)), 0\oplus 1\rangle$$

$$- \sum_{x\in X_0} |t, x, \min(f(x), f(x\oplus t)), 1\oplus 0\rangle$$

$$\left. - \sum_{x\in X_1} |t, x, \min(f(x), f(x\oplus t)), 1\oplus 1\rangle \right)$$

$$= \frac{1}{\sqrt{2^{4n+1}}} \left(\sum_{x\in X_0} |t, x, \min(f(x), f(x\oplus t)), 0\rangle \right.$$

$$+ \sum_{x\in X_1} |t, x, \min(f(x), f(x\oplus t)), 1\rangle$$

$$- \sum_{x\in X_0} |t, x, \min(f(x), f(x\oplus t)), 1\rangle$$

$$\left. - \sum_{x\in X_1} |t, x, \min(f(x), f(x\oplus t)), 0\rangle \right)$$

$$= \frac{1}{\sqrt{2^{4n}}} \left(\sum_{x\in\{0,1\}^{4n}} (-1)^{P(x,t)} |t, x, \min(f(x), f(x\oplus t)), b\rangle \right),$$

which is the desired superposition. ∎

As promised, we next investigate the classical time complexity for calculating whether an input string w belongs or not to the language L_A given by Equation(4.4). We will prove a very strong lower bound for the number of steps needed to calculate classically L_A (in the above sense). Namely, we will show that, for some function oracle A, any classical algorithm that calculates L_A needs at least $2^{\Omega(n)}$ steps on almost every input.[4] Moreover, the above property holds for an overwhelming fraction of oracles A that satisfy Simon's promise given in Definition 4.3.2. This lower bound is valid also for probabilistic classical algorithms with bounded error probability, but we consider here only deterministic classical algorithms.

We need to clarify the statement "an overwhelming fraction of oracles A." Recall that $\mathcal{A}$ is the set of all oracles that satisfy Simon's promise. One can induce

[4] We recall that a predicate $P(\cdot)$ holds almost everywhere (abbreviated a.e.) if it holds on all points in its domain except at most a finite set.

Chapter 5

One-way functions and pseudo-random generators

5.1 Chapter overview and basic definitions

Modern cryptography relies in an essential way on complexity theory. In cryptography, the typical objective is to design protocols whose functionality (e.g., the secrecy of a message, the authentication of the participating parties, etc.) cannot be altered by the malicious actions of an adversary. The most general and safest approach in cryptography it to consider as a parameter a bound for the computational power of the adversary and to ensure the functionality of the protocol against any malicious action that can be performed within this bound. This amounts to proving that any successful adversarial attack requires more computation than the assumed bound, a mission which falls in the territory of complexity theory. The cryptographical protocols need to utilize as basic primitives tasks that are computationally hard for the attacker. Moreover, having in mind the cryptographical applications, it is essential to quantify carefully the hardness of these primitives, which implies the need for a thorough quantitative analysis of the computational complexity of these tasks. In this chapter we undertake such an investigation for two such primitives, one-way functions and pseudo-random generators. We also discuss two related concepts, hard functions (which are a relaxation of one-way functions), and extractors (which are both a relaxation and a strengthening, in different aspects, of pseudo-random generators).

A one-way function is a function that is easy to compute and hard to invert. A pseudo-random generator is a function that takes a short random input, called the seed, and produces a long output that "looks" random to an adversary. These two types of functions are important per se and have numerous applications in computational complexity theory. In cryptography they play a major role and almost all cryptosystems and cryptographic protocols rely on them in a quite direct way. To illustrate, we give just two simple and familiar examples.

Example 1. Consider the way passwords are handled in a multi-user computer system. Instead of storing them explicitely, the system keeps, for each password w, the value $f(w)$, where f ideally is a one-way function. At login, the user types w and the system computes $f(w)$ and compares it with the stored value. On the other hand, since f is one-way, no one (e.g., the system administrator) having $f(w)$ is able to retrieve w.

Example 2. The one-time pad encryption scheme works by bitwise XOR-ing the message m with a random string R that acts as the private key, i.e., the ciphertext c is obtained as $c = m \oplus R$ ($\oplus$ denotes bitwise XOR). This is a perfect encryption scheme and, in fact, it is known that in any private key encryption scheme that does not leak any information to an adversary the length of the private key has to be at least as large as the message being encrypted. The drawback is that the two legal parties (usually called Alice and Bob) must share an extremely long private key. With a pseudo-random generator g, Alice and Bob need to share only a short seed r and encrypt the message m by $c = m \oplus g(r)$.

Let us now define formally the two notions of primary interest in this chapter. We start with one-way functions, and we first introduce some notation.

NOTATION. We will use more operations on binary strings and we introduce here notation that distinguishes them clearly. We recall that Σ denotes the binary alphabet $\{0, 1\}$, $|x|$ is the length of a string x, and $\|A\|$ denotes the cardinality of a set A. The concatenation of two binary strings x and y is denoted $x \odot y$.[1] This notation is extended to the concatenation of more strings and we write $x_1 \odot x_2 \odot x_3 \odot \ldots \odot x_m$ instead of $(\ldots((x_1 \odot x_2) \odot x_3) \ldots \odot x_m)$. For strings x and y having the same length (i.e., $x, y \in \Sigma^n$ for some $n \in \mathbb{N}$), we define the inner product of x and y, denoted $x \cdot y$, as follows. We view x and y as vectors over the field $GF(2)$, $x = (x_1, \ldots, x_n)$ and $y = (y_1, \ldots, y_n)$, where the x_i's and the y_j's, the bits that form x and respectively y, are identified with elements of $GF(2)$ in the natural way: If the i-th bit of x is bit 0(1), then x_i is identified with the element 0(1) in $GF(2)$. Then, the inner product is $x \cdot y = x_1 \cdot y_1 + \ldots + x_n \cdot y_n$, the arithmetical operations being done modulo 2, i.e., in $GF(2)$. Finally, we also use cartesian products of sets of strings and we use the standard tuple notation (i.e., $(y_1, y_2, \ldots, y_k)$) to denote the elements of such a cartesian products.

We consider functions $f\colon \Sigma^* \to \Sigma^*$ with the property that, for all x_1 and x_2, $|x_1| = |x_2|$ implies $|f(x_1)| = |f(x_2)|$. Such a function is called *length-regular*. This restriction is mainly a technical convenience and, moreover, appears natural for most cryptographic applications. A function $f\colon \Sigma^* \to \Sigma^*$ having the property that, for all $x \in \Sigma^*$, $|f(x)| = |x|$ is called *length-preserving*.

For any function $f : \Sigma^* \to \Sigma^*$ and $\ell \in \mathbb{N}$, f_ℓ denotes the restriction of f to Σ^ℓ. The set of functions $\{f_\ell\colon \Sigma^\ell \to \Sigma^* \mid \ell \in \mathbb{N}\}$, usually denoted $(f_\ell)_{\ell \in \mathbb{N}}$, is

---

[1]This notation for concatenation is valid in this chapter only. We felt the need for a more striking notation here so as not to confuse concatenation with the other string operations.

called the *ensemble of functions* induced by f. Conversely, given a set of functions $\{f_\ell \colon \Sigma^\ell \to \Sigma^* \mid \ell \in \mathbb{N}\}$, the function $f\colon \Sigma^* \to \Sigma^*$ defined by $f(x) = f_{|x|}(x)$, for all $x \in \Sigma^*$, is called the function induced by the ensemble $(f_\ell)_{\ell\in\mathbb{N}}$.

Note that if f is length-regular, then, for each ℓ, there is a natural number ℓ' such that $f_\ell \colon \Sigma^\ell \to \Sigma^{\ell'}$. We want f to be computable in polynomial time which implies that the length of $f(x)$ is polynomially bounded in the length of x. In other words, ℓ' is bounded by $p(\ell)$, for some polynomial p. We also want f to be hard to invert. One trivial way to achieve this is to make ℓ' much smaller than ℓ (for example, assume that f_ℓ shrinks its input by an exponential factor). In that case no polynomial-time algorithm, on input $f(x)$, has the time to print x. This kind of "hard-to-invert" function is neither useful in cryptography nor interesting in computational complexity and, consequently, to avoid this situation, we will require that ℓ is polynomially bounded in ℓ' (i.e., $\ell \le q(\ell')$, for some polynomial q) as well. If these two requierements are met for f, we say that the input and the output lengths are *polynomially related.*

We can now present the main definitions.

Definition 5.1.1 (One-way function) *Let* $\epsilon\colon \mathbb{N} \to [0,1]$ *and* $S\colon \mathbb{N} \to \mathbb{N}$ *be two functions that are considered as parameters. A length regular function* $f : \Sigma^* \to \Sigma^*$ *with polynomially related input and output lengths is a one-way function with security* (ϵ, S) *if*

(1) *There exists a deterministic polynomial-time machine* M *such that, for all* $x \in \Sigma^*$, $M(x) = f(x)$;

(2) *For all sufficiently large* ℓ *and for any circuit* C *with* $\text{size}(C) \le S(\ell)$,

$$\text{Prob}_{x\in\Sigma^\ell}(C(f_\ell(x)) \in f_\ell^{-1}(f_\ell(x))) < \epsilon(\ell).$$

A few remarks are necessary. Stating that f is easy to calculate does not raise any problem: We ask that there is a polynomial-time algorithm that calculates f.[2] Stating that f is hard to invert needs some elaboration. We want f to be resistant to inversion by an adversary endowed with some specified computational power. An adversary is represented by a circuit that attempts to invert f (at a given length) and $S(\ell)$ represents the computational power against which f is inversion resistant. The adversary is given $f_\ell(x)$ and is not looking strictly to retrieve x (since f is not necessarily 1-to-1 this would be impossible) but only some inverse of $f_\ell(x)$. We require that this happens with probability less than $\epsilon(\ell)$, where the probability is taken over x chosen uniformly at random in Σ^*.

Currently it is not known if one-way functions exist. Note that, in fact, an adversary can invert $f_\ell(x)$ *nondeterministically* in polynomial time by just guessing a value z such that $f_\ell(z) = f_\ell(x)$. Thus, the existence of one-way functions implies

[2] We could require that f is calculated by a *probabilistic* polynomial-time algoritm. All the results that we present here would remain valid with minor and obvious modifications.

P $\neq$ NP. Even under the (plausible) assumption P $\neq$ NP, it is not known whether good one-way functions exist. The main reason is that the function needs to be hard to invert on a large fraction of inputs at almost every length, while P $\neq$ NP only means that there are languages that are hard in the worst-case (perhaps on just one input per length, and not even for almost every length). However, the general opinion is that good one-way functions do exist and there are some candidates for one-way functions (e.g., based on integer factoring or on the discrete log problem) that are used with relatively high confidence in practice.

Depending on the parameters ϵ and S (in the Definition 5.1.1), we distinguish some important types of one-way functions. We mainly consider adversaries endowed with circuits whose size is larger than any polynomial function.

Definition 5.1.2 *A function $s\colon \mathbb{N} \to \mathbb{N}$ is superpolynomial if for every polynomial p, it holds that $s(\ell) \geq p(\ell)$, for all ℓ sufficiently large.*

Definition 5.1.3 *Let $C = (C_\ell)_{\ell\in\mathbb{N}}$ be a collection of circuits and let $S\colon \mathbb{N} \to \mathbb{N}$ be a function. We say that the circuits C have size at most S if, for every ℓ,* $\text{size}(C_\ell) \leq S(\ell)$.

Definition 5.1.4 (Strong one-way function) *A length-regular function $f\colon \Sigma^* \to \Sigma^*$ with polynomially related input and output lengths is a strong one-way function if for any polynomial p, f is one-way with security $(\frac{1}{p(\ell)}, p(\ell))$.*

We also consider the following particular type of a strong one-way function.

Definition 5.1.5 (Exponentially strong one-way function) *A length-regular function $f\colon \Sigma^* \to \Sigma^*$ with polynomially related input and output lengths is an exponentially strong one-way function if there is some positive constant c such that f is one-way with security $(\frac{1}{2^{c\ell}}, 2^{c\ell})$.*

In cryptography applications, one needs strong one-way functions. We will show that, in fact, it suffices if we have at hand a much weaker type of one-way function. Indeed, in the next section, we show that given a weak one-way function, as defined next, we can construct a strong one-way function.

Definition 5.1.6 (Weak one-way function) *A length-regular function $f\colon \Sigma^* \to \Sigma^*$ with polynomially related input and output lengths is a weak one-way function if there is a polynomial q such that for any polynomial p, f is one-way with security $(1 - \frac{1}{q(\ell)}, p(\ell))$.*

We move on to define formally pseudo-random generators. Intuitively, a pseudo-random generator is a function that takes random short strings and produces (much) longer strings that "look" random to an adversary. It is also desirable that the function is efficiently computable. Essential for our discussion are distributions on sets of binary strings of a given length, i.e., distributions on Σ^n, where n is some arbitrary natural number. Recall that a distribution X_n on Σ^n is a function $X_n\colon \Sigma^n \to [0, 1]$ with the property that $\sum_{\alpha\in\Sigma^n} X_n(\alpha) = 1$. A distribution will also be identified with a random variable having that distribution.

NOTATION. For each $n \in \mathbb{N}$, U_n denotes the uniform distribution on Σ^n, i.e., $U_n\colon \Sigma^n \to [0,1]$ is the function defined by $U_n(\alpha) = \frac{1}{2^n}$, for all $\alpha \in \Sigma^n$.

Suppose that in some application (e.g., a cryptography protocol, or the execution of a probabilistic algorithm) we need random strings of length n. Ideally, we would like to utilize strings in Σ^n generated by some source of randomness according to the uniform distribution. Lacking this, we are also content if the source generates strings in Σ^n according to a distribution X_n on Σ^n that is *close* to U_n. There are more ways by which two distributions can be close. They depend on the type of distance between two distributions that we consider, and there are two distances that are relevant for us, the *statistical distance* and the *computational distance*. Let us first consider the statistical distance.

Definition 5.1.7 (Statistical distance between two distributions) *Let $n \in \mathbb{N}$. Let X_n, Y_n be two distributions on Σ^n. The statistical distance between X_n and Y_n is denoted $\Delta_{\text{stat}}(X_n, Y_n)$ and is defined by*

$$\Delta_{\text{stat}}(X_n, Y_n) = \sum_{\alpha \in \Sigma^n} |X_n(\alpha) - Y_n(\alpha)|\,.$$

For example, consider the following distributions, X_3, Y_3 and Z_3 defined on Σ^3.

α	000	001	010	011	100	101	110	111
X_3	0	0	0	0	1/4	1/4	1/4	1/4
Y_3	1/16	3/16	1/16	3/16	1/16	3/16	1/16	3/16
Z_3	1/64	1/64	1/64	1/64	15/64	15/64	15/64	15/64

Note that $\Delta_{\text{stat}}(X_3, Y_3) = 1$ and $\Delta_{\text{stat}}(X_3, Z_3) = 1/8$ which agrees with the intuition that X_3 and Z_3 resemble each other more than X_3 and Y_3.

One way to estimate the closeness of two distributions X_n and Y_n is to take some subset $A \subseteq \Sigma^n$ and to compare $\text{Prob}_{X_n}(A)$ and $\text{Prob}_{Y_n}(A)$. Such a subset is called (in this context) a *statistical test* or, simply, a test. To illustrate, let us take, in the above example, the test $A_1 = \{011, 100\}$. Then $\text{Prob}_{X_3}(A_1) = \text{Prob}_{Y_3}(A_1) = \frac{1}{4}$. Thus the test A_1 is not able to distinguish between the distributions X_3 and Y_3. If we take the test $A_2 = \{000, 001, 010, 011\}$, then $\text{Prob}_{X_3}(A_2) = 0$ and $\text{Prob}_{Y_3}(A_2) = 1/2$. The test A_2 "sees" a quite significant difference between X_3 and Y_3. It is not hard to observe that the test A_2 and its complement $A_2' = \{100, 101, 110, 111\}$ "see" the largest difference between X_3 and Y_3 among all tests. In fact, the following lemma holds.

Lemma 5.1.8 *Let $n \in \mathbb{N}$, and let X_n, Y_n be two distributions on Σ^n. Then*

$$\Delta_{\text{stat}}(X_n, Y_n) = 2 \max_{A \subset \Sigma^n} |\text{Prob}_{X_n}(A) - \text{Prob}_{Y_n}(A)|\,.$$

Proof. Let $A = \{\alpha \in \Sigma^n \mid X_n(\alpha) \geq Y_N(\alpha)\}$. It is easy to see that A is the set for which $|\text{Prob}_{X_n}(A) - \text{Prob}_{Y_n}(A)|$ is maximum. Then

$$\begin{aligned}
\Delta_{\text{stat}}(X_n, Y_n) &= \sum_{\alpha \in \Sigma^n} |X_n(\alpha) - Y_n(\alpha)| \\
&= \sum_{\alpha \in A} (X_n(\alpha) - Y_n(\alpha)) + \sum_{\alpha \in \Sigma^n - A} (Y_n(\alpha) - X_n(\alpha)) \\
&= \text{Prob}_{X_n}(A) - \text{Prob}_{Y_n}(A) + (1 - \text{Prob}_{Y_n}(A)) - (1 - \text{Prob}_{X_n}(A)) \\
&= 2\,|\text{Prob}_{X_n}(A) - \text{Prob}_{Y_n}(A)| = 2 \max_A |\text{Prob}_{X_n}(A) - \text{Prob}_{Y_n}(A)|,
\end{aligned}$$

and the lemma is proved. ∎

The statistical distance has the standard properties of a distance.

Lemma 5.1.9 *Let $n \in \mathbb{N}$, and let X_n, Y_n and Z_n be distributions over Σ^n. Then*

(1) $\Delta_{\text{stat}}(X_n, Y_n) = 0 \Leftrightarrow X_n = Y_n$,

(2) $\Delta_{\text{stat}}(X_n, Y_n) = \Delta_{\text{stat}}(Y_n, X_n)$,

(3) $\Delta_{\text{stat}}(X_n, Z_n) \leq \Delta_{\text{stat}}(X_n, Y_n) + \Delta_{\text{stat}}(Y_n, Z_n)$ *(triangle inequality).*

Proof. All three properties follow immediately from the definition of statistical distance. ∎

Let us consider now a function f that maps short strings into longer strings, as a pseudo-random generator is supposed to do. For concreteness, let us suppose that f maps binary strings of length n (i.e., Σ^n) into binary strings of length $2n$ (i.e., Σ^{2n}). The function f induces naturally a distribution X_{2n} on Σ^{2n} defined by $X_{2n}(\alpha) = \text{Prob}_{x \in \Sigma^n}(f(x) = \alpha)$. We would like X_{2n} to be statistically close to U_{2n}. Unfortunately, this is not possible because if we take the test A to be the image of f, then $\text{Prob}_{X_{2n}}(A) = 1$ and $\text{Prob}_{U_{2n}}(A) = \frac{2^n}{2^{2n}} = \frac{1}{2^n}$, and thus, according to Lemma 5.1.8, $\Delta_{\text{stat}}(X_{2n}, U_{2n}) \geq 2(1 - \frac{1}{2^n})$. Consequently, the distributions X_{2n} and U_{2n} are statistically far apart. However, it is possible that any test that is able to distinguish the two distributions, such as the above set A, is very complex and, in particular, beyond the capabilities of an adversary. In other words, it may happen that for every subset $A \subseteq \Sigma^{2n}$ that is computable by a circuit of size $S(n)$, $\text{Prob}_{X_{2n}}(A) \approx \text{Prob}_{U_{2n}}(A)$. In this case, the distributions X_{2n} and U_{2n} do look similar to an adversary that is endowed with computational power $S(n)$. This justifies the following definition.

Definition 5.1.10 (Computational distance between two distributions) *Let $n, S \in \mathbb{N}$. Let X_n, Y_n be two distributions on Σ^n. The computational distance between X_n and Y_n relative to size S is denoted $\Delta_{\text{comp},S}(X_n, Y_n)$ and is defined by*

$$\Delta_{\text{comp},S}(X_n, Y_n) = \max |\text{Prob}(C(X_n) = 1) - \text{Prob}(C(Y_n) = 1)|,$$

where the maximum is taken over all circuits C with inputs of size n and having size at most S.

Definition 5.1.11 *Let $n, S \in \mathbb{N}$ and $\epsilon > 0$. Let X_n, Y_n be two distributions on Σ^n. We say that the distributions X_n and Y_n are computationally ϵ-close relative to size S if $\Delta_{\text{comp},S}(X_n, Y_n) \leq \epsilon$.*

Does the computational distance retain the three properties listed in Lemma 5.1.9? We will see in Proposition 5.1.17 that property (1) fails in a dramatic way: There are distributions that are far apart in the statistical sense, but very close in the computational sense. In fact, it is this failure that allows the very notion of a pseudo-random generator. On the other hand, properties (2) and (3) remain valid for the computational distance.

Lemma 5.1.12 *Let $n, S \in \mathbb{N}$, and let X_n, Y_n and Z_n be distributions over Σ^n. Then*

(1) $\Delta_{\text{comp},S}(X_n, Y_n) = \Delta_{\text{comp},S}(Y_n, X_n)$,

(2) $\Delta_{\text{comp},S}(X_n, Z_n) \leq \Delta_{\text{comp},S}(X_n, Y_n) + \Delta_{\text{comp},S}(Y_n, Z_n)$ *(triangle inequality).*

Proof. The two properties follow immediately from the definition of computational distance. ∎

Finally, we can define formally the notion of a pseudo-random generator.

Definition 5.1.13 (Pseudo-random generator) *Let $\ell, L, S \in \mathbb{N}$ and $\epsilon > 0$ be parameters. A length-regular function $g\colon \Sigma^\ell \to \Sigma^L$ is a pseudo-random generator with security (ϵ, S) if*

$$\Delta_{\text{comp},S}(g(U_\ell), U_L) \leq \epsilon.$$

The value $(L - \ell)$ is called the extension *of g.*

In other words, by unwrapping all these definitions, $g\colon \Sigma^\ell \to \Sigma^L$ is a pseudo-random generator with security (ϵ, S) if for every circuit C on inputs of length L and with size$(C) \leq S$,

$$\left|\text{Prob}_{x\in\Sigma^\ell}(C(g(x)) = 1) - \text{Prob}_{y\in\Sigma^L}(C(y) = 1)\right| \leq \epsilon.$$

In general, we are interested in producing pseudo-random generators for many input lengths ℓ (ideally for any input length $\ell \in \mathbb{N}$).

Definition 5.1.14 (Ensemble of pseudo-random generators) *Let $\epsilon\colon \mathbb{N} \to [0, 1]$ and $S\colon \mathbb{N} \to \mathbb{N}$ be two functions. An ensemble of pseudo-random generators with security (ϵ, S) is a family of functions $(g_\ell)_{\ell\in\mathbb{N}}$ such that*

(1) *for some function $L\colon \mathbb{N} \to \mathbb{N}$, for all $\ell \in \mathbb{N}$, $g_\ell\colon \Sigma^\ell \to \Sigma^{L(\ell)}$, and*

(2) *for each $\ell \in \mathbb{N}$, g_ℓ is a pseudo-random generator with security $(\epsilon(\ell), S(\ell))$.*

Abusing terminology, when the context is clear, an ensemble of pseudo-random generators is called a pseudo-random generator itself.

Depending on the parameters ϵ and S and similarly to the taxonomy of one-way functions that we have introduced earlier, we distingush two types of good pseudo-random generators.

Definition 5.1.15 (Strong pseudo-random generator) *An ensemble of pseudo-random generators $(g_\ell)_{\ell \in \mathbb{N}}$ with security (ϵ, S) is called strong if $1/\epsilon$ and S are both superpolynomial functions.*

Definition 5.1.16 (Exponentially strong pseudo-random generator) *An ensemble of pseudo-random generators $(g_\ell)_{\ell \in \mathbb{N}}$ with security (ϵ, S) is called exponentially strong if there is a positive constant c such that, for almost every ℓ, $1/\epsilon(\ell) > 2^{c\ell}$ and $S(\ell) > 2^{c\ell}$.*

A first observation is that good pseudo-random generators exist. We will show this for generators of the type $g_\ell \colon \Sigma^\ell \to \Sigma^{2\ell}$, but, from the demonstration, it will be clear that the assertion can be made more general.

Proposition 5.1.17 *There exists an ensemble of functions $(g_\ell)_{\ell \in \mathbb{N}}$, of type $g_\ell \colon \Sigma^\ell \to \Sigma^{2\ell}$, for all sufficiently large $\ell \in \mathbb{N}$, that is an exponentially strong pseudo-random generator.*

Proof. Let us fix $\ell \in \mathbb{N}$. We pick a function g_ℓ at random among the functions mapping strings of length ℓ into strings of length 2ℓ, and we show that, if ℓ is sufficiently large, the probability that there is a function having the property asserted in the statement is positive. It follows that such a function g_ℓ exists.

Thus, for each $x \in \Sigma^\ell$, $g_\ell(x)$ is defined to be a string in $\Sigma^{2\ell}$ picked uniformly at random. Let C be a fixed circuit on inputs in $\Sigma^{2\ell}$ and let $p = \text{Prob}_{x \in \Sigma^{2\ell}}(C(x) = 1)$. We enumerate Σ^ℓ as $\{\alpha_1, \ldots, \alpha_{2^\ell}\}$, and, for each $i \in \{1, \ldots, 2^\ell\}$, we define the random variable X_i to be 1 if $C(g(\alpha_i)) = 1$ and 0 if $C(g(\alpha_i)) = 0$. The random variables X_i, $i \in \{1, \ldots, 2^\ell\}$, are independent and the expected value of each of them is p. Therefore, by the additive Chernoff bounds (see Appendix A),

$$\text{Prob}_{g_\ell}\left(\left|\frac{1}{2^\ell}\sum X_i - p\right| \geq 2^{-\ell/4}\right) \leq 2e^{-(1/3)(2^{-\ell/4})^2 \cdot 2^\ell} < \frac{1}{2^{(1/3)2^{\ell/2}}}.$$

Since $\left|\frac{1}{2^\ell}\sum X_i - p\right|$ is $|\text{Prob}_{\alpha \in \Sigma^\ell}(C(g(\alpha)) = 1) - \text{Prob}(C(x) = 1)|$, it follows that, for a fixed circuit C, the latter value is $\geq 2^{-\ell/4}$ with probability of g_ℓ less than $2^{-(1/3)2^{\ell/2}}$.

The number of circuits C with size$(C) \leq 2^{\ell/4}$ is bounded by $2^{O(\ell \cdot 2^{\ell/4})}$ (this is shown in Section 1.1.2). Therefore the probability of the event "There exists some circuit C of size at most $2^{\ell/4}$ such that $|\text{Prob}_{\alpha \in \Sigma^\ell}(C(g(\alpha)) = 1) - \text{Prob}(C(x) = 1)|$ is $\geq 2^{-\ell/4}$." is less than $2^{O(\ell \cdot 2^{\ell/4})} \cdot 2^{-(1/3)2^{\ell/2}} < 1$. Thus, the probability of the complementary event is positive, from which, as we have discussed, the conclusion follows. ∎

The foregoing proof is non-constructive and, therefore the result has only theoretical value. Even the theoretical merit is quite limited because what we need are pseudo-random generators that are efficiently computable. For example, it would be desirable that the pseudo-random generator $g_\ell \colon \Sigma^\ell \to \Sigma^{2\ell}$, whose existence is asserted above, is computable in polynomial time. Unfortunately, the above proof does not say anything about the complexity of g_ℓ. In fact, proving the existence of an efficiently computable pseudo-random generator is beyond the current state of complexity theory. Indeed, an efficiently computable pseudo-random generator is also a one-way function (the existence of which, as we fave argued earlier, implies P $\neq$ NP). To keep the notation simple, we prove the foregoing assertion for a particular setting of some of the parameters.

Proposition 5.1.18 INFORMAL STATEMENT: *An efficiently computable pseudo-random generator is a one-way function.*

FORMAL STATEMENT: *Suppose there exists an ensemble of functions $(g_\ell)_{\ell \in \mathbb{N}}$ with the following properties:*

(1) *For some functions $\epsilon\colon \mathbb{N} \to [0,1]$ and $S\colon \mathbb{N} \to \mathbb{N}$, the ensemble $(g_\ell)_{\ell\in\mathbb{N}}$ is a pseudo-random generator with security (ϵ, S),*

(2) *for all $\ell \in \mathbb{N}$, $g_\ell\colon \Sigma^\ell \to \Sigma^{2\ell}$,*

(3) *there is a polynomial q such that, for all $\ell \in \mathbb{N}$, g_ℓ is computable in time $q(\ell)$.*

Then the function $g\colon \Sigma^* \to \Sigma^*$ induced by the ensemble $(g_\ell)_{\ell\in\mathbb{N}}$ is a one-way function with security $(\epsilon + \frac{1}{2^\ell}, S - p(\ell))$, for some polynomial p.

Proof. Let us fix $\ell \in \mathbb{N}$ sufficiently large (so that the following arguments are valid). Suppose there is a circuit C_ℓ that inverts g_ℓ with probability at least $\epsilon(\ell) + \frac{1}{2^\ell}$, i.e.,

$$\mathrm{Prob}_{x\in\Sigma^\ell}\left(C_\ell(g_\ell(x)) \in g_\ell^{-1}(g_\ell(x))\right) \geq \epsilon(\ell) + \frac{1}{2^\ell}. \tag{5.1}$$

We define $A = \{y \in \Sigma^{2\ell} \mid g_\ell(C_\ell(y)) = y\}$. Let $C_{\ell,A}$ be a circuit that calculates A (i.e., $C_{\ell,A}(y) = 1$ if and only if $y \in A$). There is a polynomial p, such that for all ℓ, $C_{\ell,A}$ can be taken with $\mathrm{size}(C_{\ell,A}) \leq \mathrm{size}(C_\ell) + p(\ell)$. Let us assume that $\mathrm{size}(C_\ell) \leq S(\ell) - p(\ell)$. Thus, $\mathrm{size}(C_{\ell,A}) \leq S(\ell)$. Since A is a subset of the image of g_ℓ, it follows that $\|A\| \leq \|\Sigma^\ell\| = 2^\ell$. Therefore,

$$\mathrm{Prob}_{y\in\Sigma^{2\ell}}\left(C_{\ell,A}(y) = 1\right) \leq \frac{2^\ell}{2^{2\ell}} = \frac{1}{2^\ell}.$$

On the other hand,

$$\mathrm{Prob}_{x\in\Sigma^\ell}\left(C_{\ell,A}(g_\ell(x)) = 1\right) \geq \epsilon(\ell) + \frac{1}{2^\ell}.$$

It follows that

$$\mathrm{Prob}_{x\in\Sigma^\ell}\left(C_{\ell,A}(g_\ell(x)) = 1\right) - \mathrm{Prob}_{y\in\Sigma^{2\ell}}\left(C_{\ell,A}(y) = 1\right) \geq \epsilon(\ell).$$

Since $\mathrm{size}(C_{\ell,A}) \leq S(\ell)$, this contradicts the fact that g_ℓ is a pseudo-random generator with security $(\epsilon(\ell), S(\ell))$. Thus, the relation (5.1) is false. ∎

On the other hand, a one-way function f is not necessarily a pseudo-random generator. Indeed, suppose a length-regular function $f\colon \Sigma^* \to \Sigma^*$, with

$f_\ell \colon \Sigma^\ell \to \Sigma^{2\ell}$, for all $\ell \in \mathbb{N}$ (the extension $\ell \to 2\ell$ has been taken arbitrarily), is one-way with security (ϵ, S), for some functions $\epsilon \colon \mathbb{N} \to [0,1]$ and $S \colon \mathbb{N} \to \mathbb{N}$. Consider the functions $g_\ell \colon \Sigma^\ell \to \Sigma^{2\ell+1}$ defined, for all $\ell \in \mathbb{N}$, by $g_\ell(x) = 0 \odot f_\ell(x)$. Then, it is easy to see that the ensemble of functions $(g_\ell)_{\ell \in \mathbb{N}}$ continues to be one-way. However, $g_\ell(x)$ does not look random at all since it always (i.e., for all ℓ and for all x) starts with 0. In particular, a small circuit that accepts an input string if and only if it starts with 0 will distinguish the distribution $g_\ell(U_\ell)$ from U_{2l+1} and, therefore, g_ℓ is not a pseudo-random generator.

This example and Proposition 5.1.18 suggest that the requirements in the definition of a pseudo-random generator are much more exacting than what a one-way function provides. In spite of this, a remarkable result of Håstad, Impagliazzo, Levin, and Luby [HILL99b] (building on the work of many other researchers) shows that, given a strong one-way function f, one can construct a polynomial-time computable strong pseudo-random generator. The proof of this result is extremely complex and beyond the scope of this book. We prove in this chapter a weaker result by assuming that f in addition to being a strong one-way function is also a permutation at each length (i.e., for each ℓ, $f_\ell \colon \Sigma^\ell \to \Sigma^\ell$ is a bijection). We show that given such a function f one can construct a polynomial-time computable strong pseudo-random generator with polynomial extension (superpolynomial extension is discussed below).

This construction, as well as many other ones in this chapter, follows a pattern: Given one function f_1 with certain properties, there is an effective procedure that computes some other function f_2. This concept is formalized in the following definition.

Definition 5.1.19 *Let $f_1 \colon \Sigma^* \to \Sigma^*$ and $f_2 \colon \Sigma^* \to \Sigma^*$. We say that f_2 is effectively computed from f_1 if there is an algorithm A that (a) has oracle access to the function f_1, and (b) on input x, calculates $f_2(x)$. The function f_1 is called the building block of the construction. In case the algorithm A runs in time $t(|x|)$ on input x, for all $x \in \Sigma^*$, we say that f_2 is effectively computed from f_1 in time $t(\cdot)$.*

The transformation of the one-way function f into a pseudo-random generator is done in two steps. In Section 5.3, using the function f, we build a polynomial-time computable pseudo-random generator that has an extension of just one bit. The second step is done in Section 5.4, where the extension is enlarged to more significant values. How large an extension can one achieve depends on the quality (i.e., the security parameters ϵ and S) of the initial one-way permutation. In particular, if the one-way permutation is strong or exponentially strong, then the extension can be made superpolynomial or, respectively, exponential. Obviously, if the pseudo-random generator has superpolynomial extension then it cannot be computed in polynomial time. Therefore, it is desirable that each bit of the randomly looking output can be computed in polynomial time independently of the other bits of the input. Such an object can be viewed as a function which on input an index i returns the i-th bit of the string produced by the pseudo-random generator and indeed it is called a *pseudo-random function*. In Section 5.5, we show how to build a pseudo-random function given, as a building block, a strong pseudo-random generator $(g_\ell)_{\ell \in \mathbb{N}}$ that doubles the length of the seed (i.e., for each ℓ, $g_\ell \colon \Sigma^\ell \rightarrow \Sigma^{2\ell}$).

We have considered efficiently computable pseudo-random generators and it seems natural to require that the output produced by such generators should look random to an adversary that is endowed with large computational resources, in particular, with computational resources that are superior to the ones needed to calculate the pseudo-random generator. Indeed our discussion so far has concentrated on pseudo-random generators with security (ϵ, S) that are computable in time significantly shorter than S. Let us call these type I pseudo-random generators. Håstad, Impagliazzo, Levin, and Luby have shown that such pseudo-random generators can be constructed given a one-way function (we only present the construction that uses a one-way permutation). Let us relax the efficiency requirement of a pseudo-random generator and allow it to be computable in time that is larger than the security parameter S. Let us call this a type II pseudo-random generator. Can we construct such pseudo-random generators under a relaxed assumption regarding the building block function? The answer is positive. Indeed, we show in Section 5.8 that if we use as a building block a hard function (in fact a hard predicate) then we can build type II pseudo-random generators. This is an interesting result for two reasons. First, hard functions do exist (even though it is not known how to actually obtain such hard functions), while the assumption needed for constructing type I pseudo-random generators, namely the existence of one-way functions, is only a conjecture (which, furthermore, implies $\mathrm{P} \neq \mathrm{NP}$). Secondly, type II pseudo-random generators, under some conditions, can be utilized to derandomize any polynomial-time probabilistic computation with bounded 2-sided error (this is usually called a BPP computation). More precisely, the alluded conditions require that (1) the pseudo-random generator has exponential extension, (2) the pseudo-random generator is computable in time polynomial in the output length, and (3) the pseudo-random generator is secure against adversaries that can spend time that is a fixed polynomial in the output length. In Section 5.9, we observe (using type II pseudo-random generators) that if these requirements are met, then $\mathrm{P} = \mathrm{BPP}$! We also show that, under a quite reasonable hypothesis, the above requirements are in fact satisfied. The hypothesis is that there exists a length-regular function $f \colon \Sigma^* \rightarrow \Sigma^*$ and two constants c_1 and c_2 such that (1) f is computable in time $2^{c_1 n}$, and (2) for almost every length n, no circuit of size $2^{c_2 n}$ can calculate f_n.

Since the construction of type II pseudo-random generators relies on hard functions, we need to clarify what we mean when we say that a function f is hard. The intent is to capture the idea that no adversary having some specified computational power can calculate f. The most basic definition of a hard function deals with functions defined for inputs of some fixed length.

Definition 5.1.20 (Hard function) *Let $\epsilon > 0$, $\ell \in \mathbb{N}, \ell' \in \mathbb{N}$ and $S \in \mathbb{N}$ be parameters. A function $f\colon \Sigma^\ell \to \Sigma^{\ell'}$ is (ϵ, S)-hard if for every circuit C of size S,*

$$\|\{x \in \Sigma^\ell \mid C(x) \neq f(x)\}\| \geq \epsilon \cdot 2^\ell.$$

We move to functions that are defined at all lengths. In this case, the adversary is represented by a collection of deterministic circuits $(C_\ell)_{\ell \in \mathrm{N}}$, where each C_ℓ calculates a function whose domain is Σ^ℓ. Abusing notation, we use C_ℓ to also denote the function computed by the circuit C_ℓ.

As we have proceeded earlier, to prevent the trivial and uninteresting case of functions being hard simply because their output is too long, we will consider only length-regular hard functions $f\colon \Sigma^* \to \Sigma^*$ with $f_\ell\colon \Sigma^\ell \to \Sigma^{\ell'(\ell)}$ for which there is a polynomial p such that, for all ℓ, $\ell'(\ell) \leq p(\ell)$.

Also, as in the case of one-way functions and pseudo-random generators, we mainly consider adversaries endowed with circuits whose size is bounded by some superpolynomial function. Intuitively, f is hard for an adversary represented by a collection of circuits $(C_\ell)_{\ell \in \mathrm{N}}$ if, for all sufficiently large ℓ, C_ℓ fails to calculate f_ℓ. The failure can be more or less severe and correspondingly we have different degrees of hardness for a function.

Definition 5.1.21 (Worst-case hard function) *A length-regular function $f\colon \Sigma^* \to \Sigma^*$ is worst-case hard if there is a superpolynomial function S so that the following holds:*

For any family of circuits $(C_\ell)_{\ell \in \mathrm{N}}$ of size at most S,

$$\|\{x \in \Sigma^\ell \mid C_\ell(x) \neq f_\ell(x)\}\| \geq 1,$$

for all sufficiently large ℓ.

Definition 5.1.22 (Constant-rate hard function) *Let k be a positive integer. A length-regular function $f\colon \Sigma^* \to \Sigma^*$ is k-constant-rate hard if there is a superpolynomial function S so that the following holds:*

For any family of circuits $(C_\ell)_{\ell \in \mathrm{N}}$ of size at most S,

$$\|\{x \in \Sigma^\ell \mid C_\ell(x) \neq f_\ell(x)\}\| \geq \frac{1}{k} \cdot 2^\ell,$$

for all sufficiently large ℓ.

Definition 5.1.23 (Crypto hard function) *A length-regular function $f\colon \Sigma^* \to \Sigma^*$ is cryptographically hard (or, in short, crypto-hard) if there is a superpolynomial function S so that the following holds:*

For any polynomial p and for any family of circuits $(C_\ell)_{\ell \in \mathrm{N}}$ of size at most S,

$$\|\{x \in \Sigma^\ell \mid C_\ell(x) \neq f_\ell(x)\}\| \geq \left(1 - \frac{1}{p(\ell)}\right) \cdot 2^\ell,$$

for all sufficiently large ℓ.

Definition 5.1.24 (Exponentially hard function) *A length-regular function $f: \Sigma^* \to \Sigma^*$ is exponentially hard if there is a constant $c > 0$ so that the following holds:*

For any family of circuits $(C_\ell)_{\ell \in \mathbb{N}}$ of size at most $2^{c\ell}$,

$$\|\{x \in \Sigma^\ell \mid C_\ell(x) \neq f_\ell(x)\}\| \geq (1 - 2^{-c\ell}) \cdot 2^\ell,$$

for all sufficiently large ℓ.

Of course, we can use probability to express the relations in Definition 5.1.21, Definition 5.1.22, and Definition 5.1.23. For example, in Definition 5.1.22, we can say

$$\text{Prob}(C_\ell(x) \neq f_\ell(x)) \geq \frac{1}{k}, \tag{5.2}$$

where the probability is taken over x chosen uniformly at random in Σ^ℓ. This formulation has the advantage that it can be extended easily to probabilistic algorithms. For instance, we can require that, for all probabilistic circuits of size at most S, the probability in (5.2) holds over x chosen uniformly at random in Σ^ℓ *and* over the random choices r made by the circuit. Note that here the fact that the adversary can utilize probabilistic algorithms does not give him much additional power. Indeed, if there is a probabilistic algorithm C_ℓ so that,

$$\text{Prob}_{x,r}(C_\ell(x,r) \neq f_\ell(x)) \geq \frac{1}{k},$$

(where x denotes the input and r denotes the random bits), then there has to be one fixed r_0 so that

$$\text{Prob}_x(C_\ell(x,r_0) \neq f_\ell(x)) \geq \frac{1}{k}.$$

By embedding r_0 into the circuit C_ℓ, this becomes deterministic and its size increases by only $|r_0|$ additional gates (needed to store r_0). Keeping in mind this observation, we restrict our attention to the case of adversary deterministic circuits.

Clearly, the property "f is crypto hard" is much stronger than "f is worst-case hard," with "f is constant-rate hard" falling in the middle.

Interestingly, hardness can be amplified, and, furthermore, the amplification can be accomplished effectively. Indeed, we show in Section 5.6 that using a worst-case function as a building block, one can construct a crypto hard function. The construction has two phases. First, we build a constant hard function from a worst-case hard function, and, in the second phase, a constant hard function is used to produce a crypto hard function.

In Section 5.7, we consider hard predicates, i.e., functions f of the form $f: \Sigma^* \to \{0,1\}$. Since the range of a predicate has only two elements, there always is a small circuit that calculates a predicate on at least half of the inputs in Σ^n. Therefore, a predicate is considered hard if no circuit of some respectable size can calculate it on a fraction of inputs in Σ^n that is signicantly larger (by, say,

1/poly(n)) than 1/2 (see Definition 5.7.1). We present an efective construction of a hard predicate using a crypto hard function as a building block.

Section 5.10 is dedicated to extractors. An extractor is a function that resembles a pseudo-random generator in that its output has to pass some randomness tests. It is used to remedy sources of randomness that are in a sense weak. More precisely, suppose that there exists a device that generates random binary strings of length n. Ideally, to obtain perfect randomness, each string should be generated with probability 2^{-n}. Suppose instead that each string is generated with probability at most 2^{-k}, for some $k < n$ (k is a parameter, called min-entropy, that together with n characterizes the source, which in this case is called a (n, k)-*weak source*). Intuitively, the strings generated have k bits of randomness (out of the total length of n). An extractor is a function that depends on five parameters n, k, d, m, and ϵ: It takes as input a string produced by a (n, k)-weak source and a perfectly random but short additional string of length d, and produces a string of length m such that the distribution of the output is at a statistical distance at most ϵ from the uniform distribution on $\{0, 1\}^m$. We show that the construction, given in Section 5.8, of a type II pseudo-random generator using as a starting primitive a hard predicate f, can also be used to build an extractor function computable in polynomial time. The key idea is to view the truth-table of f as simply being a string produced by a weak source with min-entropy k and carry on the construction from Section 5.8. In the construction of a pseudo-random generator, we argue that if the output does not have a distribution within a short computational distance from the uniform distribution, then the starting predicate f is not hard. A similar argument works in the case of extractors: If the output distribution is not within a short statistical distance from the uniform distribution, then the starting truth-table belongs to a small family of strings which contains the set of strings generated by the weak source. This contradicts the fact that the min-entropy of the source is at least k (which implies that the number of generated strings is at least 2^k). The method leads to the construction of a polynomial-time computable extractor that can remedy $(n, \gamma n)$-weak sources, for arbitrarily small constant γ, using an additional number of random bits $d = O(\log n)$ and with output length $m = k^{1-\alpha}$, for arbitrarily small positive α (for simplicity, the given proof obtains a shorter output length).

From our brief overview, it should be clear that most of the major results that are presented in this chapter are of the following form: Using a function f_1 of type T_1 as a building block, one can effectively construct a function f_2 of type T_2 (where type T_2 functions satisfy more demanding requirements than type T_1 functions). Table 5.1 gives a "road map" for these results.

Table 5.1: Summary of effective constructions. Results are of the type: Using f_1 as a building block, there is an effective construction of f_2. (Note: An extender is a pseudo-random generator with extension 1.)

building block f_1	result f_2	where
weak one-way function	strong one-way function	Theorem 5.2.1 Section 5.2
one-way permutation	extender	Theorem 5.3.11 Section 5.3
extender	type I pseudo-random generator	Theorem 5.4.1 Section 5.4
type I pseudo-random generator	pseudo-random function	Theorem 5.5.1 Section 5.5
worst-case hard function	const.- rate hard function	Theorem 5.6.3(a) Section 5.6
$(\frac{1}{2^n}, 2^{cn})$-hard function	$(\text{const.}, 2^{cn})$-hard function	Theorem 5.6.3(b) Section 5.6
const.- rate hard function	crypto hard function	Theorem 5.6.12 (a) Section 5.6
$(\text{const.}, 2^{cn})$-hard function	exp. hard function	Theorem 5.6.12 (b) Section 5.6 (no proof)
crypto hard function	crypto hard predicate	Theorem 5.7.4 Section 5.7
exp. hard function	exp. hard predicate	Theorem 5.7.4 Section 5.7
exp. hard predicate	type II pseudo-random generator	Corollary 5.8.7 Section 5.8

5.2 From weak to strong one-way functions

IN BRIEF: The "hard-to-invert" property of a one-way function can be amplified from a fraction of $1/\text{poly}(n)$ of the inputs to the fraction $(1 - (1/\text{superpoly}(n)))$ of the inputs.

Weak one-way functions and strong one-way functions have been defined in Section 5.1. Intuitively, in the case of a weak one-way function, any polynomial-time algorithm fails to invert at least a $(1/\text{poly}(n))$ fraction of the multiset $\{f(x) \mid x \in \Sigma^n\}$. It may very well be the case that some polynomial-time algorithm inverts a large majority of inputs. In the case of a strong one-way function, any polynomial-time algorithm fails much more drastically, i.e., it fails on a $(1 - 1/\text{superpoly}(n))$ fraction of the multiset $\{f(x) \mid x \in \Sigma^n\}$. Obviously, if we need a one-way function, it better be a strong one. Note that the common candidates for one-way functions that have been considered are not strong. For example, let us take the case of the integer factoring problem. This problem offers a good basis for building a one-way function because, given two integers p and q, it is easy to calculate $n = p \cdot q$ and, on the other hand, there is no known polynomial-time algorithm for inverting the process, i.e., to find p and q given n. However, half of the integers n are even and for them it is very easy to find a prime factor (namely 2). Therefore factoring certainly does not provide a strong one-way function.

Thus, while the existence of weak one-way functions looks plausible, there seems to be little direct evidence in support of strong one-way function. The "hard-to-invert" requirement in the definition of a strong one-way function appears to be much more demanding than in the case of a weak one-way function. Perhaps surprisingly, in fact, strong one-way functions exist if and only if weak one-way functions exist. Since a strong one-way function obviously is also a weak one-way function, we only need to prove the following theorem.

Theorem 5.2.1 INFORMAL STATEMENT: *If weak one-way functions exist, then strong one-way functions exist.*

FORMAL STATEMENT: *Assume* $f: \Sigma^* \rightarrow \Sigma^*$ *is a weak one-way function. Then there is* $g: \Sigma^* \rightarrow \Sigma^*$ *a strong one-way function. Moreover,* g *is effectively computable from* f *in polynomial time.*

Proof. The assumption formally means that there is a polynomial q such that, for almost every n, for any polynomial p, and for any circuit A on inputs of length n with $\text{size}(A) \leq p(n)$,

$$\text{Prob}_{x \in \Sigma^*}(A(f(x)) \notin f^{-1}(f(x))) \geq \frac{1}{q(n)}. \tag{5.3}$$

Let us fix a sufficiently large n for which the assumption holds and for which the following arguments are valid. We focus on the restriction of f on Σ^n. Let $m = n \cdot q(n)$ and consider the function $g_{m \cdot n}: \Sigma^{m \cdot n} \rightarrow \Sigma^*$ defined by

$$g_{m \cdot n}(x_1 \odot \ldots \odot x_m) = f(x_1) \odot \ldots \odot f(x_m),$$

where each substring x_i, $i = 1, \ldots, m$, of the input has length n. In other words, $g_{m \cdot n}$ is the concatenation of m independent copies of f. The function $g_{m \cdot n}$ is still computable in polynomial time and, intuitively, it is harder to invert than f because one has to invert each "chunk" $f(x_i)$, $i = 1, \ldots, m$. For example, if we indeed follow the strategy of finding the inverse of each $f(x_i)$ one at a time, the probability of success in inverting $g_{m \cdot n}(x_1 \odot \cdots \odot x_m)$ is at most $(1 - (1/q(n)))^m = (1 - (1/q(n)))^{nq(n)} < \mathrm{e}^{-n}$. We will show that no other strategy inverts $g_{m \cdot n}$ significantly better than this simple one.

The proof is by contradiction. Thus we assume that there is a polynomial p and a circuit B (running the presumed better inverting strategy) on inputs of size $m \cdot n$, with size$(B) \leq p(m \cdot n)$ (which is polynomial in n) such that

$$\text{Prob}\left(B(g_{m\cdot n}(x_1 \odot \ldots \odot x_m)) \in g^{-1}_{m\cdot n}(g_{m\cdot n}(x_1 \odot \ldots \odot x_m))\right) \geq \frac{1}{p(m \cdot n)} \tag{5.4}$$
$$> \mathrm{e}^{-n} + \frac{1}{2p(m \cdot n)}.$$

Using B, we construct a polynomial-time probabilistic algorithm D that successfully inverts $f(x)$ on more than a fraction of $(1 - (1/q(n))$ of the strings x in Σ^n. This essentially contradicts the hypothesis (5.3) (the only small problem being that D is probabilistic).

The algorithm D proceeds as follows.

Input: $y = f(x)$, for some $x \in \Sigma^n$.
Repeat the following $2n \cdot m \cdot p(n \cdot m) = 2n^2 \cdot q(n) \cdot p(n^2 \cdot q(n))$ times.
  Pick random $i \in \{1, \ldots, m\}$.
  Pick $m - 1$ random strings in Σ^n denoted $x_1, \ldots, x_{i-1}, x_{i+1}, \ldots, x_m$.
  Calculate $Y = f(x_1) \odot \ldots \odot f(x_{i-1}) \odot f(x) \odot f(x_{i+1}) \odot \ldots \odot f(x_m)$.
  Call the circuit B to invert Y. The result is $x'_1, \ldots, x'_i, \ldots, x'_m$.
  If $f(x'_i) = y$, then the algorithm outputs x'_i, reports SUCCESS, and stops.
End Repeat
If there was no SUCCESS, output arbitrarily 0^n.

We next estimate the success probability of the above algorithm. Let INV $\subseteq \Sigma^{n\cdot m}$ be the elements of $\Sigma^{n\cdot m}$ that B inverts. By our assumption,

$$\|\text{INV}\| \geq \left(\mathrm{e}^{-n} + \frac{1}{2p(n \cdot m)}\right) \cdot \|\Sigma^{n\cdot m}\|.$$

For each $x \in \Sigma^n$, let $N(x)$ be the multiset consisting of m-tuples $(x_1, \ldots, x_m)$, with each $x_i \in \Sigma^n$, such that for some $i \in \{1, \ldots, m\}$, $x = x_i$. The multiplicity of such an m-tuple is the number of entries that are equal to x. Note that, for each $x \in \Sigma^n$, $\|N(x)\| = m \cdot 2^{n(m-1)}$. For any set $T \subseteq \Sigma^n$, $N(T)$ denotes $\bigcup_{x\in T} N(x)$.

It is useful to view the algorithm D from a different angle. On input $f(x)$, the algorithm calculates $Y = f(x_1) \odot \ldots \odot f(x) \odot \ldots \odot f(x_m)$, where $X = (x_1, \ldots, x_m)$ is uniformly at random chosen in $N(x)$. The circuit B is next invoked to invert Y. If $X \in$ INV then the algorithm D is successful. Therefore, we would like to show that for a large fraction (more precisely for a fraction of $(1 - \frac{1}{q(n)}))$ of x in Σ^n, $N(x) \cap \text{INV}$ is a large set, so that the success probability of D is large in inverting the input $f(x)$. This will lead to the desired contradiction of relation (5.3).

Claim 5.2.2 *Let* $V = \left\{x \in \Sigma^n \mid \frac{\|N(x) \cap \mathrm{INV}\|}{\|N(x)\|} \geq \frac{1}{2 \cdot p(n \cdot m) \cdot m}\right\}$. *Then* $\|V\| > \left(1 - \frac{1}{q(n)}\right) \cdot \|\Sigma^n\|$.

Proof. The complementary set of V is

$$\overline{V} = \left\{x \in \Sigma^n \mid \frac{\|N(x) \cap \mathrm{INV}\|}{\|N(x)\|} < \frac{1}{2 \cdot p(m \cdot m) \cdot m}\right\}.$$

Thus, for each $x \in \overline{V}$, $\|N(x) \cap \mathrm{INV}\| < \frac{1}{2 \cdot p(n \cdot m) \cdot m} \cdot (m \cdot 2^{n(m-1)})$, and consequently,

$$\begin{aligned} \|N(\overline{V}) \cap \mathrm{INV}\| &\leq \sum_{x \in \overline{V}} \|N(x) \cap \mathrm{INV}\| \\ &< 2^n \cdot \frac{1}{2 \cdot p(n \cdot m) \cdot m} \cdot (m \cdot 2^{n(m-1)}) \\ &= \frac{1}{2 \cdot p(n \cdot m)} \cdot \|(\Sigma^n)^m\|. \end{aligned}$$

We show that this is possible only if $\|\overline{V}\| < \frac{1}{q(n)} \cdot \|\Sigma^n\|$. To this aim, let $S \subseteq \Sigma^n$ be a set with $\|S\| \geq \frac{1}{q(n)} \cdot \|\Sigma^n\|$. The essential observation is that $N(S)$ covers an overwhelming fraction of $(\Sigma^n)^m$. Indeed, note that the probability that a tuple $(x_1, \ldots, x_m)$ is not in $N(S)$ is equal to the probability of the event "$x_1 \notin S \wedge \ldots \wedge x_m \notin S$," which is bounded by

$$\left(1 - \frac{1}{q(n)}\right)^m = \left(1 - \frac{1}{q(n)}\right)^{n \cdot q(n)} < \mathrm{e}^{-n}.$$

Therefore, the complementary set of $N(S)$, denoted $\overline{N(S)}$, satisfies

$$\|\overline{N(S)}\| < \mathrm{e}^{-n} \cdot \|(\Sigma^n)^m\|.$$

Then,

$$\begin{aligned} \|N(S) \cap \mathrm{INV}\| &= \|\mathrm{INV}\| - \|\mathrm{INV} \cap \overline{N(S)}\| \\ &\geq \|\mathrm{INV}\| - \|\overline{N(S)}\| \\ &> \left(\mathrm{e}^{-n} + \frac{1}{2p(n \cdot m)}\right) \|(\Sigma^n)^m\| - \mathrm{e}^{-n} \cdot \|(\Sigma^n)^m\| \\ &= \frac{1}{2p(n \cdot m)} \cdot \|(\Sigma^n)^m\|. \end{aligned}$$

Thus, necessarily, $\|\overline{V}\| < \frac{1}{q(n)} \cdot \|\Sigma^n\|$ and, therefore, $\|V\| > \left(1 - \frac{1}{q(n)}\right) \cdot \|\Sigma^n\|$. ∎

Claim 5.2.2 implies that if $x \in V$, then the success probability of D in inverting $f(x)$ in one iteration is at least $(1/(2m \cdot p(n \cdot m)))$. Since the algorithm D makes $n \cdot (2m \cdot p(n \cdot m))$ iterations, the overall failure probability is at most $\left(1 - \frac{1}{2m \cdot p(n \cdot m)}\right)^{n \cdot (2m \cdot p(n \cdot m))} \leq \mathrm{e}^{-n}$. An inspection of the algorithm D reveals that it can be implemented by a probabilistic circuit C'_D with size polynomial in n. Using a standard technique, the circuit C'_D can be converted into a *deterministic* circuit C_D with size still polynomial in n that inverts $f(x)$ for all x in V. Namely, observe that since, for any x in V, Prob(C'_D fails to invert x) $\leq e^{-n}$, we can conclude that Prob(C'_D fails to invert some x in V) $\leq 2^n \cdot e^{-n} < 1$. The probabilities are taken over the random strings used by C'_D. Then there is such a string r, so that C'_D acting according to r, will successfully invert $f(x)$ for all x in V. We can embed r into the wiring of C'_D, obtaining the deterministic circuit C_D which simulates the circuit C'_D replacing the utilization of the random string with queries from r. The size of C_D increases by only $|r|$ bits and thus it is still polynomial in n.

To conclude, we have obtained a circuit (i.e., C_D) of polynomial size that inverts f on more than a fraction of $1 - \frac{1}{q(n)}$ of strings in the multiset $\{f(x) \mid x \in \Sigma^n\}$. This contradicts the relation (5.3) and, therefore, our assumption (5.4) is false. Thus, for any polynomial p and for any circuit B with size(B) $\leq p(n \cdot m) = p(n^2 \cdot q(n))$, B inverts $g_{n^2 \cdot q(n)}$ on less than a fraction $\frac{1}{p(n^2 \cdot q(n))}$ of the multiset $\{g_{n^2 \cdot q(n)}(y) \mid y \in \Sigma^{n^2 \cdot q(n)}\}$. This assertion holds for almost all n.

This is almost the desired conclusion. The only problem is that we have only obtained the ensemble of functions $\{g_{n^2 \cdot q(n)}\}_{n \in \mathbb{N}}$, which does not cover all possible input lengths (as required in the definition of a one-way function). Fortunately, it is easy to extend the ensemble with functions $g_{n'}: \Sigma^{n'} \to \Sigma^{n'}$, for n' not of the form $n^2 \cdot q(n)$. Namely, for such an n', we find the largest n such that $n^2 \cdot q(n) \leq n'$ and define $g_{n'}$ on input x to be $g_{n^2 \cdot q(n)}$ on the prefix of x of length $n^2 \cdot q(n)$. It is easy to check that the full ensemble $(g_n)_{n \in \mathbb{N}}$ is a strong one-way function. ∎

5.3 From one-way permutations to extenders

IN BRIEF: For each one-way function f, there exists a small variation f' that admits a hard-core predicate b, i.e., a predicate $b(x)$ that is hard to compute given $f'(x)$.

Given a one-way permutation, one can effectively construct a pseudo-random generator with an extension of one bit.

An *extender* is a pseudo-random generator whose output is one bit longer than its input.

Definition 5.3.1 (Extender) (a) *Let $\epsilon > 0$ and $S \in \mathbb{N}$. An extender with security (ϵ, S) is a function f such that (1) for some $n \in \mathbb{N}$, $f: \Sigma^n \to \Sigma^{n+1}$, and (2) f is a pseudo-random generator with security (ϵ, S).*

(b) *Let $\epsilon: \mathbb{N} \to [0,1]$ and $S: \mathbb{N} \to \mathbb{N}$ be two functions. An ensemble of extenders with security (ϵ, S) is a family of functions $(f_n)_{n \in \mathbb{N}}$ such that, for each n, $f_n: \Sigma^n \to \Sigma^{n+1}$ and f_n is an extender with security $(\epsilon(n), S(n))$.*

As before, abusing the terminology, when the context is clear, an ensemble of extenders will usually be called an extender as well.

Our goal is to build an extender starting from a one-way permutation. This is a worthy operation because we will see in the next section that, using as a building block an extender, we can obtain pseudo-random generators with more impressive extensions. The decisive step in the construction is the production of a *hard-core predicate* for the one-way function. Intuitively, a hard-core predicate $h\colon \Sigma^n \to \{0,1\}$ for a function $f\colon \Sigma^n \to \Sigma^{n'}$ provides a bit $h(x)$ that cannot be *predicted* by an adversary that knows $f(x)$ with probability significantly larger than 1/2 (which is what anyone can obtain by simply guessing the value of the bit without any knowledge of $f(x)$). For this reason, we also say that a hard-core predicate provides a *hidden bit* for f. We prefer however to give the formal definition of a hard-core predicate by saying that the adversary cannot *distinguish* between the distributions $f(U_n) \odot h(U_n)$ and $f(U_n) \odot U_1$. The connection between predictor and distinguisher adversaries will be established shortly in Theorem 5.3.3. The merit of the next definition is that it involves distributions that are computationally close and this is useful for our goal of building pseudo-random generators.

Definition 5.3.2 (Hard-core predicate) *The values $n, n', S \in \mathbb{N}$ and $\epsilon > 0$ are parameters. Let $f\colon \Sigma^n \to \Sigma^{n'}$ and $h\colon \Sigma^n \to \{0,1\}$. The function h is a hard-core predicate for f with security (ϵ, S) if the distributions $(f(U_n) \odot h(U_n))$ and $(f(U_n) \odot U_1)$ are computationally ϵ-close for circuits of size S.*

The next theorem states the promised relation between predicting a hard-core predicate $h(x)$ given $f(x)$ and distinguishing between the distributions $f(U_n) \odot h(U_n)$ and $f(U_n) \odot U_1$.

Theorem 5.3.3 (Predictors vs. Distinguishers)

INFORMAL STATEMENT: *A predicate $h(x)$ can be predicted from a function $f(x)$ with probability at least $1/2 + \epsilon$ by an adversary circuit of some given size if and only if an adversary circuit of essentially the same size can distinguish between the distributions $f(U_n) \odot h(U_n)$ and $f(U_n) \odot U_1$ with bias at least ϵ.*

FORMAL STATEMENT: *Let n and n' be two integer parameters and $\epsilon > 0$. Let $f\colon \Sigma^n \to \Sigma^{n'}$ and $h\colon \Sigma^n \to \{0,1\}$. There exists a constant c with the following properties:*

(1) *If there is a circuit D such that $\mathrm{Prob}_x(D(f(x)) = h(x)) \geq \frac{1}{2} + \epsilon$, then there is a circuit C of size at most $\mathrm{size}(D) + c$ so that*

$$\mathrm{Prob}_{x\in\Sigma^n}(C(f(x) \odot h(x)) = 1) - \mathrm{Prob}_{x\in\Sigma^n, u\in\Sigma}(C(f(x) \odot u) = 1) \geq \epsilon.$$

(2) *If there is a circuit C such that*

$$|\mathrm{Prob}_{x\in\Sigma^n}(C(f(x) \odot h(x)) = 1) - \mathrm{Prob}_{x\in\Sigma^n, u\in\Sigma}(C(f(x) \odot u) = 1)| \geq \epsilon,$$

then there is a circuit D of size at most $\mathrm{size}(C) + c$ so that

$$\mathrm{Prob}_x(D(f(x)) = h(x)) \geq \frac{1}{2} + \epsilon.$$

Proof. (1) The circuit C on input $f(x) \odot y$, where $f(x) \in \Sigma^n$ and $y \in \Sigma$, calculates $D(f(x))$ and outputs 1 if $D(f(x)) = y$, and 0, otherwise. Now it is immediate to check that the asserted inequality holds because $\text{Prob}_{x,u}(C(f(x) \odot u) = 1) = 1/2$ and $\text{Prob}_x(C(f(x) \odot h(x)) = 1) \geq 1/2 + \epsilon$.

(2) We can eliminate the absolute value and assume that in fact

$$\text{Prob}_x(C(f(x) \odot h(x)) = 1) - \text{Prob}_{x,u}(C(f(x) \odot u) = 1) \geq \epsilon,$$

because otherwise we can just consider the circuit that flips (i.e., negates) the output of C. Thus, the circuit C is more likely to accept the string $f(x)$ followed by $h(x)$ than the string $f(x)$ followed by a random bit u. Based on this fact, the circuit D on input $f(x)$ does the following: It chooses a random bit b and calculates $C(f(x) \odot b)$. If the result is 1, it outputs b, otherwise it outputs $1 - b$. Let E be the event that $D(f(x)) = h(x)$ when x and b are chosen uniformly at random in Σ^n and, respectively, Σ. From the description of D, we observe that

$$\begin{aligned}\text{Prob}(E) = {} & \text{Prob}(h(x) = b) \cdot \text{Prob}(C(f(x) \odot h(x)) = 1) \\ & + \text{Prob}(h(x) = 1 - b) \cdot \text{Prob}(C(f(x) \odot (1 - h(x))) = 0).\end{aligned}$$

Let

$$\begin{aligned} p &\stackrel{\text{def}}{=} \text{Prob}_x(C(f(x) \odot h(x)) = 1), \qquad \text{and} \\ q &\stackrel{\text{def}}{=} \text{Prob}_{x,u}(C(f(x) \odot u) = 1).\end{aligned}$$

We have

$$\begin{aligned} q &= \text{Prob}(h(x) = u) \cdot \text{Prob}(C(f(x) \odot u) = 1 \mid h(x) = u) \\ &\quad + \text{Prob}(h(x) \neq u) \cdot \text{Prob}(C(f(x) \odot u) = 1 \mid h(x) \neq u) \\ &= \frac{1}{2}\text{Prob}(C(f(x) \odot h(x)) = 1) + \frac{1}{2}\text{Prob}(C(f(x) \odot (1 - h(x))) = 1) \\ &= \frac{1}{2}p + \frac{1}{2}\text{Prob}(C(f(x) \odot (1 - h(x))) = 1).\end{aligned}$$

It follows that

$$\text{Prob}(C(f(x) \odot (1 - h(x))) = 1) = 2q - p,$$

and, thus,

$$\text{Prob}(C(f(x) \odot (1 - h(x))) = 0) = 1 - 2q + p.$$

Therefore

$$\text{Prob}(E) = \frac{1}{2} \cdot p + \frac{1}{2}(1 - 2q + p) = \frac{1}{2} + (p - q) \geq \frac{1}{2} + \epsilon.$$

This ends the proof of (2). ∎

The relation between hard-core predicates and extenders is stated in the following lemma.

Lemma 5.3.4 INFORMAL STATEMENT: *If g is a permutation and h is a hard-core predicate for g, then the function $f(x) = g(x) \odot h(x)$ is an extender.*

FORMAL STATEMENT: *Let $n, S \in \mathbb{N}$ and $\epsilon > 0$ be some parameters. Let $g\colon \Sigma^n \to \Sigma^n$ and $h\colon \Sigma^n \to \{0,1\}$ be two functions such that g is a permutation and h is a hard-core predicate for g with security (ϵ, S). Then*

$$\Delta_{\mathrm{comp},S}(g(U_n) \odot h(U_n), U_{n+1}) < \epsilon.$$

Proof. By the triangle inequality,

$$\begin{aligned}&\Delta_{\mathrm{comp},S}(g(U_n) \odot h(U_n), U_{n+1})\\ &\qquad \leq \Delta_{\mathrm{comp},S}(g(U_n) \odot h(U_n), g_n(U_n) \odot U_1) + \Delta_{\mathrm{comp},S}(g_n(U_n) \odot U_1, U_{n+1}).\end{aligned}$$

Since g is a permutation, the distribution $g(U_n) \odot U_1$ is in fact the uniform distribution U_{n+1}. Thus the second term in the right hand side is zero. By assumption, the first term in the right hand side is less than ϵ. ∎

Roughly speaking, we will show that any one-way function has a hard-core predicate. Note first that, if f is a one-way function with security (ϵ, S) (and, for concreteness, let us also assume that f is a permutation) then, given $f(x)$, no adversary circuit of size S can find x but with probability ϵ. This does not mean that any given bit of x (such as, say, the first bit) is hidden given $f(x)$. For example, if $g\colon \Sigma^n \to \Sigma^n$ is one-way with security $(\epsilon/2, S)$, then the function $f\colon \Sigma^{n+1} \to \Sigma^{n+1}$ defined by $f(0 \odot x) = (0 \odot g(x))$ and $f(1 \odot x) = (1 \odot g(x))$ is one-way with security (ϵ, S); however, clearly, an adversary, even if modestly endowed with computational power, that is seeing $f(y)$ can immediately retrieve the first bit of y. Nevertheless, the fact that x is globally hard to retrieve given $f(x)$ provides the basis for obtaining a hidden bit.

Actually, in our construction, the hard-core predicate is not built directly from a one-way function $f\colon \Sigma^n \to \Sigma^{n'}$, but from a small modification of it, $g\colon \Sigma^{2n} \to \Sigma^{n+n'}$, given by $g(x \odot r) = f(x) \odot r$ (where $|x| = |r| = n$). The hard-core predicate is obtained through the utilization of error-correcting codes.

We recall that an error-correcting code is a length-regular function c over strings written in a given alphabet with a strong injectivity property: Any two strings in the image (such strings are called *codewords*) are guaranteed to be far apart. More precisely, the Hamming distance between two strings α and β of equal length is the number of positions in which α and β differ, and it is denoted by $\mathrm{dist}(\alpha, \beta)$. An error correcting code $c\colon \Sigma^n \to \Sigma^{n'}$, for some $n, n' \in \mathbb{N}$, has the property that, for some parameter d called the *distance* of the code, and for any $x_1 \neq x_2 \in \Sigma^n$, $\mathrm{dist}(c(x_1), c(x_2)) \geq d \cdot n'$. The *Hamming distance* has the triangle property and it follows that for any $x_1 \neq x_2$ in Σ^n, the balls $B_{(d/2)n'}(c(x_1))$ and $B_{(d/2)n'}(c(x_2))$ are disjoint, where by definition $B_r(u) = \{v \mid \mathrm{dist}(u, v) < r\}$. Thus, a string y that agrees with some codeword $c(x)$ in $> (1 - \frac{d}{2})n'$ bits, defines x uniquely, and, in particular, x can be retrieved effectively from y (provided c is a computable function). This is the decoding property of error-correcting codes, which is a very strong invertibility feature of c. For many good error-correcting code, the decoding can be done efficiently.

In our construction, we take an error-correcting code Code : $\Sigma^n \to \Sigma^{\overline{n}}$ and define the hard-core predicate $P\colon \Sigma^n \times \Sigma^{\log(\overline{n})} \to \{0,1\}$ by $P(x,r) = (\text{Code}(x))(r)$ (i.e., the r-th bit of Code(x)). Then, if we assume that there is a circuit C so that $C(f(x) \odot r) = (\text{Code}(x))(r)$ for a big fraction α of r (and for many x), then the bits $(C(f(x) \odot r))_{r \in \Sigma^{\log(\overline{n})}}$ make a string that is at distance at most $(1-\alpha)\cdot\overline{n}$ from Code(x), and, this, by the property of the error-correcting code, should allow us to retrieve x. This will contradict the fact that f is one-way. One problem is that in our assumption, α can not be taken to be $1 - d/2$ (where d is the distance of the code), but much smaller. Therefore we cannot retrieve x directly, as we have claimed, and we will be content to recover only a relatively small list of elements which contains x and whose cardinality is polynomial in n. This operation is called the *list-decoding* of an error-correcting code. List-decoding is here good enough because we can try all the elements in the list and check which one maps via f into $f(x)$. Thus we are able to invert f, still obtaining the contradiction we need.

For concreteness, we will be using the *Hadamard error-correcting code*, Had: $\Sigma^n \to \Sigma^{2^n}$, defined as follows: For all $y \in \{0,\dots,2^n - 1\}$, the y-th bit of Had(x) is the inner product $x \cdot y$ (in the vector space $(\text{GF}(2))^n$.) In other wordes, we write y in base 2 as $y_1y_2\dots y_n$ and the y-th bit of Had(x), where $x = x_1x_2\dots x_n$, is $x_1y_1 + x_2y_2 + \dots + x_ny_n \pmod 2$. Since $|\text{Had}(x)| = 2^n$, it follows that $|r| = n$ (r is the string from the previous paragraph) and, thus, we can not afford to calculate $C(f(x), r)$ for all $r \in \Sigma^n$ to construct a list of elements in Σ^n one of which is x. Fortunately, such a list can be obtained by looking at only a few elements of $(C(f(x), r))_{r\in\Sigma^n}$. This property of Hadamard codes is stated in the next theorem. In the statement, we make reference to a circuit that takes as inputs a string $x \in \Sigma^n$ and a string $y \in \Sigma^{2^n}$. Since y is very long and we do not want the circuit to have that many input gates, access to y is provided via the so called *oracle access* mechanism. This means that we use a different type of a circuit, called an *oracle circuit*, and the string y plays the role of an oracle set. An oracle circuit, in addition to the usual AND, OR, and NOT gates, also has *oracle gates*. Each oracle gate has n inputs and an instance $\alpha_1\dots\alpha_n$ of these inputs, the gate outputs the bit of y located at address $\alpha_1\dots\alpha_n$. The number of queries is the number of oracle gates. It will be the case that the oracle circuit only accesses a few of the bits of y and the locations of these bits are determined by the input x.

Theorem 5.3.5 (List decoding for Hadamard codes) *Let $n \in \mathbb{N}$ and $\epsilon > 0$ be two parameters. Consider the Hadamard error correcting code* Had: $\Sigma^n \to \Sigma^{2^n}$. *There is a probabilistic circuit A that has oracle access to a string of length 2^n so that:*

(a) *If $x \in \Sigma^n$ and $y \in \Sigma^{2^n}$ are two strings so that* $\text{dist}(y, \text{Had}(x)) \le (\frac{1}{2} - \epsilon)2^n$, *then, with probability at least $\frac{3}{4}$, A on input 1^n and with oracle access to y outputs a list of $\left(n \cdot \frac{1}{\epsilon^2} + 1\right)$ strings which includes x;*

(b) *The circuit A makes $n^2 \cdot \frac{1}{\epsilon^2}$ queries to y and has size $O(n^3 \cdot \frac{1}{\epsilon^4})$.*

Proof. For $y \in \Sigma^{2^n}$ and $r \in \Sigma^n$, $y(r)$ denotes as usual the r-th bit of y. For i, $1 \le i \le n$, let e_i be the vector $(0, \ldots, 0, 1, 0, \ldots 0)$, where the single 1 is in position i. Let $\oplus$ denote the addition of vectors in $(\mathrm{GF}(2))^n$. One first observation is that if for some i we have the r and the $(r \oplus e_i)$ bits of $\mathrm{Had}(x)$, then we can retrieve $x(i)$, the i-th bit of x. This is so because

$$(\mathrm{Had}(x))(r) + (\mathrm{Had}(x))(r \oplus e_i) = \sum x_i \cdot r_i + \sum x_i(r_i \oplus e_i) = x(i),$$

where the addition $+$ is in GF(2) (i.e., modulo 2). Thus, if we pick at random r and we read the bits in positions r and $r \oplus e_i$ of y and if we are lucky and $y(r) = (\mathrm{Had}(x))(r)$ and $y(r \oplus e_i) = (\mathrm{Had}(x))(r \oplus e_i)$, then we can retrieve $x(i)$. However, the probability that "we are lucky" is only guaranteed to be at least $1 - 2(\frac{1}{2} - \epsilon) = 2\epsilon$, which may be less than $\frac{1}{2}$. Note that $\frac{1}{2}$ is the probability that we obtain if we just guess $x(i)$ directly. Therefore, if $\epsilon < 1/4$, it does not help to look at both $y(r)$ and $y(r \oplus e_i)$. To understand the idea of the algorithm, suppose that we know $(\mathrm{Had}(x))(r)$ for some random r. Then it is enough to look at $y(r \oplus e_i)$ hoping that it is equal to $(\mathrm{Had}(x))(r \oplus e_i)$. This would allow us to calculate $x(i)$ correctly with the probability that our hope comes true, i.e., with probability at least $1/2 + \epsilon$. However, we do not know $(\mathrm{Had}(x))(r)$ for any r. This obstacle can be overcome by taking a sample set S of $r's$, assigning all possible values to $((\mathrm{Had}(x))(r))_{r \in S}$, looking at the corresponding $y(r \oplus e_i)$, and taking our guess for $x(i)$ to be given by the majority vote over all r in S. Each assigned value will give one candidate for x, and, with good probability, the correct assignment gives us x. The sample set is constructed by taking the random choices of r in Σ^n to be only pairwise independent. This allows us to obtain only a small number of possible candidates (i.e., the LIST) for x.

The complete algorithm is as follows.

Let $m = \log\left(n \cdot \frac{1}{\epsilon^2} + 1\right)$.

Step 1. Select a random binary $n \times m$ matrix T. (Comment: T is used for the pairwise independent choices of various r.)

For each vector $\tau \in \Sigma^m$, we produce a string $z_\tau \in \Sigma^n$ as shown below in Steps 2, 3, and 4. In the end, LIST $= \{z_\tau \mid \tau \in \Sigma^m\}$. So, let $\tau \in \Sigma^m$.

Step 2. For each $j \in \Sigma^m - \{(0, \ldots, 0)\}$, calculate

$$r^j = T \cdot j \pmod 2.$$

Step 3. For every $i \in \{1, \ldots, n\}$ and every $j \in \Sigma^m - \{(0, \ldots, 0)\}$, calculate

$$z_i^j = (\tau \cdot j) + y(r^j \oplus e_i) \pmod 2.$$

(Comment:

We hope that $y(r^j \oplus e_i)$ is $(\mathrm{Had}(x))(r^j \oplus e_i)$ and that $\tau \cdot j$ is $(\mathrm{Had}(x))(r^j)$.)

Set z_i to the value of the majority of z_i^j (the majority is taken over $j \in \Sigma^m - \{(0,\ldots,0)\}$).

Step 4. We set $z_\tau = z_1 z_2 \ldots z_n$.

Claim 5.3.6 *Suppose* $\tau \cdot j = (\mathrm{Had}(x))(r^j)$, *for all* $j \in \Sigma^m - \{(0,\ldots,0)\}$. *Let* $i \in \{1,\ldots,n\}$. *Then*

$$\mathrm{Prob}(z_\tau(i) = x(i)) > 1 - \frac{1}{4n}.$$

Proof. For every vector $j \in \Sigma^m - \{(0,\ldots,0)\}$, we define the random variable

$$X^j = \begin{cases} 1, & \text{if } (\mathrm{Had}(x))(r^j) + y(r^j \oplus e_i) = x(i) \pmod 2, \\ 0, & \text{otherwise.} \end{cases}$$

The random variable $r^j = T \cdot j$ is uniformly distributed in Σ^n (when T is selected uniformly at random as a binary $n \times m$ matrix). Recalling that $(\mathrm{Had}(x))(r) + (\mathrm{Had}(x))(r \oplus e_i) = x(i) (\mathrm{mod}\ 2)$, it follows that

$$\mathrm{Prob}(X^j = 1) \geq \mathrm{Prob}(y(r^j \oplus e_i) = (\mathrm{Had}(x))(r \oplus e_i)) \geq \frac{1}{2} + \epsilon$$

(the last inequality follows from the Theorem's hypothesis about $\mathrm{dist}(y, \mathrm{Had}(x))$). Let $M = 2^m - 1$. The probability that the majority of z_i^j is not equal to $x(i)$ is equal to $\mathrm{Prob}\big(\sum_j X^j \leq \frac{1}{2} \cdot M\big)$. The next observation is that the variables r^j are pairwise independent. To check this, let T_i denote the i-th row of the matrix T, $i = 1,\ldots,n$, let b_1 and b_2 be two bits, and let j and k be two distinct vectors in $(\mathrm{GF}(2))^n$. It is easy to see that, for all i, $\mathrm{Prob}_T(T_i \cdot j = b_1 \text{ and } T_i \cdot k = b_2) = \frac{1}{4}$. Then, for any two vectors u, v in $(\mathrm{GF}(2))^n$,

$$\begin{aligned} \mathrm{Prob}_T(r^j = u \text{ and } r^k = v) &= \mathrm{Prob}_T(T \cdot j = u \text{ and } T \cdot k = v) \\ &= \mathrm{Prob}_T(\forall i, T_i \cdot j = u_i \text{ and } T_i \cdot k = v_i) = \\ &= \prod_{i=1}^{n} \mathrm{Prob}_T(T_i \cdot j = u_i \text{ and } T_i \cdot k = v_i) = \frac{1}{4^n} \\ &= \mathrm{Prob}_T(r^j = u) \cdot \mathrm{Prob}_T(r^k = v). \end{aligned}$$

It follows that the random variables X^j are pairwise independent as well. Thus, by Chebyshev's inequality, denoting expectation as $\mathrm{E}(\cdot)$ and variation as $\mathrm{Var}(\cdot)$,

$$\begin{aligned} &\mathrm{Prob}\Big(\sum X^j \leq \frac{1}{2} \cdot M\Big) \\ &\quad = \mathrm{Prob}\Big(\sum (1 - X^j) \geq \frac{1}{2} \cdot M\Big) \\ &\quad = \mathrm{Prob}\Big(\sum (1 - X^j) - M \cdot \mathrm{E}(1 - X^1) \geq M\Big(\frac{1}{2} - \mathrm{E}(1 - X^1)\Big)\Big) \end{aligned}$$

$$
\begin{aligned}
&\leq \text{Prob}\Big(\Big|\sum(1-X^j) - \text{E}(1-X^1)\cdot M\Big| \geq M\cdot\epsilon\Big) \\
&\leq \frac{\text{Var}\,(\sum(1-X^j))}{M^2\cdot\epsilon^2} && \text{(by Chebyshev's inequality)} \\
&= \frac{M\cdot\text{Var}\,(1-X^1)}{M^2\cdot\epsilon^2} && \text{(since } X^{j'} \text{ are pairwise independent)} \\
&\leq \frac{\text{Var}\,(1-X^1)}{\epsilon^2\cdot n\cdot\frac{1}{\epsilon^2}} \\
&\leq \frac{\frac{1}{4}}{n} = \frac{1}{4n}
\end{aligned}
$$

(we took into account that $M = 2^m - 1 = n(\epsilon^{-2})$ and that the variance of any boolean random variable is at most $1/4$). ∎

(Proof of Theorem 5.3.5 continued.) Hence, if $\tau\cdot j = (\text{Had}(x))(r^j)$ for all $j \in \Sigma^m - \{(0,\dots,0)\}$, then the string z_τ is equal to x with probability at least $1 - n\cdot\frac{1}{4n} = \frac{3}{4}$. Note that

$$
\begin{aligned}
& \tau\cdot j = (\text{Had}(x))(r^j) \\
\iff & \tau\cdot j = (\text{Had}(x))(T\cdot j) && \text{(by definition of } r^j) \\
\iff & \tau\cdot j = x\cdot(T\cdot j) && \text{(by definition of Had}(x)) \\
\iff & \tau\cdot j = (x\cdot T)\cdot j.
\end{aligned}
$$

Since we try all $\tau \in \Sigma^m$, for $\tau = x\cdot T$ it holds that $\tau\cdot j = (\text{Had}(x))(r^j)$ for all $j \in \Sigma^m - \{(0,\dots,0)\}$. Therefore with probability at least $\frac{3}{4}$, the string x is z_τ, and, thus, with probability at least $\frac{3}{4}$ it appears in LIST.

By inspecting the algorithm, we see that the number of queries to y is $n(2^m-1) = (\frac{n}{\epsilon})^2$ and that the number of elementary operations is $O(2^m\cdot n\cdot(2^m-1)) = O(n^3\cdot\epsilon^{-4})$. This concludes the proof of Theorem 5.3.5. ∎

Lemma 5.3.7 *Let $f\colon \Sigma^n \to \Sigma^{n'}$ be a function computable by a circuit of size $p(n)$. Let G be a circuit that on input $f(x)\odot r$, with $|r| = n$, attempts to calculate the function $P(x,r) \stackrel{\text{def}}{=} (\text{Had}(x))(r)$ and let*

$$\epsilon = \text{Prob}_{x,r}(G(f(x)\odot r) = P(x,r)) - \frac{1}{2}.$$

Then there is a circuit A with size bounded by $O\big(n^3\cdot\frac{1}{\epsilon^4} + n\cdot p(n)\cdot\frac{1}{\epsilon^2} + n^2\cdot\frac{1}{\epsilon^2}\cdot\text{size}(G)\big)$ so that

$$\text{Prob}_x(A(f(x)) \in f^{-1}(f(x))) \geq \frac{3}{4}\cdot\epsilon.$$

Proof. We can assume that $\epsilon > 0$ (the case $\epsilon \leq 0$ is trivial and not interesting). For each $x \in \Sigma^n$, let

$$s(x) = \text{Prob}_r(G(f(x)\odot r) = P(x,r)).$$

Let $\text{GOOD} = \{x \in \Sigma^n \mid s(x) \geq 1/2 + \epsilon/2\}$. We first observe the following fact.

Claim 5.3.8 $\|\text{GOOD}\| \geq \epsilon \cdot 2^n$.

Proof. From the hypothesis, we know that $2^{-n}\sum_x(s(x) - 1/2) = \epsilon$. Also, for all $x \in \Sigma^n$, $s(x) - 1/2 \leq 1/2$. Then, $2^n \cdot \epsilon \leq (2^n - \|\text{GOOD}\|) \cdot \frac{\epsilon}{2} + \|\text{GOOD}\| \cdot \frac{1}{2}$. After some simple calculations, we obtain $\|\text{GOOD}\|/2^n \geq \epsilon$. ∎

For each $x \in \Sigma^n$, let $y(x)$ be the string obtained by concatenating in the lexicographical order of the strings $r \in \Sigma^n$, the bits $G(f(x) \odot r)$. Then, for each $x \in \text{GOOD}$, $\text{dist}(y(x), \text{Had}(x)) \leq \frac{1}{2} - \frac{\epsilon}{2}$. We apply Theorem 5.3.5 for $y(x)$ and x; however, instead of querying the bits of $y(x)$ we calculate them using the circuit G on input $f(x)$ and various r. We obtain the set of strings LIST having $(n \cdot (\frac{\epsilon}{2})^2 + 1)$ strings and which, with probability at least $\frac{3}{4}$, contains the string x. By checking all the elements in LIST, we find one which maps via f into the input $f(x)$. This entire procedure can be implemented by a probabilistic circuit A' of size $O(n^3 \cdot \frac{16}{\epsilon^4} + \frac{4n^2}{\epsilon^2} \cdot \text{size}(G) + \frac{4n}{\epsilon^2} \cdot p(n))$. Note that

$$\begin{aligned}&\text{Prob}(A'(f(x)) \in f^{-1}(f(x)))\\ &\qquad \geq \text{Prob}(x \in \text{GOOD}) \cdot \text{Prob}(A(f(x)) \in f^{-1}(f(x)) \mid x \in \text{GOOD}) \geq \epsilon \cdot \frac{3}{4},\end{aligned}$$

where the probability is taken over x chosen uniformly at random in Σ^n and over the random choices of A'. We can convert A' to a deterministic circuit A by observing that there has to be a random choice r_0 so that the above relation holds when x is chosen in Σ^n and r is fixed to r_0 (i.e., the circuit A does what A' is doing only that it is using the fixed r_0 to simulate the random choices of A'). The size of A is equal to $\text{size}(A') + |r_0| \leq 2\text{size}(A')$. The proof of the lemma is now complete. ∎

Theorem 5.3.9 INFORMAL STATEMENT: *Given a one-way function f, we can build a hard-core predicate for a small variation of f.*

FORMAL STATEMENT: *Let $\epsilon \colon \mathbb{N} \to [0,1]$ and $S \colon \mathbb{N} \to \mathbb{N}$ be two functions such that $S(n)$ is superpolynomial and $\epsilon(n) \geq n \cdot (S(n))^{-1/4}$ for all sufficiently large n. Let $(f_n)_{n\in\mathbb{N}}$ be an (ϵ, S) one-way function. Let $P_n \colon \Sigma^n \times \Sigma^n \to \{0,1\}$ be defined by $P_n(x,r) = (\text{Had}(x))(r)$ (i.e., $P_n(x,r) = x \cdot r$.) Then there is a constant c so that, for any family of circuits $(G_n)_{n\in\mathbb{N}}$ with* $\text{size}(G_n)$ *at most $c \cdot \frac{\epsilon(n)^2}{n^2} \cdot S(n)$, and for all sufficiently large n,*

$$\text{Prob}_{x,r}(G_n(f_n(x) \odot r) = P_n(x,r)) < \frac{1}{2} + \frac{4}{3} \cdot \epsilon(n). \tag{5.5}$$

Proof. This is an easy consequence of Lemma 5.3.7. For some constant c, if $\text{size}(G_n) = c \cdot \frac{\epsilon(n)^2}{n^2} \cdot S(n)$, then the size of the circuit A that results from G_n in Lemma 5.3.7 is at most $S(n)$. Assume that the opposite of inequality (5.5) holds for a family of circuits $(G_n)_{n\in\mathbb{N}}$ with size as above and for infinitely many n. It follows that a circuit of size $S(n)$ (namely, the circuit A) can invert the function f_n with a bias that is at least $\epsilon(n)$. Since this happens for infinitely many n, we have reached a contradiction to the fact that $(f_n)_{n\in\mathbb{N}}$ is an (ϵ, S) one-way function. ∎

Note that $(\text{Had}(x))(r)$ is simply the inner product of x and r, which we denote by $x \cdot r$ (x and r are viewed here as vectors in $\text{GF}(2)^n$). Taking into account Theorem 5.3.3, the above result says that, essentially, the inner product of x and r is a hard-core predicate of $f_n(x) \odot r$, whenever $(f_n)_{n\in\mathbb{N}}$ is a one-way function. One small problem is that this hard-core predicate is defined only for inputs of even length, but this can be remedied easily. For $y \in \Sigma^*$, let y_{left} be the left half of y and y_{right} be the right half of y with the provision that if $|y|$ is odd then $y_{\text{left}} = y(1 : \lfloor |y|/2 \rfloor)$ and $y_{\text{right}} = y(\lfloor |y|/2 \rfloor + 1, |y|)$.

Lemma 5.3.10 *Let $\epsilon\colon \mathbb{N} \to [0,1]$ and $S\colon \mathbb{N} \to \mathbb{N}$ be two functions such that $S(n)$ is superpolynomial and $\epsilon(n) \geq n \cdot (S(n))^{-1/4}$ for all sufficiently large n. Let $(f_n)_{n\in\mathbb{N}}$ be an (ϵ, S) one-way function.*

(a) *For each $n \in \mathbb{N}$, let g_{2n} and h_{2n} be the functions defined on strings of length $2n$ by $g_{2n}(y) \stackrel{\text{def}}{=} f_n(y_{\text{left}}) \odot y_{\text{right}}$ and, respectively, by $h_{2n}(y) \stackrel{\text{def}}{=} y_{\text{left}} \cdot y_{\text{right}}$ Then, for some constant d and for all sufficiently large n, h_{2n} is a hard-core predicate for g_{2n} with security $(\frac{4}{3}\epsilon(n), d \cdot \frac{\epsilon(n)^2}{n^2} S(n))$.*

(b) *For each $n \in \mathbb{N}$ let g_{2n+1} and h_{2n+1} be the functions defined on strings of length $2n+1$ by $g_{2n+1}(y) \stackrel{\text{def}}{=} f_n(y_{\text{left}}) \odot y_{\text{right}}$ and, respectively, by $h_{2n+1}(y) \stackrel{\text{def}}{=} y_{\text{left}} \cdot y_{\text{right}}(1:n)$ Then, for some constant d and for all sufficiently large n, h_{2n+1} is a hard-core predicate for g_{2n+1} with security $(\frac{4}{3}\epsilon(n), d \cdot \frac{\epsilon(n)^2}{n^2} S(n))$.*

Proof. (a) By Theorem 5.3.9, any circuit of size at most $c \cdot \frac{\epsilon(n)^2}{n^2} \cdot S(n)$ on input $f_n(y_{\text{left}}) \odot y_{\text{right}}$ computes $y_{\text{left}} \cdot y_{\text{right}}$ on less than a fraction $\frac{1}{2} + \frac{4}{3}\epsilon(n)$ of all $y \in \Sigma^{2n}$. Theorem 5.3.3 implies that the computational distance between the distributions $g_{2n}(u_{2n}) \odot h_{2n}(u_{2n})$ and $g_{2n}(u_{2n}) \odot u_1$ is less than $\frac{4}{3}\epsilon(n)$ relative to circuit size $d \cdot \frac{\epsilon(n)^2}{n^2} \cdot S(n)$, for some constant d. ∎

(b) It is easy to check that Theorem 5.3.9 implies that any circuit of size at most $c \cdot \frac{\epsilon(n)^2}{n^2} \cdot S(n)$ on input $f_n(y_{\text{left}}) \odot y_{\text{right}}$ computes $y_{\text{left}} \cdot y_{\text{right}}(1:n)$ on less than a fraction $\frac{1}{2} + \frac{4}{3}\epsilon(n)$ of all $y \in \Sigma^{2n+1}$ (the last bit of y_{right} cannot be too helpful; if it were it could be fixed to some advantageous value and build a circuit that computes $y_{\text{left}} \cdot y_{\text{right}}(1:n)$ on input $f_n(y_{\text{left}}) \odot y_{\text{right}}(1:n)$ on at least a fraction of $\frac{1}{2} + \frac{4}{3}\epsilon(n)$ of all strings $y_{\text{left}} \odot y_{\text{right}}(1:n)$). The rest is as in (a). ∎

All the preparations are ready for producing an extender, provided we are given a one-way permutation.

Theorem 5.3.11 INFORMAL STATEMENT: *Given a one-way permutation, we can construct an extender.*

FORMAL STATEMENT: *Let $\epsilon\colon \mathbb{N} \to [0,1]$ and $S\colon \mathbb{N} \to \mathbb{N}$ be two functions such that $S(n)$ is superpolynomial and $\epsilon(n) \geq n \cdot (S(n))^{-1/4}$ for all sufficiently large n. Suppose that $(f_n)_{n\in\mathbb{N}}$ is a one-way function with security (ϵ, S) such that, for each n, $f_n\colon \Sigma^n \to \Sigma^n$ is a permutation. Then there exists $(\tilde{g}_n)_{n\in\mathbb{N}}$ an extender with security $(\frac{4}{3}\epsilon(n), O(\frac{\epsilon(n)^2}{n^2} S(n)))$. Moreover, there exists a polynomial q such that, for all n, $\tilde{g}_n$ is computable by a circuit of size $q(n)$.*

Proof. We consider the family of functions $(g_n)_{n\in\mathbb{N}}$ and $(h_n)_{n\in\mathbb{N}}$ constructed in Lemma 5.3.10 from the family of functions $(f_n)_{n\in\mathbb{N}}$. It is easy to see that g_n and h_n can be calculated in time bounded by $p(n)$, for some polynomial p. It is also easy to check that, for each n, g_n is a permutation of Σ^n. Since, for some constant d and for each n, h_n is a hard-core predicate for g_n with security $(\frac{4}{3}\epsilon(n), d\cdot\frac{\epsilon(n)^2}{n^2}\cdot S(n))$, using Lemma 5.3.4, we derive that

$$\Delta_{\text{comp},d\cdot\frac{\epsilon(n)^2}{n^2}\cdot S(n)}(g_n(U_n)\odot h_n(U_n), U_{n+1}) < \frac{4}{3}\epsilon(n).$$

The conclusion follows for $\tilde{g}_n(x) \stackrel{\text{def}}{=} g_n(x)\odot h_n(x)$. ∎

5.4 From extenders to pseudo-random generators

IN BRIEF: It is shown how to enlarge the extension of a pseudo-random generator.

Given an extender, we can build a pseudo-random generator with much more significant extension. We present here the construction. The starting point is a function $g\colon \Sigma^n \to \Sigma^{n+1}$ that is an extender with security (ϵ, S). We will build a pseudo-random generator $h\colon \Sigma^n \to \Sigma^L$, for some arbitrary L[3] larger than n, as follows. First we compute $g(x)$, we retain the first bit of $g(x)$, and then we use the remaining n bits of $g(x)$ as a seed for a new invocation of g. Since g is a pseudo-random generator, this seed is almost as good as a fresh independently chosen random string of length n. We iterate this process L times and we output all the bits that have been retained.

More formally, let us define

$$\text{head}(x) = g(x)(1),$$

and

$$\text{tail}(x) = g(x)(2 : n+1).$$

For each $k \geq 1$, let $h^k\colon \Sigma^n \to \Sigma^k$ be defined by

$$h^k(x) = \underbrace{(\text{head}(x)\odot\text{head}(\text{tail}(x))\odot\ldots\odot\text{head}(\text{tail}(\ldots\text{tail}(x)\ldots))}_{k}.$$

As explained above, we take

$$h(x) = h^L(x).$$

[3]In order to make the construction meaningful, the value of L will in fact be bounded by a function of ϵ and S.

Theorem 5.4.1 INFORMAL STATEMENT: *Given an (ϵ, S) extender with large S, we can construct a pseudo-random generator with large extension.*

FORMAL STATEMENT: *Let $\epsilon > 0$ and $S, q, L \in \mathbb{N}$. Let $g\colon \Sigma^n \to \Sigma^{n+1}$ be an extender with security (ϵ, S) that is computable by a circuit of size q. Then the function $h\colon \Sigma^n \to \Sigma^L$ defined as above is a pseudo-random generator with security $(L \cdot \epsilon, S - 2L \cdot q - O(1))$.*

Proof. For the sake of this proof it is helpful to introduce the following distributions $d^0, d^1, \ldots, d^L$, all defined on the space Σ^L. For each $k \in \{0, \ldots, L\}$, d^k is the random variable obtained by taking uniformly at random a k-bits long binary string followed by the $(L-k)$-bits long binary string that results from applying h^{L-k} to a random binary string of length n. Formally,

$$d^k = U_k \odot h^{L-k}(U_n).$$

Observe that $d^0 = h(U_n)$ and $d^L = U_L$. The distributions $d^0, \ldots, d^L$ are sometimes called *hybrid distributions* because they are obtained via a mixed usage of pure random strings and of the extender g. Note the fine gradual passage from d^0 to d^L realized by the intermediate distributions $d^1, \ldots, d^{L-1}$. What is important is that the computational distance between any two consecutive d^i and d^{i+1} is, as we will prove shortly, at most ϵ for circuits of size not much smaller than S, which implies that the computational distance between the two extremes d^0 and d^L cannot be too large. This proof technique is an instance of the so-called *hybrid method* which we are going to see later as well.

Suppose that the computational distance between the distributions $(h(U_n))$ and (U_L) is greater than $L \cdot \epsilon$ for circuits of size $S - 2L \cdot q(n)$. In other words, suppose that there is a circuit C of size $S - 2L \cdot q(n)$ such that

$$|\text{Prob}(C(d^0) = 1) - \text{Prob}(C(d^L) = 1)| \geq L \cdot \epsilon.$$

Clearly,

$$\begin{aligned} &|\text{Prob}(C(d^0) = 1) - \text{Prob}(C(d^L) = 1)| \\ &\quad \leq \; |\text{Prob}(C(d^0) = 1) - \text{Prob}(C(d^1) = 1)| \\ &\qquad + |\text{Prob}(C(d^1) = 1) - \text{Prob}(C(d^2) = 1)| \\ &\qquad\quad \vdots \\ &\qquad + |\text{Prob}(C(d^{L-1}) = 1) - \text{Prob}(C(d^L) = 1)|. \end{aligned}$$

Therefore, there is $k \in \{0, \ldots, L-1\}$ such that

$$|\text{Prob}(C(d^k) = 1) - \text{Prob}(C(d^{k+1}) = 1)| \geq \epsilon. \tag{5.6}$$

An inspection of d^k and d^{k+1} reveals that the difference between them stems from the fact that d^{k+1} is obtained by applying a deterministic easy-to-compute function to a random $y \in \Sigma^{n+1}$, while d^k is obtained by applying the same deterministic function to $g(x)$, with x randomly chosen in Σ^n. Thus, the relation (5.6) implies that a minor modification of the circuit C is able to distinguish with bias greater than ϵ between the distributions $g(U_n)$ and U_{n+1}, and this contradicts the hypothesis.

Let us formalize this argument. For $z \in \Sigma^{n+1}$, let

$$\text{first}(z) = z(1), \text{ and}$$

$$\text{last}(z) = z(2 : n+1).$$

Thus, $\text{head}(x) = \text{first}(g(x))$ and $\text{tail}(x) = \text{last}(g(x))$. Let $f \colon \Sigma^{n+1} \to \Sigma^{L-k}$ be defined by

$$f(z) = \text{first}(z) \odot h^{L-k-1}(\text{last}(z)).$$

It can be checked that f can be computed by a circuit of size $L \cdot q(n) + O(1)$. Now, d^k is

$$U_k \odot \underbrace{\text{head}(U_n) \odot \text{head}(\text{tail}(U_n)) \odot \ldots \odot \text{head}(\text{tail}(\ldots(\text{tail}(U_n)\ldots)))}_{L-k}$$

and d^{k+1} is

$$U_k \odot U_1 \odot \underbrace{\text{head}(U_n) \odot \text{head}(\text{tail}(U_n)) \odot \ldots \odot \text{head}(\text{tail}(\ldots(\text{tail}(U_n)\ldots)))}_{L-k-1}.$$

Thus,

$$d^k = U_k \odot f(g(U_n))).$$

Observe that d^{k+1} can be viewed as the concatenation of U_k with

$$\underbrace{\text{first}(U_{n+1}) \odot \text{head}(\text{last}(U_{n+1})) \odot \ldots \odot \text{head}(\text{tail}(\ldots(\text{tail}(\text{last}(U_{n+1}))\ldots)))}_{L-k},$$

and thus,

$$d^{k+1} = U_k \odot f(U_{n+1})).$$

Therefore, the relation (5.6) can be rewritten as

$$\Big|\text{Prob}_{U_k,U_n}(C((U_k \odot f(g(U_n)))) = 1) \\ -\text{Prob}_{U_k,U_{n+1}}(C((U_k \odot f(U_{n+1}))) = 1)\Big| \geq \epsilon.$$

Clearly, there is some fixed string u_k^0 of length k such that

$$\Big|\text{Prob}_{U_n}(C((u_k^0 \odot f(g(U_n)))) = 1) \\ -\text{Prob}_{U_{n+1}}(C((u_k^0 \odot f(U_{n+1}))) = 1)\Big| \geq \epsilon. \quad (5.7)$$

Now we can define a circuit D that is able to distinguish between the distributions $(g(U_n))$ and U_{n+1}. The circuit D has u_k^0 hardwired in its circuitry, and on input

$z \in \Sigma^{n+1}$ it simulates the circuit C on input $u_k^0 \odot f(z)$. It is easy to see that the size of D is bounded by size$(C) + L \cdot q(n) + O(1) + L \leq$ size$(C) + 2L \cdot q(n) \leq S$ (because we need to "store" u_k^0 and to calculate $f(z)$). By the definition of D,

$$\text{Prob}_{U_n}(D(g(U_n)) = 1) = \text{Prob}_{U_n}(C((u_k^0 \odot f(g(U_n)))) = 1)$$

and

$$\text{Prob}_{U_{n+1}}(D(U_{n+1}) = 1) = \text{Prob}_{U_{n+1}}(C((u_k^0 \odot f(U_{n+1}))) = 1).$$

Therefore the relation (5.7) implies

$$\left|\text{Prob}_{U_n}(D(g(U_n))) = 1) - \text{Prob}_{U_{n+1}}(D(U_{n+1})) = 1)\right| \geq \epsilon,$$

and this contradicts the fact that g is an (ϵ, S)-secure extender. ∎

We have finally built the pseudo-random generator. It is the moment to contemplate the entire construction. We have started with an ensemble $(f_n)_{n\in\mathbb{N}}$ of one-way permutations, $f_n\colon \Sigma^n \to \Sigma^n$, having security (ϵ, S). We have next obtained an ensemble of extenders $(g_n)_{n\in\mathbb{N}}$ with security (ϵ', S'), where $\epsilon'(n) = \frac{4}{3}\epsilon(n)$ and $S'(n) = O(1) \cdot \frac{\epsilon^2}{n^2}S(n)$ (provided S is a superpolynomial function and $\epsilon(n) \geq n \cdot (S(n))^{-1/4}$). There is a polynomial $q(n)$ such that each g_n can be calculated in time $q(n)$. In the last construction stage, we have produced the ensemble $(h_n)_{n\in\mathbb{N}}$, $h_n\colon \Sigma^n \to \Sigma^{L(n)}$, of pseudo-random generators with security (ϵ'', S''), where $\epsilon''(n) = L(n) \cdot \epsilon'(n)$ and $S''(n) = S'(n) - 2L(n)q(n) - O(1)$. Thus, $\epsilon''(n) = L(n) \cdot \frac{4}{3}\epsilon(n)$ and $S''(n) = O(1) \cdot \frac{\epsilon^2}{n^2}S(n) - 2L(n)q(n) - O(1)$.

Naturally, the quality of the pseudo-random generator depends on the quality of the one-way permutation. In particular, the following theorem holds.

Theorem 5.4.2 (a) *Suppose there exists a strong one-way function $(f_n)_{n\in\mathbb{N}}$ such that for all n, $f_n\colon \Sigma^n \to \Sigma^n$ is a permutation. Then there exists a strong pseudo-random generator $(g_n)_{n\in\mathbb{N}}$, where, for each n, $g_n\colon \Sigma^n \to \Sigma^{L(n)}$ and $L(n)$ is a superpolynomial function.*

(b) *Suppose there exists an exponentially strong one-way function $(f_n)_{n\in\mathbb{N}}$ such that for all n, $f_n\colon \Sigma^n \to \Sigma^n$ is a permutation. Then there exists an exponentially strong pseudo-random generator $(g_n)_{n\in\mathbb{N}}$, where, for each n, $g_n\colon \Sigma^n \to \Sigma^{L(n)}$ and $L(n)$ is an exponential function (i.e., $L(n) = 2^{cn}$ for some constant c).*

Proof. (a) We are using the notation from the previous paragraph. We can assume that $\epsilon(n) \geq n \cdot (S(n))^{-1}$. (If the opposite inequality holds, we can substitute $\epsilon(n)$ with the largerr $\epsilon_1(n) = n \cdot (S(n))^{-1/4}$ and of course $(f_n)_{n\in\mathbb{N}}$ is one-way with security $(\epsilon_1(n), S(n)^{1/4})$ and $1/\epsilon_1$ is still superpolynomial.) We take $L(n) = \min((\epsilon(n))^{-1/2}, (S(n))^{1/3})$. One can check that $1/\epsilon''(n)$ and $S''(n)$ are superpolynomial.

(b) Similar to (a). ∎

5.5 Pseudo-random functions

IN BRIEF: An exponentially strong pseudo-random generator can be converted into a pseudo-random function.

We have seen in the previous sections that, given a one-way permutation, we can build a pseudo-random generator. One drawback of the method is that the computation of the pseudo-random generator is strictly sequential. A rapid inspection of the construction reveals that the calculation of the $(i+1)$-th bit can be done only after the bits $1, \ldots, i$ are already known.

We present here an alternative construction which avoids this problem. The starting point is already a pseudo-random generator $g\colon \Sigma^n \to \Sigma^{2n}$, for some $n \in \mathbb{N}$, but we aim for superpolynomial or even exponential extension. Such a pseudo-random generator g, i.e., one that doubles the input length, can be obtained from a one-way permutation by the method that we have already seen. At the end of the construction that we present in this section, we will obtain a a pseudo-random generator of the following type. The input (i.e., the seed) will be a string x of length n and the output will be a string $f(x)$ of length $n \cdot 2^m$, for some $n, m \in \mathbb{N}$. The string $f(x)$ can be interpreted in a natural way as a function mapping strings of length m into strings of length n: We divide $f(x)$ into 2^m blocks, each of length n, and we view the i-th block (where $i = 0, \ldots, 2^{m-1}$) as the output of the function $f(x)$ on input i (formally, the input is the binary encoding of the integer i). Thus, identifying a positive integer i with the string representing i in binary, which for simplicity we denote i as well, $f(x)(i)$ denotes the i-th block of $f(x)$. Via this interpretation, when x is chosen uniformly at random in Σ^n, $f(x)$ is a random variable over the space of functions $\mathcal{F}_{m,n} = \{f \mid f\colon \Sigma^m \to \Sigma^n\}$. If $f(x)$ is computationally indistinguishable from $U_{n\cdot 2^m}$, it is natural to view $f(x)$ as a random function which is computationally indistinguishable from a function chosen uniformly at random in $\mathcal{F}_{m,n}$. This is why we call $f(x)$ a *pseudo-random function*. Furthermore, our construction will ensure that $f(x)(i)$ can be calculated efficiently and independently from $f(x)(j)$, if $i \neq j$, which is a desirable property for a function. This amounts to being able to calculate each of the 2^m blocks of $f(x)$ separately.

We next describe the construction. Let $g\colon \Sigma^n \to \Sigma^{2n}$ be a pseudo-random generator. Let

$$g_0(x) = g(x)(1:n) \text{ (the left half of } g(x)),$$

and

$$g_1(x) = g(x)(n+1:2n) \text{ (the right half of } g(x)).$$

For each $x \in \Sigma^n$ and each string $\alpha \in \Sigma^{\leq m}$, we will describe a string $f(x)(\alpha)$, and $f(x)$ will be

$$f(x)\underbrace{(00\ldots0)}_{m} \odot f(x)\underbrace{(00\ldots1)}_{m} \odot \ldots \odot f(x)\underbrace{(11\ldots1)}_{m}.$$

The strings $f(x)(\alpha)$ are defined inductively as follows:

$$f(x)(\lambda) = x,$$
$$f(x)(\alpha 0) = g_0(f(x)(\alpha)), \text{ and}$$
$$f(x)(\alpha 1) = g_1(f(x)(\alpha)).$$

Thus,

$$f(x)(\alpha_1\alpha_2\ldots\alpha_m) = g_{\alpha_m}(g_{\alpha_{m-1}}(\cdots(g_{\alpha_1}(x))\ldots)).$$

The construction can be viewed as the process of labeling the nodes of a full binary tree of height m with n-bits long strings. The root is labeled with x, and then, inductively, if a node is labeled with $y \in \Sigma^n$, then its left child is labeled with $g_0(y)$ and its right child is labeled with $g_1(y)$. The value of $f(x)$ is the concatenation of the leaves labels in order from left to right and $f(x)(i)$ is the label of the i-th leaf, $i = 0, \ldots, 2^m - 1$.

Theorem 5.5.1 *Let $\epsilon > 0$ and $n, m, q \in \mathbb{N}$. Suppose the pseudo-random generator $g\colon \Sigma^n \to \Sigma^{2n}$ has security (ϵ, S) and is computable by a circuit of size q. Then the function f described above that maps strings x of length n into strings $f(x)$ of length $n2^m$ is a pseudo-random generator with security $((2^m - 1)\cdot\epsilon, S/(4mq))$. Moreover if we partition $f(x)$ into 2^m blocks each of length n, then each block can be computed in time polynomial in m and q.*

Proof. The "moreover" part is immediate from the construction. So let us focus on the pseudo-randomness aspect. As before, for each $j \in \mathbb{N}$, let U_j denote the uniform distribution on Σ^j. The proof uses the hybrid method that we have encountered in the proof of Theorem 5.4.1. Thus, we need to design a series of hybrid distributions, the first one being $f(U_n)$ and the last one being $U_{n\cdot 2^m}$, and such that any two consecutive distributions are computationally close relative to circuit size $S/(4m\cdot q)$. The hybrid distributions will be indexed with elements from the set $\mathcal{Z} = \{\lambda\} \cup \{x1\}_{x\in\Sigma^{\leq(m-1)}}$. If we consider the full binary tree B_m of height m with nodes named with strings in $\Sigma^{\leq m}$ (i.e., the root is λ, and, inductively, the left child of node α is named $\alpha 0$ and the right child is named $\alpha 1$), $\mathcal{Z}$ consists of the names of the root and of all right children in B_m. Thus we will produce a family of distributions $(h_z)_{z\in\mathcal{Z}}$, each over the set $\Sigma^{n\cdot 2^m}$. It will hold that (a) h_λ is identical to $f(U_n)$, (b) $h_{1\ldots 1}$ is identical to $U_{n\cdot 2^m}$, and (c) the computational distance relative to circuit size $S/(4m\cdot q)$ of two consecutive distributions h_z and $h_{\text{succ}(\text{succ}(z))}$ is less than ϵ. (We have denoted by succ(z) the lexicographical successor of the string z.) Observe that if we use the lexicographical order over $\mathcal{Z}$, then two consecutive distributions are indeed of the form h_z and $h_{\text{succ}(z)}$, for $z \in \mathcal{Z} - \{1\ldots 1\}$.

Let $z \in \mathcal{Z}$. To build h_z, we consider the full binary tree B_m to whose nodes we are going to assign n-bits long binary strings obtained via a combination of independent random strings in Σ^n and applications of the pseudo-random generator g. In the end, h_z is the frontier of this tree, i.e., it is the concatenation from left to right of the labels of all leaves in the tree B_m. We need to specify the labeling process. Each node whose name is $\leq_{\text{lex}} z$ is labeled with a random n-bits long string picked independently, and each node whose name is $\geq_{\text{lex}} z$ is labeled with g_0(label of parent), if the node is a left child of its parent, or with g_1(label of parent), if the node is a right child of its parent. It is obvious that the extreme distributions h_λ and $h_{1\ldots1}$ are identical to $f(U_n)$ and $U_{n\cdot 2^m}$, as desired. It remains to evaluate the computational distance between two consecutive distributions h_z and $h_{\text{succ}(\text{succ}(z))}$.

Claim 5.5.2 *Let* $z \in \mathcal{Z} - \{1\ldots1\}$ *and let* C *be a circuit of size* $\leq S/(4m\cdot q)$. *Then*

$$\left|\text{Prob}(C(h_{\text{succ}(\text{succ}(z))}) = 1) - \text{Prob}(C(h_z) = 1)\right| < \epsilon.$$

Proof. Suppose there is a circuit of size $S/(4m\cdot q)$ such that

$$\left|\text{Prob}(C(h_{\text{succ}(\text{succ}(z))}) = 1) - \text{Prob}(C(h_z) = 1)\right| \geq \epsilon. \tag{5.8}$$

We will build a circuit D with size at most $S/(4m\cdot q)$ that is able to distinguish between the distributions $g(U_n)$ and U_{2n} with bias larger than ϵ, thus reaching a contradiction. The circuit D acts on inputs of length $2n$ and our goal is (a) to make D on input a random $y \in \Sigma^{2n}$ to behave the same as C on $h_{\text{succ}(\text{succ}(z))}$, and (b) to make D on input $g(x)$, with x random in Σ^n behave the same as C on h_z.

Description of the computation of D on input $y \in \Sigma^{2n}$. Essentially D simulates C. A problem arises when C accesses one input bit (because D has a shorter input). To handle this situation, we divide the input of C into 2^m blocks of length n (and we will speak about the blocks $0\ldots0, 0\ldots1, \ldots, 1\ldots1$, with the natural interpretation: Block $0\ldots0$ is the first block, etc.). When C reads an input bit from block α (note that $\alpha \in \Sigma^m$), C determines α' the ancestor of α located in B_m at the same level as z (formally, $\alpha' = \alpha(1 : |z|)$). The circuit D maintains a list of locations (i.e., nodes) in the full binary tree B_m to which binary strings of length n have already been assigned. We call it LIST. If there is a pair (α', γ) in LIST, then γ will be used for the next calculations. If there is no such pair in LIST, then there are several cases:

Case 1. α' is succ(z). We take γ to be the left half of y, i.e., $\gamma = y(1:n)$. (Recall that y of length $2n$ is the input of D.)

Case 2. α' is succ(succ(z)). We take γ to be the right half of y, i.e., $\gamma = y(n+1:2n)$.

Case 3. $\alpha' \leq_{\text{lex}} z$. Then D randomly chooses a string γ of length n and uses it for the next calculations. It also inserts the pair (α', γ) in LIST.

Case 4. $\alpha' >_{\text{lex}} z$ and the Cases 1 and 2 do not apply. Then $\alpha' = \alpha''0$ or $\alpha' = \alpha''1$ for some string α'' (note that α' cannot be λ in Case 4). D checks if there is a pair (α'', γ') in LIST. If yes, then D calculates $\gamma = g_0(\gamma')$ or $\gamma = g_1(\gamma')$ depending on whether $\alpha' = \alpha''0$ or, respectively, $\alpha' = \alpha''1$. If no, then D picks randomly a string γ' of length n, it inserts the pair (α'', γ') in LIST and calculates γ from γ' as above.

many irreducible factors of the form $(Y - p(X))$ and this gives the bound on the number of polynomials $p(X)$. ∎

We now are ready to attack the hardness amplification issue.

Theorem 5.6.3 (Step 1 of hardness amplification)

INFORMAL STATEMENT: *Given a worst-case hard function, we can construct a constant-rate hard function.*

FORMAL STATEMENT: (a) *Let* $g\colon \Sigma^* \to \Sigma^*$ *be a length-regular function that is worst-case hard. Assume that* g *is length-preserving. Then there is a length-preserving function* $g'\colon \Sigma^* \to \Sigma^*$ *that is* $1/100$*-constant rate hard. Moreover,* g' *can be constructed effectively from* g *in time* $2^{O(\ell)}$.

(b) *Let* $g\colon \Sigma^* \to \Sigma^*$ *be a length-regular function such that, for some positive constant* c *and for all sufficiently large* ℓ, g_ℓ *is* $(\frac{1}{2^\ell}, 2^{c\ell})$*-hard. Assume that* g *is length-preserving. Then there is a length-preserving function* $g'\colon \Sigma^* \to \Sigma^*$ *such that, for some constant* c' *and for all sufficiently large* ℓ, g'_ℓ *is* $(\frac{1}{100}, 2^{c'\ell})$*-hard. Moreover,* g' *can be constructed effectively from* g *in time* $2^{O(\ell)}$.

Note. The assumption $|g(x)| = |x|$, for all $x \in \Sigma^*$ (i.e., g is length-preserving), is a technical convenience and is not essential. The constant $1/100$ is arbitrary and can be substituted with any positive constant.

Proof. We first prove (a). We fix ℓ in $\mathbb{N}$, and we consider the restriction of g to Σ^ℓ. We identify Σ^ℓ with the finite field with 2^ℓ elements $\mathbb{F} = \mathrm{GF}(2^\ell)$. Let $L = 2^\ell$ and let us denote the elements of $\mathbb{F}$ by $\{0, 1, ..., L-1\}$. We next define the polynomial p with ℓ variables and of degree 1 in each variable, $p\colon \mathbb{F}^\ell \to \mathbb{F}$, so that

$$\begin{aligned} p(0,\ldots,0) &= g(0\ldots 0) \\ p(0,\ldots,1) &= g(0\ldots 1) \\ &\vdots \\ p(1,\ldots,1) &= g(1\ldots 1). \end{aligned} \tag{5.9}$$

Note that a polynomial with ℓ variables and of degree 1 in each one of them has 2^ℓ coefficients in $\mathbb{F}$ which can be determined from the 2^ℓ equations above. Therefore g determines uniquely the polynomial p. Furthermore, observe that $p(x)$ can be calculated in time $2^{O(\ell)}$, given oracle access to g.

The key idea of this construction is to utilize *error-correcting codes* (see the discussion in Section 5.3, in the paragraphs preceding Theorem 5.3.5).

If we consider the $2^\ell \cdot \ell$-long binary string obtained by concatenating all the outputs of g

$$\tilde{g} = g(0\ldots 0) \odot g(0\ldots 1) \odot \ldots \odot g(1\ldots 1)$$

as a message string, then the $2^{\ell^2} \cdot \ell$-long binary string

$$\tilde{p} = p(0,\ldots,0) \odot p(0,\ldots,1) \odot \ldots \odot p(L-1,\ldots,L-1)$$

is just the codeword of $\tilde{g}$ obtained using the Reed-Muller error-correcting code with certain parameters over the alphabet $\{0, 1, \ldots, L-1\}$. We will show a decoding property that is sufficient for our purposes. Namely, we prove that if somehow we are able to obtain a word $\tilde{p}'$ that is within Hamming distance $(1/100) \cdot |\tilde{p}|$ from $\tilde{p}$, then we can reconstruct the entire $\tilde{p}$. This implies immediately that we can calculate the function g in all the points of its domain. Since this contradicts the worst-case hardness of g, we conclude that in fact we cannot get a string $\tilde{p}'$ as above.

We next proceed with the formal argument. It is useful to consider probabilistic circuits. These are circuits that, for some parameters ℓ and ℓ', have ℓ standard input gates and ℓ' special input gates which are considered to hold random bits. Thus the input to such a circuit is a pair of strings $(x, r) \in \Sigma^{\ell} \times \Sigma^{\ell'}$, where x denotes the proper input and r denotes the string of random bits used by the circuit. We say that a probabilistic circuit C computes a function h if, for all $x \in \Sigma^{\ell}$,

$$\text{Prob}_{r \in \Sigma^{\ell'}}(C(x, r) = h(x)) \geq \frac{3}{4}. \tag{5.10}$$

The value $\frac{3}{4}$ is somewhat arbitrary: Any constant $\frac{1}{2}+\epsilon$ would work as well. Abusing notation, in case C is a probabilistic circuit that computes a function, we will denote this function by C as well. The utilization here of probabilistic circuits is just a technical convenience, as we can convert them to deterministic ones with only a small increase in the size.

Lemma 5.6.4 *Let C be a probabilistic circuit that computes a function $h: \Sigma^{\ell} \rightarrow \Sigma^{n}$, using ℓ' random bits, for some $\ell, \ell', n \in \mathbb{N}$. Then there is a deterministic algorithm C' of size bounded by $O(\ell\ell') \cdot \text{size}(C)$ that computes h.*

Proof. On input $x \in \Sigma^{\ell}$, we iterate $N = 432\ell$ times the run of circuit C using at iteration i the random string r_i (the strings r_i, $i = 1, \ldots, N$, are independent). Let X_i be the 0-1 random variable defined by $X_i = 1$ if $C(x, r_i) = h(x)$, and $X_i = 0$, otherwise. We have that $\text{Prob}(X_i = 1) \geq 3/4$, and, therefore, using the multiplicative form of Chernoff bounds, we obtain

$$\begin{aligned}\text{Prob}\Big(\frac{1}{N}\sum_{i}^{N} X_i < \frac{2}{3}\Big) &= \text{Prob}\Big(\frac{1}{N}\sum_{i}^{N} X_i < \Big(1 - \frac{1}{9}\Big)\frac{3}{4}\Big) \\ &\leq \mathrm{e}^{-((1/9)^2 \cdot (3/4) \cdot N)/2} = \mathrm{e}^{-2\ell} < 2^{-2\ell}.\end{aligned}$$

Thus the probability that there exists some $x \in \Sigma^{\ell}$ such that fewer than $\frac{2}{3}N$ iterations produce $h(x)$ is less than $2^{\ell} \cdot 2^{-2\ell} = 2^{-\ell}$. Therefore there exists a fixed sequence of strings $r_0 = (r_{0,1}, \ldots, r_{0,N})$, such that, for all $x \in \Sigma^{\ell}$, on at least $\frac{2}{3}N$ iterations we obtain $h(x)$, if at iteration i we use the fixed string $r_{0,i}$ as the random bits needed at that iteration. The deterministic circuit C' is built as follows. The sequence r_0 is hard-wired in the circuit. On input x, C' simulates N times C on x using in the simulation of the i-th iteration $r_{0,i}$ in the role of the needed random bits. C' outputs the value that is produced at least $\frac{2}{3}N$ times during the N simulations. By the above remarks, it holds that $C'(x) = h(x)$, for all $x \in \Sigma^\ell$. A simple inspection of the construction shows that $\text{size}(C') = O(\ell\ell') \cdot \text{size}(C)$. ∎

We resume the proof of Theorem 5.6.3. We fix $\ell \in \mathbb{N}$ and p is the polynomial given by the equations (5.9).

Claim 5.6.5 *Suppose there is a circuit C of size s computing a function that maps $\mathbb{F}^\ell$ to $\mathbb{F}$ (i.e., in binary notation, it maps Σ^{ℓ^2} to Σ^ℓ) such that*

$$\|\{\overline{x} \in \mathbb{F}^\ell \mid C(\overline{x}) = p(\overline{x})\}\| \geq \frac{99}{100} \cdot 2^{\ell^2}.$$

Then there is a polynomial q such that, if ℓ is sufficiently large, there is a probabilistic circuit C' of size $s' \leq q(\ell) \cdot s$ that calculates $p(\overline{x})$, for all $\overline{x}$ in $\mathbb{F}^\ell$. This implies that there is a probabilistic circuit of size at most $s' + \ell \leq q(\ell) \cdot s + \ell$ that calculates g for all $x \in \Sigma^\ell$.

Proof. Let us fix $\overline{x}$ in $\mathbb{F}^\ell$. We first pick at random and independently $\overline{y}$ and $\overline{z}$ in $\mathbb{F}^\ell$. The elements $\overline{y}$ and $\overline{z}$ define the following set:

$$Q_{\overline{y},\overline{z}} = \{\overline{y}t^2 + \overline{z}t + \overline{x} \mid t \in \mathbb{F}\}.$$

The key properties of $Q_{\overline{y},\overline{z}}$ are: (1) It represents a sample of points from $\mathbb{F}^\ell$ that are pairwise independent (as we will see shortly), and (2) it can be parameterized using a variable from $\mathbb{F}$, namely t.

The first property of $Q_{\overline{y},\overline{z}}$ allows us to show that the agreement of p and C on the sample set is not much different from their agreement on the entire $\mathbb{F}^\ell$ (which is assumed to be 99/100), and the second property allows us to move to single variable polynomials for which reconstruction can be done using the algorithm from Theorem 5.6.2. Let $p_{(\overline{x},\overline{y},\overline{z})}$ (in some sense, the restriction of p to $Q_{\overline{y},\overline{z}}$) be the polynomial defined by

$$p_{(\overline{x},\overline{y},\overline{z})}(t) = p(\overline{y}t^2 + \overline{z}t + \overline{x}).$$

Note that $p_{(\overline{x},\overline{y},\overline{z})}$ is a single variable polynomial over $\mathbb{F}$ and that it has degree 2ℓ. Analogously, let

$$C_{(\overline{x},\overline{y},\overline{z})}(t) = C(\overline{y}t^2 + \overline{z}t + \overline{x}).$$

We show that with probability of $\overline{y}$ and $\overline{z}$ at least $1 - \frac{25}{(99/100)\cdot(\|\mathbb{F}\|-1)}$, $p_{(\overline{x},\overline{y},\overline{z})}$ and $C_{(\overline{x},\overline{y},\overline{z})}$ agree on at least a fraction of $\frac{99}{100} \cdot \frac{4}{5} = \frac{99}{125}$ of the points in $\mathbb{F}$. This results from the following Claim.

Definition 5.6.6 *Let f_1, f_2 be two functions with a common domain D, and $\epsilon \in [0,1]$. We say that f_1 and f_2 have ϵ-agreement on a set $D' \subseteq D$ if*

$$\|\{x \in D' \mid f_1(x) = f_2(x)\}\| = \epsilon \cdot \|D'\|.$$

Claim 5.6.7 *Let $\epsilon > 0$. Suppose the functions p and C mapping $\mathbb{F}^\ell$ to $\mathbb{F}$ have ϵ-agreement on $\mathbb{F}^\ell$. Choose uniformly at random and independently $\overline{y}$ and $\overline{z}$ in $\mathbb{F}^\ell$. Then the functions $p_{(\overline{x},\overline{y},\overline{z})}$ and $C_{(\overline{x},\overline{y},\overline{z})}$ defined as above have $\frac{4}{5}\epsilon$-agreement on $\mathbb{F}$ with probability of $\overline{y}$ and $\overline{z}$ at least $1 - 25/(\epsilon \cdot (\|\mathbb{F}\| - 1))$.*

Proof. For each $t \in \mathbb{F}$, define the random variable (depending on $\overline{y}$ and $\overline{z}$),

$$X_t = \begin{cases} 1, & \text{if } p(\overline{y}t^2 + \overline{z}t + \overline{x}) = C(\overline{y}t^2 + \overline{z}t + \overline{x}), \\ 0, & \text{otherwise.} \end{cases}$$

Note that for each $t \in \mathbb{F}$, we have

$$\text{Prob}_{\overline{y},\overline{z}}(X_t = 1) = \epsilon,$$

because $\overline{y}t^2 + \overline{z}t + \overline{x}$ is uniformly distributed in $\mathbb{F}^\ell$.

We next show that the random variables $X_1, \ldots, X_{L-1}$ are pairwise independent (recall that we have named the elements of the field $\mathbb{F}$, $0, 1, \ldots, L-1$, and 0 denotes the zero element in the field). Let MATCH be the points in $\mathbb{F}^\ell$ where p and C coincide and let t_1 and t_2 be two distinct elements in $\mathbb{F} - \{0\}$. We have

$$\begin{aligned} &\text{Prob}_{\overline{y},\overline{z}}(X_{t_1} = 1 \text{ and } X_{t_2} = 1) \qquad (5.11) \\ &= \text{Prob}_{\overline{y},\overline{z}}(\overline{y}t_1^2 + \overline{z}t_1 + \overline{x} \in \text{ MATCH and } \overline{y}t_2^2 + \overline{z}t_2 + \overline{x} \in \text{ MATCH}) \\ &= \sum_{\overline{u},\overline{v} \in \text{MATCH}} \text{Prob}_{\overline{y},\overline{z}}(\overline{y}t_1^2 + \overline{z}t_1 + \overline{x} = \overline{u} \text{ and } \overline{y}t_2^2 + \overline{z}t_2 + \overline{x} = \overline{v}). \end{aligned}$$

Let us focus on

$$\text{Prob}_{\overline{y},\overline{z}}(\overline{y}t_1^2 + \overline{z}t_1 + \overline{x} = \overline{u} \text{ and } \overline{y}t_2^2 + \overline{z}t_2 + \overline{x} = \overline{v}),$$

for fixed $\overline{u}$ and $\overline{v}$ in $\mathbb{F}^\ell$. At component k of the vectors ($k \in \{1, \ldots, \ell\}$), we have

$$\begin{aligned} y_k t_1^2 + z_k t_1 &= u_k - x_k \\ y_k t_2^2 + z_k t_2 &= u_k - x_k. \end{aligned}$$

The determinant of the system is

$$\begin{vmatrix} t_1^2 & t_1 \\ t_2^2 & t_2 \end{vmatrix} = t_1 t_2 (t_1 - t_2) \neq 0.$$

Therefore, the system has a unique solution, and thus,

$$\text{Prob}_{\overline{y},\overline{z}}(\overline{y}t_1^2 + \overline{z}t_1 + \overline{x} = \overline{u} \text{ and } \overline{y}t_2^2 + \overline{z}t_2 + \overline{x} = \overline{v}) = \frac{1}{\|F^\ell\| \cdot \|F^\ell\|}.$$

Returning to Equation (5.11),

$$\begin{aligned} &\sum_{\overline{u},\overline{v} \in \text{MATCH}} \text{Prob}_{\overline{y},\overline{z}}(\overline{y}t_1^2 + \overline{z}t_1 + \overline{x} = \overline{u} \text{ and } \overline{y}t_2^2 + \overline{z}t_2 + \overline{x} = \overline{v}) \\ &= \frac{\|\text{MATCH}\| \cdot \|\text{MATCH}\|}{\|\mathbb{F}^\ell\| \cdot \|\mathbb{F}^\ell\|} \end{aligned}$$

$$= \frac{\epsilon\|\mathbb{F}^\ell\| \cdot \epsilon\|\mathbb{F}^\ell\|}{\|\mathbb{F}^\ell\| \cdot \|\mathbb{F}^\ell\|} = \epsilon^2$$
$$= \text{Prob}(X_{t_1} = 1) \cdot \text{Prob}(X_{t_2} = 1).$$

Analogously, for each pair $(b_1, b_2) \in \{0,1\} \times \{0,1\}$,

$$\text{Prob}(X_{t_1} = b_1 \text{ and } X_{t_2} = b_2) = \text{Prob}(X_{t_1} = b_1) \cdot \text{Prob}(X_{t_2} = b_2).$$

Therefore, $X_1, X_2, \ldots, X_{L-1}$ are pairwise independent, which allows for an easy application of the Chebyshev's Inequality. Let

$$Z = \frac{1}{L-1} \cdot \sum_{i=1}^{L-1} X_i.$$

We have that $E[Z] = \epsilon$ and

$$Z \leq \frac{4}{5}\epsilon \Leftrightarrow (\epsilon - Z) \geq \frac{1}{5}\epsilon.$$

So, by the Chebyshev's Inequality,

$$\text{Prob}\Big(Z \leq \frac{4}{5}\epsilon\Big) \leq \text{Prob}\Big(|Z - \epsilon| \geq \frac{1}{5}\epsilon\Big) \leq \frac{\text{Var}(Z)}{\frac{1}{25}\epsilon^2}.$$

For each $i \in \{1, \ldots, L-1\}$, $\text{Var}(X_i) = E(X_i^2) - (E(X_i))^2 = \epsilon - \epsilon^2$, and taking into account the pairwise independence established before,

$$\text{Var}(Z) = \frac{1}{(L-1)^2} \cdot \sum_{i=1}^{L-1} \text{Var}(X_i) = \frac{1}{(L-1)^2} \cdot (L-1) \cdot (\epsilon - \epsilon^2) = \frac{\epsilon(1-\epsilon)}{L-1}.$$

Consequently,

$$Prob\left(Z \leq \frac{4}{5}\epsilon\right) \leq \frac{25}{\epsilon^2} \cdot \frac{\epsilon(1-\epsilon)}{L-1} \leq \frac{25}{\epsilon \cdot (L-1)}.$$

This closes the proof of Claim 5.6.7. ∎

We will assume in what follows that $\overline{y}$ and $\overline{z}$ are such that $p_{(\overline{x},\overline{y},\overline{z})}$ and $C_{(\overline{x},\overline{y},\overline{z})}$ have 99/125 agreement on $Q_{\overline{y},\overline{z}}$. Note that it is enough to determine the polynomial $p_{(\overline{x},\overline{y},\overline{z})}$, because $p_{(\overline{x},\overline{y},\overline{z})}(0) = p(\overline{x})$.

In principle, we can reconstruct $p_{(\overline{x},\overline{y},\overline{z})}$ via Theorem 5.6.2 using the pairs of points $(\overline{y}t^2 + \overline{z}t + \overline{x}, C_{(\overline{x},\overline{y},\overline{z})}(\overline{y}t^2 + \overline{z}t + \overline{x}))_{t \in F}$. However there would be at least $2^\ell/2$ pairs and the algorithm would not be efficient. Instead, we sample points from $Q_{\overline{y},\overline{z}}$ and we do the reconstruction using the pairs induced by the sample points. More precisely, we sample ℓ^2 points $u_1, \ldots, u_{\ell^2}$ in $Q_{\overline{y},\overline{z}}$ by picking ℓ^2 random points $t_1, \ldots, t_{\ell^2}$ in $\mathbb{F}$ and taking $u_i = \overline{y}t_i^2 + \overline{z}t_i + \overline{x}$. Let S be the multiset

$\{u_i \mid i = 1, \ldots, \ell^2\}$. With high probability (over the choice of S), the agreement of $p_{(\overline{x},\overline{y},\overline{z})}$ and $C_{(\overline{x},\overline{y},\overline{z})}$ on S is still high. Indeed, let

$$T = \{\overline{y}t^2 + \overline{z}t + \overline{x} \in S \mid p_{(\overline{x},\overline{y},\overline{z})}(t) = C_{(\overline{x},\overline{y},\overline{z})}(t)\}.$$

The cardinality of T is the sum of ℓ^2 independent 0-1 valued random variables each having an expected value of at least 99/125. Therefore, by the multiplicative Chernoff bounds (see Appendix A),

$$\begin{aligned}\text{Prob}\left(\frac{\|T\|}{\ell^2} < \frac{3}{4}\right) &= \text{Prob}\left(\frac{\|T\|}{\ell^2} < \left(1 - \frac{7}{132}\right) \cdot \frac{99}{125}\right) \\ &\leq \mathrm{e}^{\left(-\left(\frac{7}{132}\right)^2 \cdot \frac{99}{125} \cdot \ell^2\right)/2}.\end{aligned}$$

Since we are sampling ℓ^2 elements from $\mathbb{F}$ and $\|\mathbb{F}\| = 2^\ell$, with high probability, the ℓ^2 sampled elements in the set S are distinct. Indeed, let A_i be the event that u_i is equal to one of $u_1, \ldots, u_{i-1}$. Note that, for any $j < i$, $u_i = u_j$ if either $t_i = t_j$ or t_i is the other root (than t_j) of the equation $\overline{y}t^2 + \overline{z}t + \overline{x} = u_j$. Therefore $\text{Prob}(A_i) \leq 2(i-1)/(2^\ell)$ and Prob(at least 2 sampled points coincide) is at most (by the union bound) $\text{Prob}(A_1) + \ldots + \text{Prob}(A_{\ell^2}) = \frac{2 \cdot 0}{2^\ell} + \frac{2 \cdot 1}{2^\ell} + \ldots + \frac{2 \cdot \ell^2 - 1}{2^\ell} = \frac{\ell^2(\ell^2-1)}{2^\ell}$. Therefore, the probability that all the sampled points are distinct is at least $1 - \frac{\ell^2(\ell^2-1)}{2^\ell}$.

Thus, with probability at least $1 - \mathrm{e}^{\left(-\left(\frac{7}{132}\right)^2 \cdot \frac{99}{125} \cdot \ell^2\right)/2} - \frac{\ell^2(\ell^2-1)}{2 \cdot 2^\ell}$, $p_{(\overline{x},\overline{y},\overline{z})}$ and $C_{(\overline{x},\overline{y},\overline{z})}$ have 3/4 agreement on S and S has at least ℓ^2 distinct elements.

Thus, let us assume that the above conditions hold. The next observation is that, if $\ell > 4$, $p_{(\overline{x},\overline{y},\overline{z})}$ is the unique single variable polynomial of degree 2ℓ that has 3/4 agreement with $C_{(\overline{x},\overline{y},\overline{z})}$ on S. The reason is that if two polynomials, p_1 and p_2, have agreement at least 3/4 with $C_{(\overline{x},\overline{y},\overline{z})}$ on S, then they must have agreement at least 1/2 with each other. So p_1 and p_2 must coincide on $\frac{1}{2} \cdot \ell^2$ points. On the other hand, two distinct polynomials of degree 2ℓ can be equal on at most 2ℓ points and so $2\ell \geq \frac{1}{2} \cdot \ell^2$, which is possible only if $\ell \leq 4$. We can now use Theorem 5.6.2 to reconstruct the polynomial $p_{(\overline{x},\overline{y},\overline{z})}$. Let us consider the points $(u_i, C_{(\overline{x},\overline{y},\overline{z})}(u_i))_{i=1,\ldots,\ell^2}$. We are looking for the unique polynomial p_1 of degree 2ℓ that satisfies

$$\|\{i \in \{1, \ldots, \ell^2\} \mid p_1(u_i) = C_{(\overline{x},\overline{y},\overline{z})}(u_i)\}\| \geq \frac{3}{4}\ell^2.$$

For $\ell > 64/9$, it holds that $\frac{3}{4}\ell^2 \geq \sqrt{2 \cdot (2\ell) \cdot \ell^2}$. Therefore the conditions required by Theorem 5.6.2 are satisfied. The algorithm from Theorem 5.6.2, given the points $(u_i, C_{(\overline{x},\overline{y},\overline{z})}(u_i))_{i=1,\ldots,\ell^2}$, will return the unique polynomial of degree 2ℓ that has agreement $(3/4)\ell^2$ with this set of points. This polynomial is $p_{(\overline{x},\overline{y},\overline{z})}$ and, having $p_{(\overline{x},\overline{y},\overline{z})}$, we can calculate $p(\overline{x})$.

To summarize $p(\overline{x})$ is calculated as follows:

Choose randomly $\overline{y}$ and $\overline{z}$ in $\mathbb{F}^\ell$, which yield the set $Q_{\overline{y},\overline{z}}$.

Choose randomly ℓ^2 points in $Q_{\overline{y},\overline{z}}$ obtaining the set $S = \{u_1, \ldots, u_{\ell^2}\}$.

Run the algorithm from Theorem 5.6.2 on input $(u_i, C_{(\overline{x},\overline{y},\overline{z})}(u_i))_{i=1,\ldots,\ell^2}$ seeking the polynomials of degree 2ℓ that have $(3/4)\ell^2$ agreement with the set of points. The algorithm returns one polynomial p_1. The algorithm from Theorem 5.6.2 is probabilistic and may, with small probability, run for a long time. We stop the algorithm if it does not finish in $r^2(\ell)$ steps, where $r(x)$ is the polynomial that bounds the expected time of the algorithm. If more polynomials are returned, ignore them, as this is an erroneous, but unlikely, case. The running time for this part is $\text{poly}(\ell^2, \log\|F\|) = \text{poly}(\ell)$.

Calculate and return $p_1(0)$. This is $p(\overline{x})$ with probability at least 3/4 (if ℓ is sufficiently large).

From our analysis, the "bad" cases are: $p_{(\overline{x},\overline{y},\overline{z})}$ and $C_{(\overline{x},\overline{y},\overline{z})}$ have agreement than a fraction of 99/125 on $\mathbb{F}$, the multiset S does not have ℓ^2 distinct elements, the agreement of $p_{(\overline{x},\overline{y},\overline{z})}$ and $C_{(\overline{x},\overline{y},\overline{z})}$ on S is less than a fraction of 3/4 of S, the polynomial reconstruction algorithm does not terminate in the allotted time. They all have small probability (at most $1/\text{poly}(\ell)$) when ℓ is large. It follows that, if ℓ is sufficiently large, the error probability is, as claimed in the above summary, bounded by 1/4, and thus the relation (5.10) is satisfied. It can be seen that this algorithm can be performed by a probabilistic circuit C' of size $s' = q(\ell) \cdot s$, for some polynomial q (most of the work is in computing the ℓ^2 values $C_{(\overline{x},\overline{y},\overline{z})}(u_i)$). This concludes the proof of Claim 5.6.5 ∎

We resume once again the proof of Theorem 5.6.3. For each $\ell \in \mathbb{N}$, and for each restriction of g to Σ^ℓ, we have obtained a function p with the properties in Claim 5.6.5 (the properties hold provided ℓ is sufficiently large). Let h be the union of all these functions p taken over all $\ell \in \mathbb{N}$. Note that h is defined on $\bigcup_{\ell\in\mathbb{N}} \Sigma^{\ell^2}$. By hypothesis, there is a superpolynomial function s, so that for any collection of circuits $(C_\ell)_{\ell\in\mathbb{N}}$ of size at most s, and for any ℓ sufficiently large,

$$\|\{x \in \Sigma^\ell \mid C_\ell(x) = g(x)\}\| \le 2^\ell - 1.$$

Suppose there are a function s' and a collection of circuits $C = (C_{\ell^2})_{\ell\in\mathbb{N}}$ of size at most s' so that, for infinitely many ℓ,

$$\|\{x \in \Sigma^{\ell^2} \mid C_{\ell^2}(x) = h(x)\}\| \ge \frac{99}{100} \cdot 2^{\ell^2}.$$

Then, by Claim 5.6.5, we obtain a family of probabilistic circuits $C' = (C'_\ell)_{\ell\in\mathbb{N}}$, of size at most $\ell^c \cdot s'(\ell^2)$, for some constant c, so that for infinitely many ℓ, $C'_\ell(x) = g(x)$, for *all* $x \in \Sigma^\ell$. Using Lemma 5.6.4, we can assume that the circuits C'_ℓ are actually deterministic. We obtain a contradiction if $\ell^c \cdot s'(\ell^2) \le s(\ell)$. Therefore, if $s'(\ell^2) \le \frac{1}{\ell^c} \cdot s(\ell)$, then for any collection of circuits $C = (C_{\ell^2})_{\ell\in\mathbb{N}}$ of size at most s', and for all sufficiently large ℓ,

$$\|\{x \in \Sigma^{\ell^2} \mid C_{\ell^2}(x) = h(x)\}\| < \frac{99}{100} \cdot 2^{\ell^2}.$$

Clearly, there is a superpolynomial function s' satisfying the above inequality. The function h is almost what we need, the only problem being that it is defined only on $\bigcup_{\ell \in \mathrm{N}} \Sigma^{\ell^2}$ instead of the entire Σ^*. This can be remedied easily by extending h. Let g' be the extension of h to Σ^* defined as follows: $g'(x) = h(x)$ for all $x \in \bigcup_{\ell \in \mathrm{N}} \Sigma^{\ell^2}$, and, for $x \in \Sigma^*$ with $\ell^2 < |x| < (\ell+1)^2$ for some ℓ,

$$g'(x) = h(x(1 : \ell^2)).$$

The following claim finishes the proof.

Claim 5.6.8 g' *is* $1/100$*-constant-rate hard.*

Proof. Let s' be the superpolynomial function such that if $(C_\ell)_{\ell \in \mathrm{N}}$ is a collection of circuits of size at most $s'(\ell)$, then, for all sufficiently large ℓ,

$$\|\{x \in \Sigma^{\ell^2} \mid C_{\ell^2}(x) = h(x)\}\| < \frac{99}{100} \cdot 2^{\ell^2}. \tag{5.12}$$

Let us suppose that the above relation holds for all ℓ larger than some fixed ℓ_0. Let s'' be a superpolynomial function so that, for all ℓ,

$$s''((\ell+1)^2) + (2\ell+1) \le s'(\ell^2).$$

To reach a contradiction, suppose that for some length k, with $\ell^2 < k < (\ell+1)^2$ for some integer $\ell > \ell_0$, there is a circuit C of size $s''(k)$ so that

$$\mathrm{Prob}_{x \in \Sigma^k}(C(x) = g'(x)) \ge \frac{99}{100}.$$

Under this assumption, we will show that there is a circuit D of size less $s'(\ell^2)$ that has agreement with h on Σ^{ℓ^2} at least a fraction of 99/100 the inputs, which contradicts the relation (5.12) and thus terminates the proof of the claim. We first define a circuit D' having two inputs, $y \in \Sigma^{\ell^2}$, and $z \in \Sigma^{k-\ell^2}$. The circuit D' on inputs y and z simply simulates $C(yz)$. If Y and Z denote random variables uniformly distributed in Σ^{ℓ^2} and, respectively, $\Sigma^{k-\ell^2}$, we observe that

$$\mathrm{Prob}_{Y,Z}(D'(Y,Z) = h(Y)) = \mathrm{Prob}_{Y,Z}(C(YZ) = g'(YZ)) \ge \frac{99}{100}.$$

Since $\mathrm{Prob}_{Y,Z}(D'(Y,Z) = h(Y))$ is the average of $\mathrm{Prob}_Y(D'(Y,z) = h(Y))$ over all $z \in \Sigma^{k-\ell^2}$, there must be some fixed z_0 such that

$$\mathrm{Prob}_Y(D'(Y,z_0) = h(Y)) \ge \frac{99}{100}.$$

Consequently, if we take such a z_0 and hard-wire it into D', we obtain a circuit D that calculates correctly h on at least a $\frac{1}{100}$ fraction of Σ^{ℓ^2}. Finally, observe that the size of D is bounded by $s''(k)+(k-\ell^2) \leq s''((\ell+1)^2)+(2\ell+1) \leq s'(\ell^2)$ (we have used the fact that $k < (\ell+1)^2$). We have reached the desired contradiction. ∎

By inspecting the construction, one can see that, for all x, $|g'(x)| \leq |x|$. We can easily make the function g' length-preserving by padding $g'(x)$ with some 0s. Clearly this will not make g' any less harder. Looking back at the construction of the polynomial p (see relations (5.9)), which is the core component of the whole construction, one can see that, given oracle access to g, the function g' can be calculated in time $2^{O(\ell)}$. ∎

Proof of (b). The approach used in the proof of (a) does not fully work because the inputs of the polynomial p (see Equation 5.9) have length ℓ^2 and the circuits that fail to calculate p on at least a 1/100 fraction of the inputs would have size $2^{c\ell}$ (for some constant c), which is not exponential in the new input length. Therefore we need to adapt the proof of (a) so that the input length of the function that we construct is only larger by a constant factor. We present the modifications that need to be done.

Let j be an integer such that the polynomial that bounds the running time of the algorithm in Theorem 5.6.2 is n^j. Let $k = \lceil 6j/c \rceil$. As in the proof of (a), we encode g with a polynomial p. This time, the polynomial p has k variables and is of degree $d = 2^{\lceil \ell/k \rceil} - 1$ in each variable. The encoding is done as follows. Let $b\colon \mathbb{F}^k \to \mathbb{F}$ be some standard fast-computable injection. We define p by requiring

$$p(b(0)) = g(0), p(b(1)) = g(1), \ldots, p(b(L)) = g(L),$$

where, as before, $L = 2^\ell$ and we have identified Σ^ℓ with the field $\mathbb{F} = \mathrm{GF}(2^\ell)$. The polynomial p needs to have at least L coefficients and this is true because p has $(d+1)^k$ coefficients (because of the ceiling function used in the definition of d and k, p actually may have a few more coefficients than L and we may require p to be zero in a few additional points, so as to assure the uniqueness of p). The analogue of Claim 5.6.5 is the following.

Claim 5.6.9 *Suppose there is a circuit C of size $s = 2^{(1/3)c\ell}$ computing a function that maps $\mathbb{F}^k$ to $\mathbb{F}$ (i.e., in binary notation, it maps $\Sigma^{\ell k}$ to Σ^ℓ) such that*

$$\|\{\overline{x} \in \mathbb{F}^k \mid C(\overline{x}) = p(\overline{x})\}\| \geq \frac{99}{100} \cdot 2^{\ell k}.$$

Then there is a probabilistic circuit C' of size $O(2^{(2/3)c\ell})$ that calculates $p(\overline{x})$, for all $\overline{x}$ in $\mathbb{F}^k$.

Proof. As in the proof of Claim 5.6.5, we pick random $\overline{y}$ and $\overline{z}$ in $\mathbb{F}^k$ and define $Q_{\overline{y},\overline{z}}$, $p_{(\overline{x},\overline{y},\overline{z})}$ and $C_{(\overline{x},\overline{y},\overline{z})}$ in the same way. The proof of Claim 5.6.7 works in the new setting and we derive that, with high probability, $p_{(\overline{x},\overline{y},\overline{z})}$ and $C_{(\overline{x},\overline{y},\overline{z})}$ have 99/125 agreement on $\mathbb{F}$. The polynomial $p_{(\overline{x},\overline{y},\overline{z})}$ has a single variable and has degree $2kd$. We sample n points in $Q_{\overline{y},\overline{z}}$, where $n = 2^{(c/(3j))\ell}$, in the same way, and we obtain the multiset $S = \{u_1, \ldots, u_n\}$. With high probability, S has actually n distinct elements and $p_{(\overline{x},\overline{y},\overline{z})}$ and $C_{(\overline{x},\overline{y},\overline{z})}$ have 3/4 agreement on S. It can be shown in the same way as in (a) that $p_{(\overline{x},\overline{y},\overline{z})}$ is the unique polynomial of degree $2kd$ that has 3/4 agreement with $C_{(\overline{x},\overline{y},\overline{z})}$ on S. We want to reconstruct $p_{(\overline{x},\overline{y},\overline{z})}$ via the algorithm in Theorem 5.6.2 using the pairs $(u_i, C_{(\overline{x},\overline{y},\overline{z})}(u_i))_{i=1,\ldots,n}$. The condition in Theorem 5.6.2 is $(3/4)n \geq \sqrt{2dn}$ and it holds true. Therefore we obtain the polynomial $p_{(\overline{x},\overline{y},\overline{z})}$ from which we can calculate $p(\overline{x})$. The procedure can be performed by a circuit of size $O(n^j + n \cdot s)$ ($O(n^j)$ is needed to run the algorithm in Theorem 5.6.2, $n \cdot s$ is needed to calculate the n pairs), which is $O\big((2^{(c/(3j))\ell})^j + 2^{(c/(3j))\ell}2^{(1/3)\ell}\big) < 2^{(2/3)c\ell}$, for ℓ sufficiently large. ∎

Under the hypothesis of the last claim, there would be a circuit of size less than $2^{c\ell}$ that calculates g in every point $x \in \Sigma^\ell$. This contradicts the assumed hardness of g, and thus the hypothesis of Claim 5.6.9 does not hold, i.e., there is no circuit of size $2^{(c/(3k))\cdot k\ell}$ that calculates $p\colon \Sigma^{k\ell} \to \Sigma^\ell$ with agreement 99/100. From here, the proof continues as in (a). ∎

We move next to step two of hardness amplification, going from g', which is 1/100-constant-rate hard to g'', which is crypto hard. The function g'' is obtained essentially by concatenating multiple copies of g'. This method (together with variants of it) is called the "direct product" method. The intuition is that if a circuit succeeds to compute g' only on a fraction of at most 99/100 of the inputs, then we might expect that it succeeds to compute k copies of g' only on a fraction of at most $(99/100)^k$ of the inputs. The crux of the proof is in the following lemma which addresses the concatenation of two functions.

Lemma 5.6.10 *Let ℓ and m be integers ≥ 1, and $\rho_1(\ell), \rho_2(m) \in [0,1]$. Let $f_1\colon \Sigma^\ell \to \Sigma^{\ell'}$ be a function such that for all circuits C of size at most $s_1(\ell)$,*

$$\text{Prob}_Y(C(Y) = f_1(Y)) \leq \rho_1(\ell),$$

and $f_2\colon \Sigma^m \to \Sigma^{m'}$ be a function such that for all circuits C of size at most $s_2(m)$,

$$\text{Prob}_Z(C(Z) = f_2(Z)) \leq \rho_2(m),$$

where Y and respectively Z denote random variables uniformly distributed in Σ^ℓ and respectively Σ^m. Let $f\colon \Sigma^{\ell+m} \to \Sigma^{\ell'+m'}$ be defined by $f(y \odot z) = f_1(y) \odot f_2(z)$. Let ϵ be a real value such that $\epsilon^3 \geq 16\ell/(2^m)$. Then, for every C of size $s \leq \min\left(\frac{\epsilon^3}{48\cdot\ell\cdot m}s_1(\ell), s_2(m)\right)$,

$$\text{Prob}_{Y,Z}(C(Y \odot Z) = f(Y \odot Z)) \leq \rho_1(\ell) \cdot \rho_2(m) + \epsilon.$$

Proof. Let C be a circuit of size s that attempts to calculate f. We can assume that C outputs strings of length $\ell' + m'$ and we let $C_1(y \odot z)$ denote the first ℓ' bits of $C(y \odot z)$, and $C_2(y \odot z)$ denote the last m' bits of $C(y \odot z)$. For the forthcoming argument it is helpful to visualize a table T having 2^ℓ rows indexed in order by the strings $y_0, y_1, \ldots, y_{2^\ell-1}$ of Σ^ℓ, and 2^m columns indexed in order by the strings $z_0, z_1, \ldots, z_{2^m-1}$ of Σ^m. An entry in the table is a pair of bits (b_1, b_2) such that the entry of coordinates (i,j) (i.e., with row label y_i and column label z_j) contains the truth values of the predicates $C_1(y_i \odot z_j) = f_1(y_i)$ and $C_2(y_i \odot z_j) = f_2(z_j)$ (for example, if $C_1(y_i \odot z_j) = f_1(y_i)$ and $C_2(y_i \odot z_j) \neq f_2(z_j)$, then the entry on row i and column j will be $(1,0)$). We use capital letters such as Y and Z to denote random variables. The probabilities, unless specified otherwise, are with respect to the distributions of these random variables.

Our approach is to evaluate $\text{Prob}(C(Y \odot Z) = f(Y \odot Z))$ by decomposing it into

$$\text{Prob}(C_2(Y \odot Z) = f_2(Z)) \cdot \text{Prob}(C_1(Y \odot Z) = f_1(Y) \mid C_2(Y \odot Z) = f_2(Z)),$$

and then estimating the two factors. The first factor is easy to estimate (it is at most $\rho_2(m)$, otherwise the hypothesis about f_2 would be violated); the evaluation of the second one is more delicate and needs a certain sampling procedure. To make the sampling succeed we need to do a more refined decomposition of the event "$(C(Y \odot Z) = f(Y \odot Z)$."

To this aim, let us call $y \in \Sigma^\ell$ "good" if a fraction of more than $\epsilon/2$ of the entries on row y in T have the second component 1, i.e., y is "good" if $\text{Prob}(C_2(y \odot Z) = f_2(Z)) \geq \epsilon/2$. Let $G \subseteq \Sigma^\ell$ denote the set of good strings y.

We first take care of the case in which y is not "good."

$$\begin{aligned}\text{Prob}(C(&Y \odot Z) = f(Y \odot Z) \text{ and } Y \notin G)\\ &\leq \text{Prob}(C_2(Y \odot Z) = f_2(Z)) \text{ and } Y \notin G)\\ &= \sum_{y \in \overline{G}} \text{Prob}_{y \in \overline{G}}(Y = y) \cdot \text{Prob}(C_2(y \odot Z) = f_2(Z) \mid Y = y)\\ &\leq \frac{\epsilon}{2} \sum \text{Prob}_{y \in \overline{G}}(Y = y) \leq \frac{\epsilon}{2},\end{aligned}$$

where $\overline{G}$ is the complement of G. Thus,

$$\begin{aligned}\text{Prob}(C(Y \odot Z) = f(Y \odot Z)) &= \text{Prob}(C(Y \odot Z) = f(Y \odot Z) \text{ and } Y \in G)\\ &\quad + \text{Prob}(C(Y \odot Z) = f(Y \odot Z) \text{ and } Y \notin G)\\ &\leq \text{Prob}(C(Y \odot Z) = f(Y \odot Z) \text{ and } Y \in G) + \frac{\epsilon}{2}.\end{aligned}$$

The rest of the proof is devoted to showing that

$$\text{Prob}(C(Y \odot Z) = f(Y \odot Z) \text{ and } Y \in G) \leq \rho_1(\ell) \cdot \rho_2(m) + \frac{\epsilon}{2}, \tag{5.13}$$

which, combined with the previous relation, establishes the lemma.

The following fact is the key point of the proof.

Claim 5.6.11 *There exists a circuit C'' of size bounded by $48 \cdot \ell \cdot m \cdot (1/\epsilon)^3 \cdot \text{size}(C)$ such that*

$$\begin{aligned} &\text{Prob}(C''(Y) = f_1(Y)) \\ &\quad > \text{Prob}(C_1(Y \odot Z) = f_1(Y) \textit{ and } Y \in G \mid C_2(Y \odot Z) = f_2(Z)) - \frac{\epsilon}{2}. \end{aligned}$$

Proof. Let $t = 16\ell/(\epsilon^3)$. We choose uniformly at random and independently t strings from Σ^m. Let $S = \{Z_1, \ldots, Z_t\}$ be the set of these strings. For each y in G, let S_y be the set of $Z \in S$ such that the (y, Z) entry in the table T is of the form $(*, 1)$ (i.e., the second component is 1, which means $C_2(y \odot Z) = f_2(Z)$). For each $y \in \Sigma^\ell$,

$$p_y \stackrel{\text{def}}{=} \text{Prob}(C_1(y \odot Z) = f_1(y) \mid C_2(y \odot Z) = f_2(Z))$$

is the fraction of entries in row y of the table T that have the form $(1, 1)$ among those entries that have the form $(*, 1)$. Therefore, since the elements in S are chosen at random, we expect that

$$\frac{\|\{Z \in S_y \mid C_1(y \odot Z) = f_1(y)\}\|}{\|S_y\|}$$

is a good approximation of p_y. Of course, we need to make the last statement more precise. Let us fix y in G. We first estimate the size of S_y. For each $i \in \{1, \ldots, t\}$,

$$\text{Prob}(C_2(y, Z_i) = f_2(Z_i)) \stackrel{\text{def}}{=} \mu_y \geq \frac{\epsilon}{2}.$$

The multiplicative form of the Chernoff bounds states that if $X_1, \ldots, X_m$ are independent 0-1 random variables with expected values μ, then (see Appendix A),

$$\text{Prob}\left(\frac{1}{m}\sum_{i=1}^{m} X_i < (1 - \gamma) \cdot \mu\right) \leq e^{-\gamma^2 \mu m/2}.$$

Thus, letting X_i, $i = 1, \ldots, t$, be defined by $X_i = 1$ if $C_2(y, Z_i) = f_2(Z_i)$ and $X_i = 0$ otherwise, we obtain

$$\begin{aligned} \text{Prob}\left(\frac{1}{t}\sum_{i=1}^{t} X_i < \left(1 - \frac{1}{2}\right)\mu_y\right) &\leq e^{-(\frac{1}{2})^2 \cdot \mu_y \cdot t/2} \\ &\leq e^{-(\frac{1}{2})^2 \cdot (\epsilon/2) \cdot t/2} \\ &= e^{-(\epsilon t)/16} = e^{-(1/\epsilon)^2 \cdot \ell} \end{aligned}$$

(we took into account that $t = 16\ell(1/\epsilon^3)$). So, with probability at least $1 - e^{-(1/\epsilon)^2 \cdot \ell}$,

$$\|S_y\| \geq \frac{1}{2} \cdot t \cdot \frac{\epsilon}{2} = 4\frac{1}{\epsilon^2}\ell.$$

$$= \text{Prob}(C_2(Y \odot Z) = f_2(Z)) \cdot \text{Prob}(C_1(Y \odot Z) = f_1(Y) \text{ and } Y \in G \mid C_2(Y \odot Z) = f_2(Z))$$

$$\leq \rho_2(m)\left(\rho_1(\ell) + \frac{\epsilon}{2}\right) \leq \rho_1(\ell) \cdot \rho_2(m) + \frac{\epsilon}{2}.$$

This is exactly the desired Equation (5.13) and thus the proof of Lemma 5.6.10 is complete. ∎

Theorem 5.6.12 (Step 2 of hardness amplification)

INFORMAL STATEMENT: *Given a constant-hard function, we can construct a crypto-hard function.*

FORMAL STATEMENT: (a) *Let g' be a length-preserving function that is $1/100$-constant-rate hard. Then there is a length-preserving function g'' that is crypto-hard. Moreover, g'' can be constructed effectively from g' in polynomial time.*

(b) *Let g' be a length-preserving function such that, for some constant c' and for all sufficiently large ℓ, g'_ℓ is $(\frac{1}{100}, 2^{c'\ell})$-hard. Then there is a length-preserving function g'' such that for some constant c'' and for all sufficiently large ℓ, g''_ℓ is $(1 - 2^{-c''\ell}, 2^{c''\ell})$-hard. Moreover, g'' can be constructed effectively from g' in polynomial time.*

Proof. We prove (a). The construction is done separately for each length. So, let us fix a length $\ell \in \mathbb{N}$ and let us consider g'_ℓ the restriction of g' on Σ^ℓ. For each integer $k \geq 1$ and for all $x_1, \ldots, x_k$ strings in Σ^ℓ, we define

$$g''_{\ell,k}(x_1 \odot x_2 \odot \ldots \odot x_k) = g'_\ell(x_1) \odot g'_\ell(x_2) \odot \ldots \odot g'_\ell(x_k).$$

By hypothesis, if ℓ is sufficiently large, then g'_ℓ is $1/100$-constant-rate hard against circuits of size $s(\ell)$, for some superpolynomial function s.

We first show the following.

Claim 5.6.13 *For any integer $k > 1$, for any $\epsilon \in (0,1)$ with $\epsilon^3 > 16\ell/2^\ell$, $g''_{\ell,k}$ can have agreement at most $(\frac{99}{100})^k + 100 \cdot \epsilon$ on its domain with any circuit of size $\frac{\epsilon^3}{48\cdot(k-1)\cdot\ell^2} \cdot s(\ell)$.*

Proof. By induction on k, using Lemma 5.6.10 and the observation that for each i, $g''_{\ell,i+1}(x_1\odot\ldots\odot x_{i+1})$ is the concatenation of $g'_\ell(x_1)$ and $g''_{\ell,i}(x_2\odot\ldots\odot x_{i+1})$, it can be shown that $g''_{\ell,k}$ agrees on at most a fraction of $(\frac{99}{100})^k + 100\cdot\epsilon(1-(\frac{99}{100})^{k-1})$ of inputs with any circuit of size bounded by $\frac{\epsilon^3}{48\cdot(k-1)\cdot\ell^2} \cdot s(\ell)$. ∎

The rest of the proof is easy. We take $k = \ell + 1$ and $\epsilon = (s(\ell))^{-1/6}$. Then the function $g''_{\ell,\ell+1}$ is defined on inputs of length $\ell(\ell+1)$ and disagrees with any circuit of size $\frac{(s(\ell))^{1/2}}{48\cdot\ell^3}$, which is superpolynomial as a function of $\ell(\ell+1)$, on at least a fraction of $\left(1 - \frac{1}{p(\ell(\ell+1))}\right)$ of its domain, for any polynomial p, and for any sufficiently large ℓ. Finally we take the function g'' as the union of the functions

$g''_{\ell,(\ell+1)}$, over all ℓ. As stated now, g'' is not defined at all lengths, but this can be fixed by extending g'' similarly to the procedure in Lemma 5.6.4.

The proof of (b) is more difficult and requires techniques that go beyond the scope of this book. Consequently, we skip it. (The approach used in Theorem 5.6.12 blows up the length of the input from ℓ to $\ell(\ell+1)$ because g'' is obtained by concatenating the outputs of g' on $\ell+1$ independent inputs. The size of the adversary circuits that fail to calculate g'' cannot be made larger than the size of the adversary circuits that fail to calculate g' and therefore we do not obtain exponential hardness. One of the known proofs of (b) manages to construct g'' by taking a "direct product" of the function g' on dependent inputs, so that the input length of g'' is larger by only a multiplicative constant factor than the input length of g'.) ∎

The combination of Theorem 5.6.3 and Theorem 5.6.12 (i.e., Step 1 and Step 2 of hardness amplification) yields the desired conclusion.

Theorem 5.6.14 (a) *Let $g\colon \Sigma^* \to \Sigma^*$ be a length-preserving function that is worst-case hard. Then there is a length-preserving function $g''\colon \Sigma^* \to \Sigma^*$ that is crypto-hard. Moreover, g'' can be constructed effectively from g' in time $2^{O(\ell)}$.*

(b) *Let $g\colon \Sigma^* \to \Sigma^*$ be a length-regular function such that, for some positive constant c and for all sufficiently large ℓ, g_ℓ is $(\frac{1}{2^\ell}, 2^{c\ell})$-hard. Assume that g is length-preserving*[4]. *Then there is a length-preserving function g'' such that, for some constant c'' and for all sufficiently large ℓ, g''_ℓ is $(1-2^{-c''\ell}, 2^{c''\sqrt{\ell}})$-hard. Moreover, g'' can be constructed effectively from g' in time $2^{O(\ell)}$.*

5.7 Hard predicates

IN BRIEF: Given a hard function, one can effectively construct a hard predicate. The type of hardness (crypto hardness or exponential hardness) is preserved.

Predicate functions are particularly interesting because they model decision problems, i.e., languages, which are objects of major interest in computational complexity. Also, the construction of the type II pseudo-random generators that will be presented in the next section utilizes hard predicates as the starting point. Consequently, we focus in this section on *hard predicates.* Recall that a predicate is a function whose outputs can only be 0 or 1. Clearly, any predicate f can be calculated correctly on a fraction of $\frac{1}{2}$ of the inputs at each length by quite simple circuits. Indeed, either the circuit that outputs 1 on all inputs, or the circuit that outputs 0 on all inputs, will agree with f on least half of the inputs. Therefore, when considering the performance of a circuit that attempts to calculate a predicate only the bias from $\frac{1}{2}$ is relevant. Accordingly, we give the following definitions

[4] As noted before this property of g is not essential.

for the general form of a hard predicate as well as for two particular forms of strong hardness for predicates.

Definition 5.7.1 ((ϵ, S)-hard predicate) *A predicate $f\colon \Sigma^{\ell} \to \{0,1\}$ is (ϵ, S)-hard if for every circuit C of size S,*

$$\text{Prob}_{x\in\Sigma^{\ell}}(C(x) \neq f(x)) \geq \frac{1}{2} - \epsilon.$$

Definition 5.7.2 (Crypto-hard predicate) *A predicate $f\colon \Sigma^* \to \{0,1\}$ is crypto-hard if there is a superpolynomial function S so that the following holds. For any polynomial p and for any family of circuits $(C_\ell)_{\ell\in\mathbb{N}}$ of size at most S,*

$$\text{Prob}_{x\in\Sigma^{\ell}}(C_\ell(x) \neq f_\ell(x)) \geq \frac{1}{2} - \frac{1}{p(\ell)},$$

for all sufficiently large ℓ.

Definition 5.7.3 (Exponentially-hard predicate) *A predicate $f\colon \Sigma^* \to \{0,1\}$ is exponentially-hard if there is a constant c so that for any family of circuits $(C_\ell)_{\ell\in\mathbb{N}}$ with* size$(C_\ell) \leq 2^{c\ell}$, *for all ℓ,*

$$\text{Prob}_{x\in\Sigma^{\ell}}(C_\ell(x) \neq f_\ell(x)) \geq \frac{1}{2} - \frac{1}{2^{c\ell}},$$

for all sufficiently large ℓ.

The next result shows that a length-preserving crypto-hard (exponential hard) function can be converted into a crypto-hard (exponential hard, respectively) predicate.

Theorem 5.7.4 (a) *Let $f\colon \Sigma^* \to \Sigma^*$ be a length-preserving function which is crypto-hard. Then there exists $f'\colon \Sigma^* \to \{0,1\}$ a crypto-hard predicate. Moreover f' can be constructed effectively from f in polynomial time.*

(b) *Let $f\colon \Sigma^* \to \Sigma^*$ be a length-preserving function such that for some positive constant c and for all sufficiently large ℓ, f_ℓ is $(2^{-c\ell}, 2^{c\ell})$-hard. Then there exists $f'\colon \Sigma^* \to \{0,1\}$ an exponentially-hard predicate. Moreover f' can be constructed effectively from f in polynomial-time.*

Proof. We prove (a). Let $p(\ell)$ be an arbitrary polynomial and let $s(\ell)$ be a superpolynomial function so that for any circuit C of size $s(\ell)$, for ℓ sufficiently large,

$$\text{Prob}(C(x) = f(x)) < \frac{1}{q(\ell)},$$

where $q(\ell) = (4/3) \cdot p(\ell) \cdot (4\ell \cdot (p(\ell))^2 + 1)$. As in the previous proofs, we fix an ℓ for which the above relation holds and consider the restriction of f to Σ^{ℓ} (abusing notation, this restriction will also be called f). Thus $f\colon \Sigma^{\ell} \to \Sigma^{\ell}$.

We will be using once again error-correcting codes. This time we will utilize the Hadamard error-correcting code and we will take advantage of its list-decoding property (see Theorem 5.3.5). Namely, for $f(x)$ of length ℓ, we consider $\text{Had}(f(x))$ and define the predicate on inputs x and r to be the r-th bit of $\text{Had}(f(x))$. If a circuit can calculate correctly at least a fraction of $\frac{1}{2}+\frac{1}{p(\ell)}$ of the bits of $\text{Had}(f(x))$, then, by Theorem 5.3.5, we can produce a short list which contains $f(x)$. By picking randomly one element from this list, we have a fairly good chance of retrieving $f(x)$. This contradicts the hardness of f.

We now proceed with the formal proof. Let

$$s_1(\ell) = \frac{c}{\ell^3 \cdot (p(\ell))^4} \cdot s(\ell), \tag{5.15}$$

where the constant c will be chosen later. Recall that $\text{Had}(f(x))$ is a string of length 2^ℓ whose r bit is the inner product modulo 2 of $f(x)$ and r, denoted $f(x)\cdot r$, where $r \in \{0,1\}^\ell$. Therefore, according to our plan, we define the predicate $f'\colon \Sigma^\ell \times \Sigma^\ell \to \{0,1\}$ by

$$f'(x,r) = f(x)\cdot r.$$

Clearly, given oracle access to f, f' can be calculated in polynomial time. Assume that there is a circuit C of size $s_1(\ell)$ such that

$$\text{Prob}_{x,r}(C(x,r) = f'(x,r)) \geq \frac{1}{2} + \frac{1}{p(\ell)}.$$

Then a small variation of the above relation holds for a polynomial fraction of fixed x. Indeed, let $B = \{x \in \Sigma^\ell \mid \text{Prob}_r(C(x,r) = f'(x,r)) \geq \frac{1}{2} + \frac{1}{2p(\ell)}\}$.

Claim 5.7.5 $\text{Prob}(B) > \frac{1}{p(\ell)}$.

Proof. Let $\alpha = \text{Prob}(B)$. Since

$$\begin{aligned}\text{Prob}_{x,r}(C(x,r) = f'(x.r)) = {} & \text{Prob}(x \in B)\cdot \text{Prob}(C(x,r) = f'(x,r) \mid x \in B)\\ & + \text{Prob}(x \notin B)\cdot \text{Prob}(C(x,r) = f'(x,r) \mid x \notin B),\end{aligned}$$

it follows that $\frac{1}{2}+\frac{1}{p(\ell)} \leq \alpha\cdot 1 + (1-\alpha)\cdot(\frac{1}{2}+\frac{1}{2\cdot p(\ell)})$, which implies $\alpha > \frac{1}{p(\ell)}$. ∎

Therefore, for any $x \in B$, $C(x,r) = f(x)\cdot r$ for at least a fraction of $\frac{1}{2}+\frac{1}{2p(\ell)}$ of the r in Σ^ℓ. In other words, if $\overline{C}$ denotes the string $C(x,0\ldots0)\ldots C(x,1\ldots1)$ of length $\ell\cdot 2^\ell$, then the Hamming distance between $\overline{C}$ and $\text{Had}(f(x))$ is at most $\frac{1}{2}-\frac{1}{2\cdot p(\ell)}$. By Theorem 5.3.5, there is an oracle probabilistic circuit A' of size $O(\ell^3\cdot\frac{1}{(2p(\ell))^4})$ that makes $\ell^2\cdot(2p(\ell))^2$ queries to C (formally to the oracle string $\overline{C}$) with the following property: For every $x \in B$, with probability at least $\frac{3}{4}$, A' outputs a list of $\ell\cdot(2p(\ell))^2+1$ strings which includes $f(x)$. By embedding the circuit C into A', we get a probabilistic circuit A of size bounded by $O(\ell^3\cdot(p(\ell))^4)\cdot\text{size}(C)$

that, for all $x \in B$, outputs a list as above. We further modify the circuit A so that at the end it randomly picks one element from this list. With probability at least $\frac{1}{4\ell \cdot (p(\ell))^2+1}$, this element is $f(x)$.

Thus, the probability that the modified A on input x computes $f(x)$ is at least

$$\begin{aligned} &\text{Prob}(x \in B) \cdot \text{Prob}_{x,rand}(A(x, rand) = f(x) \mid x \in B) \\ &\geq \frac{1}{2p(\ell)} \cdot \frac{3}{4} \cdot \frac{1}{4\ell(p(\ell))^2 + 1} = \frac{1}{q(\ell)}. \end{aligned}$$

(Here, $rand$ denotes the random bits used by the circuit A). The size of A is bounded by $s(\ell)$ (for an appropriate choice of the constant c in Equation (5.15)). Since the polynomial $p(\ell)$ is arbitrary and $s_1(\ell)$ is superpolynomial, it follows that f' is a crypto-hard predicate.

The proof of (b) is virtually identical. ∎

5.8 From hard predicates to pseudo-random generators

IN BRIEF: Given an exponentially hard predicate, one can effectively construct a pseudo-random generator with exponential extension that is computable in time $2^{c_1 n}$ and is secure against adversary circuits of size $2^{c_2 n}$, for some constants c_1 and c_2.

There are two building blocks that are utilized in the construction: (1) A hard predicate, and (2) a certain combinatorial object called a *design*. It is not difficult to see why a hard predicate is useful. For concreteness, let us assume that we are given a hard predicate f that is exponentially hard. Let us consider $f_\ell \colon \Sigma^\ell \to \{0, 1\}$ (recall that f_ℓ is the restriction of f to Σ^ℓ). Then the function $g_\ell \colon \Sigma^\ell \to \Sigma^{\ell+1}$ defined by $g_\ell(x) = x \odot f_\ell(x)$ is a pseudo-random generator with quite good parameters (more precisely, it is an extender, see Definition 5.3.1). The intuition is that an adversary that is not able to calculate $f_\ell(x)$ has a hard time in distinguishing it from a random bit that is independent of x. The formal argument is as follows.

Claim 5.8.1 *Suppose there is a circuit C of size S such that*

$$|\text{Prob}_{x \in \Sigma^\ell}(C(g_\ell(x)) = 1) - \text{Prob}_{z \in \Sigma^{\ell+1}}(C(z) = 1)| > \epsilon.$$

Then there is a circuit B of size $S + c$, for some constant c (that does not depend on ℓ), such that

$$\text{Prob}(B(x) = f_\ell(x)) > \frac{1}{2} + \epsilon. \tag{5.16}$$

Proof. The claim follows from Theorem 5.3.3, by noting that $g_\ell(x) = x \cdot f_\ell(x)$, and taking in Theorem 5.3.3, $f(x) = x$ and $h(x) = f_\ell(x)$. ∎

Claim 5.8.1 immediately implies that if f_ℓ is exponentially hard, then circuits of exponential size cannot distinguish between $g_\ell(x) = x \odot f_\ell(x)$ and a pure random string $u_{\ell+1}$ of length $\ell + 1$, but with a bias that is exponentially small.

Thus, indeed, hard functions look promising as building blocks for a pseudo-random generator. Some problems arise. First, note that calculating $g_\ell(x)$ involves the calculation of $f_\ell(x)$ which is a difficult task. In particular, it requires more computational power than the one we allow our adversary to have. This is an inherent drawback of this approach. We have called such pseudo-random generators type II. However, we will see that there are circumstances when type II pseudo-random generators are still useful. Another problem is that, of course, we want larger extension (g_ℓ has extension 1). One strategy is to iterate the construction and make the pseudo-random generator on input $x_1, x_2, \ldots, x_m$ output $x_1 \odot f(x_1) \odot \ldots \odot x_m \odot f(x_m)$. We do get a larger extension, however, the ratio between the output length and the input length still tends to 1. We need a more efficient way to iterate without blowing up too much the input length.

To this aim, we introduce the second building block, the design.

Definition 5.8.2 ((weight ℓ, intersection α)-design) *The values n, t, ℓ, α are positive integer parameters. A (weight ℓ, intersection α)-design of type (m, t) is an $m \times t$ (i.e., m rows and t columns) matrix of 0s and 1s so that each row has exactly ℓ 1s and any two distinct rows have at most α common positions filled with 1s.*

Each row i of the design can be regarded as being the characteristic vector of a subset S_i of $\{1, \ldots, t\}$ (that is, the k-th entry is 1 if $k \in S_i$, and it is 0, if $k \notin S_i$) and we can say that for all $i \in \{1, \ldots, m\}$, $\|S_i\| = \ell$ and for all $i \neq j \in \{1, \ldots, m\}$, $\|S_i \cap S_j\| \leq \alpha$.

Let $x \in \Sigma^t$ and let S be a subset of cardinality ℓ of the set $\{1, \ldots, t\}$, $S = \{i_1 < i_2 < \ldots < i_\ell\}$. By $x|_S$ we denote the string $x(i_1) \odot x(i_2) \odot \ldots \odot x(i_\ell)$ (i.e., $x|_S$ is the projection of x on the positions indicated by S). Let A be a design of type (m, t) with weight ℓ and intersection α. Let $S_1, S_2, \ldots, S_m$ be the subsets of $\{1, \ldots, t\}$ whose characteristic vectors are the rows of A. The design A will allow us to iterate a function $f \colon \Sigma^\ell \to \{0, 1\}$ without increasing the input length too much. To this aim, we define the function $g_{f,A} \colon \Sigma^t \to \Sigma^m$ by

$$g_{f,A}(x) = f(x|_{S_1}) \odot f(x|_{S_2}) \odot \ldots \odot f(x|_{S_m}). \tag{5.17}$$

We will show that if f is an exponentially-hard predicate and if the design A has appropriate parameters, then $g_{f,A}$ is a pseudo-random generator.

Note that $g_{f,A}$ amounts to computing f on m inputs. However, instead of taking m independent inputs of length ℓ, we need, for $g_{f,A}$, just one input x of length t. We obtain a saving if $t < m \cdot \ell$. In order for $g_{f,A}$ to have large extension, we want $t << m$. To prove that $g_{f,A}$ is a pseudo-random generator, we also need that $\alpha = O(\log m)$. With these requirements in mind, we construct the design that we need.

Lemma 5.8.3 *Let ℓ be a sufficiently large positive integer and let c be any positive constant such that $c < \frac{1}{12}$. There is a design of type $(2^{c\ell}, \frac{1}{8} \cdot \frac{1}{c} \cdot \ell)$ with weight ℓ and intersection $18c\ell$. Moreover, such a design can be constructed in time $2^{O(\ell)}$.*

Proof. Let $m = 2^{c\ell}, t = \frac{1}{8} \cdot \frac{1}{c} \cdot \ell$. We need m subsets $S_1, \dots, S_m$ of $\{1, \dots, t\}$, each having cardinality ℓ such that the intersection of any two subsets has at most $18 \log m$ elements. Once we have them, the m subsets define characteristic vectors in $(0,1)^t$, which we take to be the rows of the design. The construction is as follows. Sequentially, we choose m subsets of $\{1, \dots, t\}$ of cardinality ℓ such that each new set intersects each of the old sets in at most $18 \log m$ elements.

Let us consider Step i ($1 \leq i \leq m$) of the construction when S_i is built. By this time, we already have the subsets $S_1, S_2, \dots, S_{i-1}$. We first construct probabilistically a subset $T \subseteq \{1, \dots, t\}$ using the following procedure:

$T = \emptyset$.
for $k = 1, \dots, t$
    insert k in T with probability $p = 12c$.

Let S_j be some fixed "old" set (i.e., $1 \leq j \leq i-1$), and let $S_j = \{s_1, \dots, s_\ell\}$. For $k \in \{1, \dots, \ell\}$, let X_k be the random variable defined by

$$X_k = \begin{cases} 1 & \text{if } s_k \in T \\ 0 & \text{if } s_k \notin T \end{cases}$$

The random variables X_k are independent and $\text{Prob}(X_k = 1) = 12c$. Note that $X_1 + \dots + X_\ell = \|S_j \cap T\|$. By the multiplicative Chernoff bounds,

$$\begin{aligned} &\text{Prob}\Big(\|S_j \cap T\| > 18 \log m\Big) \\ &= \text{Prob}\left(\frac{X_1 + \dots + X_\ell}{\ell} > 18 \frac{\log m}{\ell}\right) \\ &= \text{Prob}\left(\frac{X_1 + \dots + X_\ell}{\ell} > 18c\right) \\ &= \text{Prob}\left(\frac{X_1 + \dots + X_\ell}{\ell} > \frac{3}{2} \cdot 12c\right) \\ &\leq \mathrm{e}^{-(1/2)^2 \cdot (12c) \cdot \ell / 3} = \mathrm{e}^{-c\ell}. \end{aligned}$$

There are less than $m = 2^{c \cdot \ell}$ such "old" subsets S_j. The probability that there is some "old" subset S_j such that $\|S_j \cap T\| > 18 \log m$ is, thus, at most

$$2^{c \cdot \ell} \cdot \mathrm{e}^{-c \cdot \ell} = \left(\frac{e}{2}\right)^{-c\ell}.$$

Therefore, with high probability (if ℓ is sufficiently large), the set T intersects all the "old" subsets in $\leq 18 \log m$ elements. It remains to show that it is likely that

$\|T\| \geq \ell$ and, consequently, we can obtain S_i as desired (by taking S_i to be an arbitrary subset of T of size exactly ℓ). Note that the expected size of T is

$$\mu = t \cdot p = \left(\frac{1}{8} \cdot \frac{1}{c} \cdot \ell\right) \cdot (12c) = \frac{3}{2}\ell.$$

Therefore, using again the multiplicative Chernoff bounds,

$$\text{Prob}(\|T\| < \ell) = \text{Prob}\left(\|T\| < \left(1 - \frac{1}{3}\right) \cdot \mu\right) \leq \mathrm{e}^{-(1/9)\cdot(12c)\cdot\ell/2} = \mathrm{e}^{-(2/3)cl}.$$

For ℓ sufficiently large, $(e/2)^{-c\ell} + \mathrm{e}^{-(2/3)\cdot c\cdot\ell} < 1$. Therefore, there is a set $S_i \subseteq \{1, \ldots, t\}$ of cardinality ℓ such that $\|S_i \cap S_j\| < 18 \log m$, for all $j < i$. We can construct the set S_i deterministically by brute force. We simply try all subsets of size ℓ of $\{1, \ldots, t\}$ as candidates for being S_i and check if the current candidate intersects each of $S_1, \ldots, S_{i-1}$ in at most $18 \log m$ elements. By our calculations above, there will be a candidate that has this intersection property. Finding this subset takes time at most $O(2^t \cdot t \cdot m)$ (there are at most 2^t candidates, at most m previously constructed sets $S_1, \ldots, S_{i-1}$, and the intersection condition can be checked in $O(t)$ time). Therefore, the construction of all m subsets $S_1, \ldots, S_m$ can be done in $O(2^t \cdot t \cdot m^2) = O(2^{(1/8)(1/c)\ell} \cdot (\frac{1}{8} \cdot \frac{1}{c} \cdot \ell) \cdot 2^{2c\ell}) \leq 2^{(1/c)\cdot\ell}$. ∎

The following Lemma is the key result underlying the construction of the pseudo-random generator. It establishes the connection between the hardness of a predicate $f\colon \Sigma^\ell \to \{0,1\}$ and the pseudo-randomness of the function $g_{f,A}$ defined by the relation (5.17).

Lemma 5.8.4 INFORMAL STATEMENT: *If there is a test D that distinguishes $g_{f,A}$ from the uniform distribution with bias ϵ, then there is a small circuit C such that $D(C(\cdot))$ is equal to $f(\cdot)$ on a fraction of inputs that is $> \frac{1}{2} + \frac{\epsilon}{m}$.*

FORMAL STATEMENT: *The values ℓ, m, t, and α are positive integer parameters, $t \leq m$. Let $f\colon \Sigma^\ell \to \{0,1\}$ be a predicate and let A be a (weight ℓ, intersection α)-design of type (m,t). Let $g_{f,A}\colon \Sigma^t \to \Sigma^m$ be defined by relation (5.17). Suppose there is a function $D\colon \Sigma^m \to \{0,1\}$ such that*

$$|\text{Prob}_{r\in\Sigma^m}(D(r) = 1) - \text{Prob}_{z\in\Sigma^t}(D(g_{f,A}(z)) = 1)| > \epsilon.$$

(In other words, D is a test that distinguishes the distributions $g_{f,A}(U_t)$ and U_m with bias ϵ.) Then there is a circuit C of size $O(m \cdot 2^\alpha)$ such that

$$\left|\text{Prob}_{x\in\Sigma^\ell}(D(C(x)) = f(x)) - \frac{1}{2}\right| > \frac{\epsilon}{m}.$$

Proof. We use again the method of hybrid distributions. We will consider the hybrid distributions $H_0, H_1, \ldots, H_m$ all defined over the set Σ^m. For $i \in \{0, \ldots, m\}$, H_i consists of m random bits constructed as follows: The first i bits are the first i bits of $g_{f,A}(z)$, for random $z \in \Sigma^t$, and the remaining $m-i$ bits are independently and uniformly at random chosen bits. Observe that the extreme distributions satisfy the requirements that H_m is $g_{f,A}(U_t)$ and H_0 is U_m. We can assume that

$$\text{Prob}_{z\in\Sigma^t}(D(g_{f,A}(z)) = 1) - \text{Prob}_{r\in\Sigma^m}(D(r) = 1) > \epsilon,$$

because the other case (i.e., $\text{Prob}_{r\in\Sigma^m}(D(r) = 1) - \text{Prob}_{z\in\Sigma^t}(D(g_{f,A}(z)) = 1) > \epsilon$) can be handled similarly. Thus

$$\begin{aligned}\epsilon &< \text{Prob}(D(H_m) = 1) - \text{Prob}(D(H_0) = 1)\\ &= \sum_{i=1}^{m} \text{Prob}(D(H_i) = 1) - \text{Prob}(D(H_{i-1}) = 1).\end{aligned}$$

It follows that there is $i \in \{1, \ldots, m\}$ such that

$$\text{Prob}(D(H_i) = 1) - \text{Prob}(D(H_{i-1}) = 1) > \frac{\epsilon}{m}. \tag{5.18}$$

Let us put H_{i-1} and H_i side by side and compare them.

$$\begin{aligned}H_{i-1} &= f(z|_{S_1}) \odot f(z|_{S_2}) \ldots f(z|_{S_{i-1}}) \odot \quad r_i \quad \odot r_{i+1} \ldots r_m\\ H_i &= f(z|_{S_1}) \odot f(z|_{S_2}) \ldots f(z|_{S_{i-1}}) \odot f(z|_{S_i}) \odot r_{i+1} \ldots r_m,\end{aligned}$$

where z is a random string in Σ^t and $r_i, r_{i+1}, \ldots, r_m$ are independent random bits. The only difference between H_{i-1} and H_i is in the i-th bit. In H_{i-1} this bit is random and independent of the previous bits, while in H_i this bit is $f(z|_{S_i})$, for a random $z \in \Sigma^t$. To simplify the notation, we can consider that $S_i = \{1, 2, \ldots, \ell\}$. We denote $z_\ell \stackrel{\text{def}}{=} z(1 : \ell)$ and $z_r \stackrel{\text{def}}{=} z(\ell+1 : t)$. The relation (5.18) can be rewritten as

$$\begin{aligned}&\text{Prob}(D(f(z_\ell \odot z_r|_{S_1}) \odot \ldots \odot f(z_\ell \odot z_r|_{S_{i-1}}) \odot f(z_\ell) \odot r_{i+1} \odot \ldots \odot r_m) = 1)\\ &\quad - \text{Prob}(D(f(z_\ell \odot z_r|_{S_1}) \ldots f(z_\ell \odot z_r|_{S_{i-1}}) \odot r_i \odot r_{i+1} \odot \ldots \odot r_m) = 1)\\ &\quad > \frac{\epsilon}{m}.\end{aligned}$$

Observe that we are in a situation very similar to the one in Claim 5.8.1. With the same approach, we can calculate $f(z_\ell)$ with probability at least $\frac{1}{2} + \frac{\epsilon}{m}$ by executing the following algorithm.

Input: $z_\ell \in \Sigma^\ell$
Pick random $r_i, \ldots, r_m \in \{0, 1\}$. Pick random $z_r \in \Sigma^{t-\ell}$.
if $D(f(z_\ell \odot z_r|_{S_1}) \odot \ldots \odot f(z_\ell \odot z_r|_{S_{i-1}}) \odot r_i \odot r_{i+1} \odot \ldots \odot r_m) = 1$
  output r_i
else
  output $1 - r_i$.

Let us denote by $B(z_\ell, r_i, \ldots, r_m, z_r)$ the output of the above algorithm on input $z_\ell \in \Sigma^\ell$ and with the random choices $r_i, \ldots, r_m$ and $z_r \in \Sigma^{t-\ell}$. Not surprisingly, the proof of the following claim is almost identical to the proof of Theorem 5.3.3.

Claim 5.8.5 $\mathrm{Prob}_{z_\ell, r_i, \ldots, r_m, z_r}(B(z_\ell, r_i, \ldots, r_m, z_r) = f(z_\ell)) > \frac{1}{2} + \frac{\epsilon}{m}$.

Proof. Let E be the event that $B(z_\ell, r_i, \ldots, r_m, z_r) = f(z_\ell)$ (taken over uniform random choices of $z_\ell, r_1, \ldots, r_m$ and z_r). Let

$$G_i = f(z|_{S_i}) \odot f(z|_{S_2}) \odot \ldots \odot f(Z|_{S_{i-1}}) \odot (1 - f(z_\ell)) \odot r_{i+1} \odot \ldots \odot r_m,$$

i.e., G_i is obtained from H_i by replacing $f(z_\ell)$ with $(1 - f(z_\ell))$ in the i-th coordinate. From the algorithm, we see that

$$\begin{aligned} \mathrm{Prob}(E) &= \mathrm{Prob}(f(z_\ell) = r_i) \cdot \mathrm{Prob}(D(H_i) = 1) \\ &\quad + \mathrm{Prob}(f(z_\ell) = 1 - r_i) \cdot \mathrm{Prob}(D(G_i) = 0). \end{aligned}$$

Let $p \stackrel{\text{def}}{=} \mathrm{Prob}(D(H_i) = 1)$ and $q \stackrel{\text{def}}{=} \mathrm{Prob}(D(H_{i-1} = 1))$. Then,

$$\begin{aligned} q &= \mathrm{Prob}(f(z_\ell) = r_i) \cdot \mathrm{Prob}(D(H_{i-1})) = 1 \mid f(z_\ell) = r_i) \\ &\quad + \mathrm{Prob}(f(z_\ell) \neq r_i) \cdot \mathrm{Prob}(D(H_{i-1}) = 1 \mid f(z_\ell) \neq r_i) \\ &= \frac{1}{2}\mathrm{Prob}(D(H_i) = 1) + \frac{1}{2}\mathrm{Prob}(D(G_i) = 1) \\ &= \frac{1}{2}p + \frac{1}{2}\mathrm{Prob}(D(G_i)) = 1). \end{aligned}$$

Therefore, $\mathrm{Prob}(D(G_i) = 1) = 2q - p$, and, thus, $\mathrm{Prob}(D(G_i) = 0) = 1 - 2q + p$. Thus,

$$\mathrm{Prob}(E) = \frac{1}{2}p + \frac{1}{2}(1 - 2q + p) = \frac{1}{2} + (p - q) > \frac{1}{2} + \frac{\epsilon}{m},$$

which is the desired conclusion. ∎

From Claim 5.8.5, it follows that there are fixed bits $r_i^*, r_{i+1}^*, \ldots, r_m^*$ and a fixed string $z_r^* \in \Sigma^{t-\ell}$ such that

$$\mathrm{Prob}_{z_\ell \in \Sigma^\ell}(B(z_\ell, r_i^*, \ldots, r_m^*, z_r^*) = f(z_\ell)) > \frac{1}{2} + \frac{\epsilon}{m}.$$

It remains to evaluate the complexity of the above procedure B. Since z_r^* is fixed, for each $j \in 1, \ldots, i-1$, $f(z_\ell \odot z_r^*|_{S_j})$ depends only on the bits of z_ℓ that are in positions $S_i \cap S_j$. Taking into account the intersection property of the design A, it follows that $f(z_\ell \odot z_r^*|_{S_j})$ depends on at most α bits of z_ℓ and these positions are independent of z_ℓ (because they are given by $S_i \cap S_j$). Thus, $f(z_\ell \odot z_r^*|_{S_j})$ is computable by a circuit of size $O(2^\alpha)$.[5] We need to calculate $f(z_\ell \odot z_r^*|_{S_1}) \odot$

[5]Recall from Section 1.1.2 that any boolean function with α variables can be calculated by a circuit of size $\leq (1 + O(\frac{1}{\sqrt{\alpha}})\frac{2^\alpha}{\alpha} = O(2^\alpha)$. In brief, the circuit stores the truth table of the function. See, e.g., [Sav98, page 80].

$f(z_\ell \odot z_r^*|_{S_2}) \odot \ldots \odot f(z_\ell \odot z_r^*|_{S_{i-1}})$ and $i - 1 < m$. By the above observation this can be done by a circuit of size bounded by $(i-1) \cdot O(2^\alpha) \leq O(m \cdot 2^\alpha)$. Consequently, by hard-wiring in the circuit the bits $r*_i, \ldots, r_m^*$ and z_r^*, there is a circuit C of size $O(m \cdot 2^\alpha)$ that on input z_ℓ calculates

$$f(z_\ell \odot z_r^*|_{S_1}) \odot f(z_\ell \odot z_r^*|_{S_2}) \odot \ldots \odot f(z_\ell \odot z_r^*|_{S_{i-1}}) \odot r_i^* \odot \ldots \odot r_m^*.$$

By inspecting the if-else clause in the algorithm, we note that (a) if r_i^* is 1, then the algorithm $B(z_\ell, r_i^*, \ldots, r_m^*, z_r^*)$ outputs 1 if and only if $D(C(z_\ell)) = 1$, and (b) if r_i^* is 0, then the algorithm $B(z_\ell, r_i^*, \ldots, r_m^*, z_r^*)$ outputs 1 if and only if $D(C(z_\ell)) = 0$. Thus, in case (a),

$$\begin{aligned}\text{Prob}_z(D(C(z)) = f(z)) &= \text{Prob}_z(B(z, r_i^*, \ldots, r_m^*, z_r^*) = f(z)) \\ &> \frac{1}{2} + \frac{\epsilon}{m},\end{aligned}$$

and, in case (b),

$$\begin{aligned}\text{Prob}_z(D(C(z)) = f(z)) &= \text{Prob}_z(B(z, r_i^*, \ldots, r_m^*, z_r^*) = 1 - f(z)) \\ &= 1 - \text{Prob}_z(B(z, r_i^*, \ldots, r_m^*, z_r^*) = f(z)) \\ &< 1 - \left(\frac{1}{2} + \frac{\epsilon}{m}\right) = \frac{1}{2} - \frac{\epsilon}{m}.\end{aligned}$$

In both cases,

$$\left|\text{Prob}_z(D(C(z)) = f(z)) - \frac{1}{2}\right| > \frac{\epsilon}{m},$$

which concludes the proof. ∎

The main result of this section is obtained by combining Lemma 5.8.4 and Lemma 5.8.3.

Theorem 5.8.6 INFORMAL STATEMENT: *Given a hard predicate $f: \Sigma^\ell \to \{0,1\}$, one can build a good pseudo-random generator g that can be calculated in time $2^{O(\ell)}$ if we are provided oracle access to f.*

FORMAL STATEMENT: *The parameters ℓ, S are positive integers, and ϵ is a positive real number.*

Let $f: \Sigma^\ell \to \{0,1\}$ be a predicate that is (ϵ, S) hard. Then for some constant d and for every constant $c < \frac{1}{12}$ the following holds. Let $m = 2^{c\ell}$ and $c' = (1/8)(1/c)$. There is a pseudo-random generator $g: \Sigma^{c'\ell} \to \Sigma^m$ having the property that

$$\Delta_{\text{comp},S_1}(U_m, g(U_{c'\ell})) \leq m \cdot \epsilon,$$

where $S_1 = S - d \cdot m^{19}$. Moreover, g can be calculated effectively from f in time $2^{O(\ell)}$ (i.e., there is a procedure that has oracle access to f, runs in time $2^{O(\ell)}$, and calculates f).

Proof. By Lemma 5.8.3, there is a (weight ℓ, intersection $18 \log m$)-design A of type $(m, c'\ell)$. We define $g\colon \Sigma^{c'\ell} \to \Sigma^m$ by $g(x) = g_{f,A}(x)$ for all $x \in \Sigma^{c'\ell}$, where $g_{f,A}$ is defined as in relation (5.17) from the hard function f and the design A.

To obtain a contradiction, suppose that there is some circuit D of size $S_1 = S - d \cdot m^{19}$, for a constant d that will be specified in the next paragraph, such that

$$|\text{Prob}_{r \in \Sigma^m}(D(r) = 1) - \text{Prob}_{z \in \Sigma^t}(D(g_{f,A}(z)) = 1)| > m \cdot \epsilon.$$

Lemma 5.8.4 states that there is a circuit C of size $d \cdot m \cdot 2^{18 \log m}$, for some constant d (which is the constant we referred to above), such that

$$\left| \text{Prob}_{x \in \Sigma^\ell}(D(C(x)) = f(x)) - \frac{1}{2} \right| > \epsilon.$$

Note that the function $D(C(x))$ can be calculated by a circuit that first simulates the circuit C on input x and then passes the output of this calculation to a simulation of the circuit D. It follows that the function $D(C(x))$ can be calculated by a circuit C' of size $S_1 + d \cdot m^{19} = S$. Thus,

$$\left| \text{Prob}_{x \in \Sigma^\ell}(C'(x)) = f(x)) - \frac{1}{2} \right| > \epsilon.$$

We can assume that $\text{Prob}(C'(x) = f(x)) \geq \frac{1}{2}$ (because, otherwise, we can take the circuit C'' that flips the output of C' and the relation holds for C''). It follows that $\text{Prob}(C'(x) = f(x)) > \frac{1}{2} + \epsilon$ and $\text{size}(C') = S$. This contradicts the (ϵ, S) hardness of f.

Note that computing $g(x)$, where $x \in \Sigma^{9\ell}$, implies the construction of the design A, which by Lemma 5.8.3 takes $2^{O(\ell)}$ time, and $m = 2^{O(\ell)}$ invocations of the function f. Thus, provided we have access to an oracle for the function f, $g(x)$ can be calculated in time $2^{O(\ell)}$. ∎

As a corollary we obtain the fact that an exponentially-hard predicate yields via the above construction an exponentially strong pseudo-random generator having exponential extension.

Corollary 5.8.7 INFORMAL STATEMENT: *An exponentially-hard predicate yields an exponentially-strong pseudo-random generator with exponential extension.*

FORMAL STATEMENT: *Let $f = (f_\ell)_{\ell \in \mathbb{N}}$ be an exponentially-hard predicate. Then there exists an exponentially-strong pseudo-random generator $(g_\ell)_{\ell \in \mathbb{N}}$, where for each ℓ, $g_\ell\colon \Sigma^\ell \to \Sigma^{L(\ell)}$ and $L(\ell)$ is an exponential function (i.e., $L(\ell) = 2^{c\ell}$ for some constant c). Moreover, g can be calculated effectively from f in time $2^{O(\ell)}$.*

Proof. Since $f = (f_\ell)_{\ell \in \mathbb{N}}$ is exponentially-hard, there exists a constant c' such that, for all sufficiently large ℓ, f_ℓ is a $(\frac{1}{2^{c'\ell}}, 2^{c'\ell})$-hard predicate. We fix ℓ and take $m = 2^{(c'/38)\ell}$. For any ℓ, let $h(\ell)$ denote $(1/8)(1/c')\ell$. By Theorem 5.8.6, there is a pseudo-random generator $g'_{h(\ell)}\colon \Sigma^{h(\ell)} \to \Sigma^m$ such that

$$\Delta_{\text{comp}, S_1}(U_m, g'_{h(\ell)}(U_{h(\ell)})) < m \cdot \epsilon = 2^{-(37/38)c'\ell},$$

where $S_1 = 2^{c'\ell} - d \cdot 2^{(c'/2)\ell}$, for some constant d. Clearly, there is some c'' such that $S_1 < 2^{c''\ell}$ and $2^{(37/38)c'\ell} < 2^{c''\ell}$ (if ℓ is sufficiently large). Thus the family of functions $(g'_{(1/8)(1/c')\ell})_{\ell \in \mathbb{N}}$ is almost what we need, the only missing property being that g' is not defined for all input lengths ℓ. This is a technicality that is easy to remedy.

For any ℓ, take $\ell' = \lfloor c' \cdot 8 \cdot \ell \rfloor$ (i.e., ℓ' is the largest integer such that $(1/8)(1/c')\ell' \leq \ell$). We take $L(\ell) = 2^{(c'/38)\ell'}$. Observe that $L(\ell) = 2^{O(\ell)}$. We finally define $g_\ell \colon \Sigma^\ell \to \Sigma^{L(\ell)}$ by

$$g_\ell(x) = g'_{h(\ell')}(x(1 : h(\ell'))),$$

i.e., $g_\ell(x)$ is obtained by applying the appropriate member of the family g' to the largest prefix of x having the type of length that is permitted for the functions in the family g'. It is easy to see that for all $a \in \Sigma^{L(\ell)}$,

$$\text{Prob}(g_\ell(U_\ell) = a) = \text{Prob}(g'_{h(\ell')}(U_{h(\ell')}) = a).$$

It follows that for any circuit C,

$$\text{Prob}(C(g_\ell(U_\ell)) = 1) = \text{Prob}(C(g'_{h(\ell')}(U_{h(\ell')}) = 1)),$$

and thus

$$\Delta_{\text{comp},S_1}(U_{L(\ell)}, g_\ell(U_\ell)) = \Delta_{\text{comp},S_1}(U_{L(\ell)}, g'_{h(\ell')}(U_{h(\ell')}))) < 2^{c''\ell'}.$$

Since $\ell' = O(\ell)$, there is a constant c such that $S_1 = 2^{c\ell}$ and $2^{c''\ell'} < 2^{c\ell}$. It follows that indeed $(g_\ell)_{\ell \in \mathbb{N}}$ is an exponentially-strong pseudo-random generator, where $g_\ell \colon \Sigma^\ell \to \Sigma^{L(\ell)}$ and $L(\ell) = 2^{O(\ell)}$. ∎

Taking into account the hardness amplification results from Section 5.6 and Section 5.7, the assumptions in Corollary 5.8.7 can be considerably relaxed.

Corollary 5.8.8 INFORMAL STATEMENT: *A function that is worst-case hard with respect to circuits of exponential size, yields an exponentially strong pseudo-random generator with exponential expansion.*

FORMAL STATEMENT: *Let $f \colon \Sigma^* \to \Sigma^*$ be a length-preserving function such that for some positive constant c and for all sufficiently large ℓ, f_ℓ is $(\frac{1}{2^\ell}, 2^{c\ell})$-hard. Then there exists an exponentially-strong pseudo-random generator $(g_\ell)_{\ell \in \mathbb{N}}$, where for each ℓ, $g_\ell \colon \Sigma^\ell \to \Sigma^{L(\ell)}$ and $L(\ell)$ is an exponential function (i.e., $L(\ell) = 2^{c\ell}$ for some constant c). Moreover, g can be calculated effectively from f in time $2^{O(\ell)}$.*

Proof. We basically need to show that from a function f as in the above statement we can build a function of the type required in Corollary 5.8.7. This follows immediately by combining Theorem 5.6.14 (b) and Theorem 5.7.4(b). ∎

5.9 BPP = P?

IN BRIEF: If there is a function $f\colon \Sigma^* \to \Sigma^*$ computable in time $2^{O(n)}$ and a constant c such that, for almost all n, no circuit of size 2^{cn} calculates f_n, then BPP = P.

Let us first observe that, by Theorem 5.6.1, a function as the one required in Corollary 5.8.8 exists. A related, stronger, still-quite-reasonable assumption yields the very interesting fact that any problem in BPP can actually be solved *deterministically* in polynomial time. The assumption is that there exists a hard function as in the hypothesis of Corollary 5.8.8 having the additional property that is computable in exponential time (that is in $2^{O(n)}$ time). More precisely, the assumption is that there exists a length-regular function $f\colon \Sigma^* \to \Sigma^*$ computable in exponential time such that for some positive constant c and for all sufficiently large ℓ, f_ℓ is $(\frac{1}{2^\ell}, 2^{c\ell})$-hard. This assumption looks very plausible since by the classical time-hierarchy theorem (see [GHS91]), there are functions $f\colon \Sigma^* \to \Sigma^*$ such that $f(x)$ is computable in time $2^{2|x|}$ and not computable in time $2^{|x|}$ for almost all x. "Computable" in the time-hierarchy theorem is taken in the uniform sense, i.e., the computation is performed by machines that work on inputs of all lengths. However, it seems very believable that a function f such as the one above cannot either be calculated by circuits of size $2^{c|x|}$, for some positive constant c.

By looking at Corollary 5.8.8, it is not hard to see how to simulate deterministically in polynomial time any BPP computation, provided that the above assumption holds. Indeed, by Corollary 5.8.8, we would get an exponentially-strong pseudo-random generator $(g_\ell)_{\ell \in \mathrm{N}}$ such that g_ℓ can be calculated in time $2^{O(\ell)}$. Note that any BPP algorithm M on input x can be simulated deterministically by running it on all possible choices of strings that can be used as random strings by M on x, and by taking at the end the majority vote (that is, if more than half of these simulations accept x, then our verdict will be "accept," and if more than half of the simulations reject x, then the verdict will be "reject"). This simulation takes exponential time because, in general, there are exponentially many choices as substitutes for the random string.

Using a pseudo-random generator g_ℓ with exponential extension, we can avoid this, by considering only all strings generated by an appropriate g_ℓ as possible substitutes for the random string needed by M on input x. Because of the exponential extension, $\ell = O(\log |x|)$, and thus there are only polynomially many (in $|x|$) simulations to do. Also, a substitute string $g_\ell(y)$ is calculated in $2^{O(\ell)}$ time which is polynomial in $|x|$. The fraction of acceptances (or rejections) will be very close to the corresponding fraction in the simulation from the previous paragraph because the output of the pseudo-random generator looks random to computations that are performed in time that is a fixed polynomial in $|x|$, and, thus, the majority vote still works.

We present next the precise result and the proof which formalizes the above intuitive argument.

Theorem 5.9.1 *Suppose that there exists a function $f\colon \Sigma^* \to \Sigma^*$ that (a) is computable in time $2^{O(n)}$, (b) is length-regular, and (c) for some positive constant c and for all sufficiently large ℓ, f_ℓ is $(\frac{1}{2^\ell}, 2^{c\ell})$-hard. Then* P = BPP.

Proof. Applying Corollary 5.8.8 to the function f, we obtain an exponentially strong pseudo-random generator g having exponential expansion.

Let A be in a set in BPP. This means that there is a polynomial-time probabilistic machine M that on input x uses $p(|x|)$ random bits, for some polynomial p such that, for all x,

$$\text{Prob}_{r\in\Sigma^{p(|x|)}}(M(x,r) = A(x)) \geq \frac{3}{4}.$$

Let us fix $x \in \Sigma^*$ sufficiently long for the argument below to work, and let us assume that $x \in A$, i.e., $A(x) = 1$ (the case $A(x) = 0$ is similar). Let

$$\text{GOOD} = \{r \in \Sigma^{p(|x|)} \mid M(x,r) = 1\}.$$

The machine M runs in time that is polynomial in $|x|$. Hence, for some polynomial q, there is a circuit C of size $q(|x|)$ that accepts GOOD, i.e., $C(x) = 1 \Leftrightarrow r \in$ GOOD. Our goal is to replace the pure random strings r used by M, by strings produced by the pseudo-random generator g. There are positive constants c_1 and c_2 such that for ℓ sufficiently large,

$$g_\ell\colon \Sigma^\ell \to \Sigma^{L(\ell)},$$

where $L(\ell) = 2^{c_1\ell}$ and for all circuits D of size at most $2^{c_2\ell}$,

$$\left|\text{Prob}_{z\in\Sigma^{L(\ell)}}(D(z) = 1) - \text{Prob}_{y\in\Sigma^\ell}(D(g_\ell(y)) = 1)\right| < 2^{-c_2\ell}.$$

We need a value of ℓ such that $2^{c_1\ell} \geq p(|x|)$ (so that the generator produces at least as many bits as M needs on input x) and $2^{c_2\ell} \geq q(|x|)$ (so that the generator produces outputs that look random to the test indiced by the set GOOD). Clearly, there is a positive constant c such that $\ell = c\log|x|$ satisfies both conditions.

Without loss of generality, we can assume that g_ℓ produces exactly $p(|x|)$ bits, i.e., $L(\ell) = p(|x|)$ (if $L(\ell)$ is larger, the unnecessary bits are discarded). Since the set GOOD can be calculated by a circuit C of size $q(|x|) \leq 2^{c_2\ell}$, it follows that

$$\left|\text{Prob}_{z\in\Sigma^{L(\ell)}}(z \in \text{GOOD}) - \text{Prob}_{y\in\Sigma^\ell}(g_\ell(y) \in \text{GOOD})\right| < \frac{1}{2^{c_2\ell}}.$$

Therefore, taking into account that $\text{Prob}_z(z \in \text{GOOD}) \geq \frac{3}{4}$,

$$\text{Prob}_{y\in\Sigma^\ell}(g_\ell(y) \in \text{GOOD}) \geq \frac{3}{4} - \frac{1}{2^{c_2\ell}} > \frac{2}{3},$$

for ℓ sufficiently large. In other words, we have shown that if $x \in A$, then for a fraction $> \frac{2}{3}$ of strings $y \in \Sigma^\ell$, $M(x, g_\ell(y)) = 1$. Similarly, we can show that if $x \notin A$, then for a fraction $> \frac{2}{3}$ of strings $y \in \Sigma^\ell$, $M(x, g_\ell(y)) = 0$. Therefore, we can simulate $M(x, g_\ell(y))$ for all strings in Σ^ℓ and see if we obtain more 1s than 0s or vice versa. This tells whether $x \in A$ or $x \notin A$. The simulation takes time $2^{O(\ell)} = \text{poly}(|x|)$ because there are 2^ℓ strings y in Σ^ℓ, $g_\ell(y)$ can be calculated in time $2^{O(\ell)}$ and the simulation of $M(x, g_\ell(y))$ takes time polynomial in $|x|$. Consequently, in deterministic polynomial time, we can decide whether $x \in A$ or $x \notin A$, which means that $A \in \text{P}$. ∎

5.10 Extractors

IN BRIEF: An extractor is function that corrects a source of randomness that emits strings containing imperfect randomness. We present an explicit construction of a polynomial-time extractor with good parameters utilizing techniques first seen in the construction of pseudo random generators from hard functions.

Extractors are functions that produce good-quality randomness from low-quality randomness. They can be used to remedy a source of randomness whose outputs are not perfectly random. In some aspects, extractors are similar to pseudo-random generators. Like pseudo-random generators, an extractor uses a short random seed to produce a long output string that is random in some sense. There is however a drastic difference: In the case of pseudo-random generators, the output is random in a computational sense, i.e., the computational distance between the output's distribution (when the seed is chosen uniformly at random) and the uniform distribution is less than a small parameter ϵ. In the case of extractors, the output is random in the absolute sense, i.e., the statistical distance between the output's distribution (implied by the uniformly distributed seed and by the source's distribution) and the uniform distribution is less than a small parameter ϵ. An extractor's objective is therefore more ambitious. In exchange, they have at their disposal more than what is given to a pseudo-random generator. In addition to the seed, they are given an extra input, which already contains "some" randomness. In fact, in most applications of extractors, it is the extra input who is the main one. This input can be viewed as being produced by a source of randomness that generates strings having partial randomness. The seed is used as a short catalyst that helps to extract the randomness existing in the string produced by the source. An immediate task is to introduce tools by which we can gauge the quality of randomness. The classical way to measure the amount of randomness in a distribution is to calculate its Shannon entropy. For a random variable X taking values in the set $\{0,1\}^n$, the Shannon entropy is defined by

$$H(X) = \sum_a \text{Prob}(X = a) \log(1/\text{Prob}(X = a)),$$

Table 5.2: Extractors vs. pseudo-random generators

	Input		Output
pseudo-random $g(\cdot)$ generator		random seed y	$g(y)$, compututationally close to U_m
extractor $E(\cdot,\cdot)$	string x with "some" randomness	random seed y	$E(x, y)$, statistically close to U_m

where the sum is taken over $a \in \{0,1\}^n$ with $\text{Prob}(X = a) \neq 0$. This definition is based on the intuition that the amount of randomness in the outcome a of a random variable X is $\log(1/\text{Prob}(X = a))$. Thus, $H(X)$ is just the average amount of randomness. For example, for U_3, the uniform distribution on $\{0,1\}^3$, $H(U_3) = 3$, while for a distribution X on $\{0,1\}^3$ defined by $\text{Prob}(X = 100) = \text{Prob}(X = 101) = \text{Prob}(X = 110) = \text{Prob}(X = 111) = 1/4$ (and the probability that X is any of the other strings is 0), we have $H(X) = 2$. This corresponds to our intuition that all the three bits of the outcome of U_3 are random, and only two of the bits in X's outcome are random. The Shannon entropy does not always capture well our intuition of randomness. Consider, for example, a random variable on $\{0,1\}^n$ that "puts" $1 - 1/\sqrt{n}$ of its probability mass on the string $0\ldots0$ (i.e., $\text{Prob}(X = 0^n) = 1 - 1/\sqrt{n}$), and distributes the remaining probability mass equally among all the other strings. The Shannon entropy $H(X)$ is close to $\sqrt{n}$, which is quite large, however, X will almost always be 0^n, and thus seems to be far from random. Sometimes (and this will be the case in this section), we want to say that X has "good" randomness if we are guaranteed that no possible outcome has too much probability mass, which implies that the probability mass is alloted in a balanced way. This leads to the definition of the *min-entropy* of a random variable X.

Definition 5.10.1 (Min-entropy) *Let $n \in \mathbb{N}$ be a parameter. The min-entropy of a random variable taking values in $\{0,1\}^n$ is given by*

$$\min\left\{ \log \frac{1}{\text{Prob}(X = a)} \;\middle|\; a \in \{0,1\}^n, \text{Prob}(X = a) \neq 0 \right\}$$

Thus, if X has min-entropy $\geq k$, then for all a in the range of X, $\text{Prob}(X = a) \leq 1/2^k$. We are now ready to define an extractor formally.

Definition 5.10.2 *The values n, k, d, m are integer parameters, and $\epsilon > 0$ is a real number parameter. A function $E\colon \{0,1\}^n \times \{0,1\}^d \to \{0,1\}^m$ is a (k,ϵ)-extractor if for every distribution X on $\{0,1\}^n$ with min-entropy at least k, the distribution $E(X, U_d)$ is ϵ-close to the uniform distribution U_m in the statistical sense, i.e.,*

$$\Delta_{\text{stat}}(E(X, U_d), U_m) \leq \epsilon.$$

Thus, an extractor has as input (a) a string x produced by an imperfect distribution X, where the defect of the distribution is measured by $k = \text{min-entropy}(X)$, and (b) a random seed y of length d. The output is $E(x, y)$, a string of length m. The key property is that, for every subset $W \subseteq \Sigma^m$,

$$|\text{Prob}_{x \in X \{0,1\}^n, y \in \{0,1\}^d}(E(x,y) \in W) - \text{Prob}_{z \in \Sigma^m}(z \in W)| \leq \epsilon. \tag{5.19}$$

Typically, we view the input x as being generated by a source of imperfect randomness. We need to add d bits of perfect randomness (the seed) and the extractor will produce from the mix m bits having almost perfect randomness.

An extractor $E\colon \{0,1\}^n \times \{0,1\}^d \to \{0,1\}^m$ can also be viewed as a regular bipartite graph where the set of "left" nodes is $V_{\text{left}} = \{0,1\}^n$ and the set of "right" nodes is $V_{\text{right}} = \{0,1\}^m$. The degree of each node in V_{left} is 2^d, and two nodes $x \in V_{\text{left}}$ and $z \in V_{\text{right}}$ are connected if there is $y \in \{0,1\}^d$ such that $E(x,y) = z$. We can imagine that each $x \in V_{\text{left}} = \{0,1\}^n$ is throwing 2^d arrows at $V_{\text{right}} = \{0,1\}^m$.

An extractor is characterized by five parameters: n, the input length, k, the min-entropy of the source, d, the seed length, m, the output length, and ϵ, the output's statistical closeness to the uniform distribution. For simplicity, we have defined individual extractors. However, implicitely we think of a family of extractors indexed by n and with the other parameters being uniform functions of n. In this way we can talk about efficient constructions of extractors.

How should the parameters be so that we can say that we have a good extractor? If we consider n and k as given (these are the parameters of the source), it is desirable that d is small, m is large, and ϵ is small. Furthermore, we want that the family of extractors is computable in polynomial time.

The following lower bounds have been proved: If $m \geq d + 1$ (i.e., the extractor outputs more bits than are input through the seed), then

(a) $d \geq \log(n-k) + 2\log(1/\epsilon) - O(1)$ (provided $\epsilon \leq 1/2$), and
(b) $m \leq d + k - 2\log(1/\epsilon) + O(1)$.

The lower bound (b) says that the extractor $E(x,y)$ cannot output more random bits (i.e., m bits) than the randomness existing in the input which consists of the randomness in x (i.e., the min-entropy k) and the random bits of y (i.e., d bits). There is an inherent loss of $2\log(1/\epsilon) - O(1)$ bits. The lower bound (a) says that if, for example, k is at most a constant fraction of n and ϵ is constant, then the seed length has to be at least $\Omega(\log n)$.

In what follows, we will construct an extractor that (a) is computable in polynomial time, (b) works for sources (i.e., distributions) on $\{0,1\}^n$ having min-entropy an arbitrarily small constant fraction of n, (c) uses a seed of length $O(\log n)$, and (d) outputs $m = n^\beta$ bits for some constant β (we will get $\beta = 18/(19 \cdot 20)$, however, with a more careful analysis, the constant β can be made arbitrarily close to 1). Ideally, the output length should be equal to the min-entropy of the source (i.e., $m = k$ or, even better, $m \approx k + d$) in which case the extractor extracts the entire randomness existing in the source. The extractor that we build falls short in this respect. At some point it is using the design given in Lemma 5.8.3. The same construction using a more elaborate type of design can achieve $m = \Omega(k)$ (however that extractor has $d = O(\log^2 n)$). Section 5.11 contains references for other types of extractors.

To understand better Equation (5.19), let us look deeper into the structure of an extractor. We fix parameters n, d, m and ϵ and a function $E\colon \{0,1\}^n \times \{0,1\}^d \to \{0,1\}^m$. Let us consider an arbitrary set $W \subseteq \{0,1\}^m$ and a string $x \in \{0,1\}^n$. We say that x hits W ϵ-correctly if the fraction of outgoing edges from x that land in W is ϵ-close to the fraction $\|W\|/\|\{0,1\}^m\|$, i.e.,

$$\left|\frac{\|\{E(x,y) \mid y \in \{0,1\}^d\} \cap W\|}{\|\{0,1\}^d\|} - \frac{\|W\|}{\|\{0,1\}^m\|}\right| \leq \epsilon.$$

If we look at a fixed x, it cannot hold that for every $W \subseteq \{0,1\}^m$, x hits W ϵ-correctly (for example, take $W = \{E(x,y) \mid y \in \{0,1\}^d\}$). Fortunately, for E to be an extractor, all we need is that any $W \subseteq \{0,1\}^m$ is hit ϵ-correctly by most $x \in \{0,1\}^n$.

Lemma 5.10.3 *Let $E\colon \{0,1\}^n \times \{0,1\}^d \to \{0,1\}^m$ and $\epsilon > 0$. Suppose that for every $W \subseteq \{0,1\}^m$, the number of $x \in \{0,1\}^n$ that do not hit W ϵ-correctly is at most 2^t, for some t. Then E is a $(t + \log(1/\epsilon), 2\epsilon)$-extractor.*

Proof. Let X be a distribution on $\{0,1\}^n$ with min-entropy at least $t + \log(1/\epsilon)$ and let W be a subset of $\{0,1\}^m$. There are at most 2^t x's that do not hit W ϵ-correctly and the distribution X allocates to these x's a mass probability of at most $2^t \cdot 2^{-(t+\log(1/\epsilon))} = \epsilon$. We have,

$$\begin{aligned}&\text{Prob}_{x\in_X\{0,1\}^n, y\in\{0,1\}^d}(E(x,y) \in W)\\ &\quad = \text{Prob}_{x\in_X\{0,1\}^n, y\in\{0,1\}^d}(E(x,y) \in W \text{ and } x \text{ hits } W\ \epsilon\text{-correctly})\\ &\qquad + \text{Prob}_{x\in_X\{0,1\}^n, y\in\{0,1\}^d}(E(x,y) \in W \text{ and } x \text{ does not hit } W\ \epsilon\text{-correctly}).\end{aligned}$$

The first term in the right hand side is between $\frac{\|W\|}{\|\{0,1\}^m\|} - \epsilon$ and $\frac{\|W\|}{\|\{0,1\}^m\|} + \epsilon$, because for each x that hits W ϵ-correctly,

$$\text{Prob}_{y\in\{0,1\}^d}(E(x,y) \in W) \in \left[\frac{\|W\|}{\|\{0,1\}^m\|} - \epsilon, \frac{\|W\|}{\|\{0,1\}^m\|} + \epsilon\right].$$

The second term is bounded by

$$\text{Prob}_{x\in_X\{0,1\}^n, y\in\{0,1\}^d}(x \text{ does not hit } W\ \epsilon\text{-correctly}),$$

which is, as we have seen, between 0 and ϵ. Plugging these estimates in the above equation, we obtain that

$$\left|\text{Prob}_{x\in_X\{0,1\}^n, y\in z0^d}(E(x,y) \in W) - \frac{\|W\|}{\|\{0,1\}^m\|}\right| \leq 2\epsilon.$$

Thus, E is a $(t + \log\frac{1}{\epsilon}, 2\epsilon)$-extractor. ∎

Consequently, to show that $E: \{0,1\}^n \times \{0,1\}^d \to \{0,1\}^m$ is an extractor, we need to bound, for an arbitrary $W \subseteq \{0,1\}^m$, the number of x's in $\{0,1\}^n$ that do not hit W ϵ-correctly. Our strategy to produce such bounds will be as follows. We will show that if some x does not hit W ϵ-correctly, then, with the help of a few additional bits of information, we can reconstruct x. More formally, this means that for "small" t there is an injective function $\text{Rec}: \{0,1\}^t \to \{0,1\}^n$ such that the set of x's that do not hit W ϵ-correctly is included in the range of Rec. This implies that the cardinality of the set of such "bad" x's is bounded by 2^t.

With this set-up in mind, we now have to do the real work and build the extractor function $E: \{0,1\}^n \times \{0,1\}^d \to \{0,1\}^m$ and the reconstruction function Rec. It is handy to recall the similarity between extractors and pseudo-random generators and, in particular, to look in this light at the construction of the pseudo-random generator given in Lemma 5.8.4. This lemma shows that if we start with a function $f: \{0,1\}^\ell \to \{0,1\}$ and with a design A of type (m,d) with weight ℓ and intersection α, we can construct a function $g_{f,A}: \{0,1\}^d \to \{0,1\}^m$, such that if there is a function $D: \{0,1\}^m \to \{0,1\}$ with

$$|\text{Prob}_{z\in\{0,1\}^d}(D(g_{f,A}(z)) = 1) - \text{Prob}_{r\in\{0,1\}^m}(D(r) = 1)| > \epsilon, \tag{5.20}$$

then there is a circuit C of size $O(m2^\alpha)$ so that $D(C(\cdot))$ agrees with $f(\cdot)$ on at least a fraction of $\frac{1}{2} + \frac{\epsilon}{m}$ of the 2^ℓ positions. In Section 5.8, we have used Lemma 5.8.4 to argue that if the circuit D is computable by a circuit of size S, then there is a circuit of size $S + O(m \cdot 2^\alpha)$ that agrees with f on a fraction of $\frac{1}{2} + \frac{\epsilon}{m}$ of the inputs. Taking the contrapositive, we derived that if f is $(\frac{1}{2} - \frac{\epsilon}{m}, S + O(m \cdot 2^\alpha))$-hard, then no circuit D of size S can satisfy the relation (5.20), and therefore $g_{f,A}$ is a pseudo-random generator. If we discard the property that D is computable by some circuit of bounded size, we can use the same argument to show that, to some extent, we can carry over the reconstruction of f as needed by our plan. Let us be more specific. The design A can be constructed algorithmically as in Lemma 5.8.3. We can consider the function f as being the first input of an extractor, and the strings $(g_{f,A}(z))_{z\in\{0,1\}^d}$, the points where f "hits" $\{0,1\}^m$. The relation (5.20) shows that if there is a set $D \subseteq \{0,1\}^m$ that is not hit by f ϵ-correctly, then, using the additional information given in the relatively small circuit C, we can reconstruct f with some approximation (namely, of $\frac{1}{2} + \frac{\epsilon}{m}$). This resembles our goal, but is not good enough because we need to reconstruct f perfectly (without any approximation).

Therefore, we revise the plan a bit. Using an error-correcting code, we would like to encode the function f into a codeword $\hat{f}$ such that if we are able to produce a function that agrees with $\hat{f}$ on a fraction of $\frac{1}{2} + \frac{\epsilon}{m}$ positions, then we can reconstruct $\hat{f}$ perfectly. Because the agreement parameter $\frac{1}{2} + \frac{\epsilon}{m}$ is so low, we are not able to achieve this. Instead, by list-decoding, we will produce a relatively short list of functions that includes $\hat{f}$. This is good enough, because if we are given as additional information the rank of $\hat{f}$ in the list of functions (ordered, say, lexicographically), then we can reconstruct $\hat{f}$ perfectly.

We proceed to implement this plan. We first fix an error-correcting code. This will be given by the following theorem, whose proof we defer for the moment. For $y \in \{0,1\}^n$ and $\epsilon > 0$, the *ball* $B(y, \epsilon)$ is the set of binary strings $z \in \{0,1\}^n$ with $\mathrm{dist}(y, z) \leq \epsilon$.

Theorem 5.10.4 INFORMAL STATEMENT: *There exists an error-correcting code* $\mathrm{ECC}\colon \{0,1\}^n \rightarrow \{0,1\}^{\mathrm{poly}(n)}$*, computable in polynomial time, such that every ball* $B(y, (1/2 - \delta)|y|)$ *has at most* $O(1/\delta^3)$ *codewords.*

FORMAL STATEMENT: *For every* $n \in \mathbb{N}$ *and* $\delta > 0$*, there exist* $\overline{n}$ *and a function* $\mathrm{ECC}\colon \{0,1\}^n \rightarrow \{0,1\}^{\overline{n}}$*, such that, for all* $y \in \{0,1\}^{\overline{n}}$*, there are at most* $2\sqrt{2}(1/\delta)^3$ *strings* $x \in \{0,1\}^n$ *with* $\mathrm{dist}(y, \mathrm{ECC}(x)) \leq (1/2 - \delta)\overline{n}$*. The output length* $\overline{n}$ *of* ECC *is bounded by* $16 \cdot n^2 \cdot (1/\delta)^8$*. Furthermore, the function* ECC *is computable in polynomial time.*

We are at this point ready to construct the extractor. We fix some parameter ℓ and start with a function $f\colon \{0,1\}^\ell \rightarrow \{0,1\}$. The function f is represented by its truth-table, a 2^ℓ-bit string, which, abusing notation, is called f as well. This will be the first input of the extractor. We use the function ECC given in Theorem 5.10.4 with the choice of parameters $n = 2^\ell$ and $\delta = 2^{-2\ell}$, and we take $\hat{f} \stackrel{\mathrm{def}}{=} \mathrm{ECC}(f)$. The length of $\hat{f}$ is bounded by $16 \cdot n^2 \cdot (1/\delta)^8 = 16 \cdot 2^{2\ell} \cdot 2^{16\ell} = 2^{18\ell+4}$. Note that $\hat{f}$ (if necessary, after some padding) can be considered the truth-table of a predicate function mapping $\{0,1\}^{18\ell+4}$ to $\{0,1\}$. This predicate will be denoted $\hat{f}$ as well. Using Lemma 5.8.3, we construct in time polynomial in 2^ℓ a design A that, for an arbitrary constant $c < 1/12$ (which will be fixed later), has the following parameters: The number of rows is $m = 2^{c\cdot(18\ell+4)}$, the number of columns is $d = (1/8)\cdot(1/c)\cdot(18\ell+4)$, the weight of each row is $18\ell + 4$, and the intersection parameter is $\alpha = 18 \cdot c \cdot (18\ell+4)$. We consider the function $g_{\hat{f},A}\colon \{0,1\}^d \rightarrow \{0,1\}^m$ defined as in relation (5.17), i.e.,

$$g_{\hat{f},A}(y) = \hat{f}(y|_{S_1}) \odot \hat{f}(y|_{S_2}) \odot \ldots \odot \hat{f}(y|_{S_m}),$$

where $S_1, \ldots, S_m$ are the rows of the design A and $y|_{S_i}$ is the string obtained by projecting y into the positions where S_i is 1 (for details see Section 5.8).

Finally, we define the extractor function $E\colon \{0,1\}^{2^\ell} \times \{0,1\}^d \rightarrow \{0,1\}^m$, by

$$E(f, y) = g_{\mathrm{ECC}(f),A}(y) \quad (= g_{\hat{f},A}(y)).$$

We denote $n \stackrel{\mathrm{def}}{=} 2^\ell$ and take $c = \frac{1}{19\cdot 20}$. Then, $E\colon \{0,1\}^n \times \{0,1\}^{\gamma\cdot\log n} \rightarrow \{0,1\}^{n^\beta}$, where $\gamma = (1/160)$ and $\beta = 18/(19\cdot 20)$.

Theorem 5.10.5 INFORMAL STATEMENT: *There exists an extractor that takes as input a distribution on* $\{0,1\}^n$ *with min-entropy* γn *(for arbitrarily small constant* γ*) and a seed of length* $O(\log n)$ *and outputs* $n^{\Omega(1)}$ *bits that are statistically close to the uniform distribution.*

FORMAL STATEMENT: *Let $\gamma > 0$ and $\epsilon > 0$. The function E defined above is a $(\gamma n, 2\epsilon)$-extractor, provided n is sufficiently large and $1/\epsilon < n$. Furthermore, the function E is computable in time polynomial in n.*

Proof. Let $D \subseteq \{0,1\}^m$ and let f be a 2^ℓ-bit string (viewed as the truth-table of a function $f\colon \{0,1\}^\ell \to \{0,1\}$). Suppose that f does not hit D ϵ-correctly, i.e.,

$$|\text{Prob}_{y\in\{0,1\}^t}(\hat{f}(y) \in D) - \text{Prob}_{z\in\{0,1\}^m}(z \in D)| > \epsilon.$$

By Lemma 5.8.4, there is a circuit C of size $c_2 \cdot m \cdot 2^\alpha$ for some constant c_2 such that either

$$\text{Prob}(D(C(y)) = \hat{f}(y)) > \frac{1}{2} + \frac{\epsilon}{m}, \tag{5.21}$$

or

$$\text{Prob}(\overline{D}(C(y)) = \hat{f}(y)) > \frac{1}{2} + \frac{\epsilon}{m},$$

where $\overline{D}$ is the complement of D. We view the truth-tables of the functions $D(C(\cdot))$ (or $\overline{D}(C(\cdot))$) and $\hat{f}(\cdot)$ as $2^{18\ell+4}$ long binary strings. Relation (5.21) implies that the Hamming distance between the two strings is at most $\frac{1}{2} - \frac{\epsilon}{m}$. Note that $m < 2^\ell$ and, by hypothesis, $1/\epsilon \le n = 2^\ell$. Thus, $\epsilon/m > 2^{-2\ell}$ (recall that $\delta = 2^{-2\ell}$ is the parameter used to define ECC). Therefore, by Theorem 5.10.4, the set B of codewords of the error-correcting code ECC that are in the ball of radius $\frac{1}{2} - \frac{\epsilon}{m}$ centered in the truth-table of $D(C(\cdot))$ (or $\overline{D}(C(\cdot))$) has cardinality at most $2\sqrt{2} \cdot (2^{2\ell})^3 = 2\sqrt{2} \cdot (2^{6\ell})$. The truth-table of $\hat{f}$ is one of these codewords. Note that this set of codewords can be obtained if we are given D and C (for example, by trying all codewords and retaining those that have $(\frac{1}{2} + \frac{\epsilon}{m})$-agreement with the truth-table of $D(C(\cdot))$). It follows that we can reconstruct $\hat{f}$ if we are given D, C, one extra bit which indicates whether relation (5.21) holds for D or $\overline{D}$, and the rank of $\hat{f}$ in the set B. The number of circuits having the same size as C is $2^{O(m\cdot 2^\alpha \cdot \log(m\cdot 2^\alpha))}$, and therefore, a description of C takes in binary $O(m \cdot 2^\alpha(\log m + \alpha)) = c_3 \cdot m \cdot \log m \cdot 2^\alpha$, for some constant c_3. The rank of $\hat{f}$ in the set is at most $2\sqrt{2}(2^{6\ell}) = O(n^6)$.

The reconstruction function takes as input the index of a circuit C in the list of circuits of size $c_2 \cdot m \cdot 2^\alpha$ and an index in the list of codewords that are within distance $\frac{1}{2} - \frac{\epsilon}{m}$ from the truth-table of $D(C(\cdot))$, and produces the function $\hat{f}$. It follows that the number of functions f such that $\hat{f}$ (which is ECC(f)) does not hit D ϵ-correctly is bounded by $O(1) \cdot (2^{c_3\cdot m\cdot \log m\cdot 2^\alpha}) \cdot n^6$. By Lemma 5.10.3, the function $E\colon \{0,1\}^n \times \{0,1\}^{\gamma\cdot\log n} \to \{0,1\}^{n^\beta}$ is a $(O(1) + c_3(m \log m \cdot 2^\alpha + 6\log n) + \log(1/\epsilon), 2\epsilon)$-extractor. Recall that $m = 2^{c\cdot(18\ell+4)} < 2^{1/(19\cdot 20)\cdot 19\ell} = n^{(1/20)}$ (for $\ell \ge 4$) and $\alpha = 18 \cdot 1/(19/20) \cdot (18\ell + 4) \le (18/20)\ell = (18/20)\log n$ (again for $\ell \ge 4$). Taking into account that $1/\epsilon < n$, it follows that if n is sufficiently large, the min-entropy parameter is less than γn. Consequently, for n sufficiently large, $E\colon \{0,1\}^n \times \{0,1\}^{\gamma \log n} \to \{0,1\}^{n^\beta}$ is a $(\gamma n, 2\epsilon)$ extractor. The function $E(f,y) = g_{\hat{f},A}(y)$ is computable in polynomial time because (a) $\hat{f} = \text{ECC}(f)$ is

that has the coefficients $x_1, \ldots, x_n$. Let $y_1, \ldots, y_{2^q}$ be the elements of GF(2^q) and let $z_i \stackrel{\text{def}}{=} p_x(y_i)$, for each $i = 1, \ldots, 2^q$. Note that the string $z_1 \odot z_2 \odot \ldots \odot z_{2^q}$ is the Reed-Solomon codeword encoding x. Each z_i has q bits when written in binary. We use the Hadamard error-correcting code, Had: $\{0,1\}^q \to \{0,1\}^{2^q}$, to further encode each z_i, $i = 1, \ldots, 2^q$. Finally, we define

$$\text{ECC}(x) = \text{Had}(p_x(y_1)) \odot \ldots \odot \text{Had}(p_x(y_{2^q})).$$

Note that, for each $x \in \{0,1\}^n$,

$$|\text{ECC}(x)| = 2^q \cdot 2^q < (4n \cdot (1/\delta^4))^2 = 16 \cdot n^2 \cdot (1/\delta)^8.$$

Let us take a string $u \in \{0,1\}^{2^{2q}}$, $u = u_1 \odot \ldots \odot u_{2^q}$, with each u_i of length 2^q. We want to evaluate the size of the set of ECC codewords that are within Hamming distance $(1/2 - \delta)2^{2q}$ from u. Let $x \in \{0,1\}^n$ be such that $u = u_1 \odot \ldots \odot u_{2^q}$ has agreement $\geq (1/2 + \delta) \cdot 2^{2q}$ with $\text{ECC}(x) = \text{Had}(p_x(y_1)) \odot \ldots \odot \text{Had}(p_x(y_{2^q}))$.[6] Let

$$A \stackrel{\text{def}}{=} \{i \in \{1, \ldots, 2^q\} \mid \text{agreement}(u_i, \text{Had}(p_x(y_i))) \geq (1/2 + \delta/2) \cdot 2^q\}.$$

Claim 5.10.7 $\|A\| \geq \delta \cdot 2^q$.

Proof. Let $\alpha = \|A\|$. The agreement of u and ECC(x) is less than $\alpha \cdot 2^q + (2^q - \alpha) \cdot (1/2 + \delta/2) \cdot 2^q = 2^{2q}(\alpha/2^q + (1 - \alpha/2^q)(1/2 + \delta/2))$. Since the agreement between u and ECC(x) is at least $(1/2 + \delta) \cdot 2^{2q}$, it follows that $\alpha/2^q + (1 - \alpha/2^q)(1/2 + \delta/2) \geq 1/2 + \delta$, which, after some simple calculations, implies $\alpha \geq \delta \cdot 2^q$. ∎

By Theorem 5.10.6, there are at most $1/(4(\delta/2)^2) = 1/\delta^2$ Hadamard codewords at distance $1/2 - \delta/2$ from any u_i. By the definition of A, if $u_i \in A$, then the set of codewords at distance $\leq 1/2 - \delta/2$ from u_i includes Had$(p_x(y_i))$. For each u_i, we define the set of pairs

$$L_i = \{(y_i, z) \mid \text{agreement}(u_i, \text{Had}(z)) \geq (1/2 + \delta/2)\}.$$

We take L to be the union of all the sets L_i, $i = 1, \ldots, 2^q$. By the above observations, $\|L\| \leq 2^q \cdot (1/\delta^2)$ and L contains at least $\|A\| \geq \delta \cdot 2^q$ pairs of the form $(y_i, p_x(y_i))$. We are now ready to use Theorem 5.6.2. We are looking for all the polynomials of degree $n - 1$ that pass through at least $\delta \cdot 2^q$ of the points in L. The condition is $\delta \cdot 2^q \geq \sqrt{2(n-1) \cdot (2^q(1/\delta^2))}$, which holds true for our choice of q. Therefore, by Theorem 5.6.2, it follows that there are at most

$$\sqrt{\frac{2(2^q(1/\delta^2))}{n-1}} \leq \sqrt{\frac{2(4(n-1)/\delta^4)(1/\delta^2)}{n-1}} = 2\sqrt{2} \cdot (1/\delta^3)$$

polynomials p of degree $n-1$ such that p passes through at least $\delta \cdot 2^q$ points in L.

---

[6] We recall that the agreement of two binary strings of the same length is the number of positions in which the two strings coincide.

Since each $x \in \{0,1\}^n$ such that ECC(x) is within distance $(1/2 - \delta) \cdot 2^{2q}$ from u defines in an injective way a polynomial p_x (i.e., two distinct x's define two distinct polynomials) of degree $n-1$ that passes through $\delta \cdot 2^q$ of the points in L, it follows that the number of such x's is also bound by $2\sqrt{2} \cdot (1/\delta^3)$. ∎

In addition to their ability to correct defects in a source of randomness, extractors have numerous other applications. We will present just one (see Section 5.11 for references to papers presenting other applications). Namely, an extractor can be used to reduce the error probability of a BPP algorithm in a randomness efficient way.

A BPP algorithm for a predicate function $f\colon \Sigma^* \to \{0,1\}$ is performed by a polynomial-time probabilistic Turing machine M that on input x of length ℓ uses a random string r of length $m = m(\ell)$ and such that, for all $x \in \Sigma^*$,

$$\text{Prob}_{r \in \{0,1\}^{m(|x|)}}(M(x,r) = f(x)) \geq \frac{2}{3}.$$

In this definition the error probability can be as large as 1/3 and we would like to reduce it to 2^{-t}, for some function t of ℓ. The standard way to do this is to iterate the algorithm k times using independent random strings $r_1, \ldots, r_k$ at each iteration and to take the output (0 or 1) that appears a majority of times. Using the Chernoff bounds, it can be easily seen that with $k = O(t)$ iterations, the error probability is reduced to 2^{-t}. Note that in this scheme the total number of random bits is $O(t \cdot m)$.

Using a polynomial-time computable extractor $E\colon \{0,1\}^n \times \{0,1\}^d \to \{0,1\}^m$ of type $(\gamma n, 1/100)$, with $d = O(\log n)$, $m = O(n)$ and γ a positive constant,[7] the error probability can be reduced to 2^{-t} using only $O(t+m)$ random bits.

The algorithm is as follows. Let us fix $x \in \Sigma^*$ of some length ℓ. The goal is to calculate $f(x)$ using the Turing machine M that uses m random bits on inputs of length ℓ. We choose z randomly in $\{0,1\}^n$; we calculate $M(x, E(z,y))$ for all $y \in \{0,1\}^d$ and take the majority output. Since m is polynomial in ℓ, it follows that $n = \text{poly}(\ell)$ and $d = \log(\ell)$, and, therefore, the above scheme can be implemented in time polynomial in ℓ.

Let us evaluate the error probability. Let

$$W_x = \{r \in \{0,1\}^m \mid M(x,r) = f(x)\},$$

i.e., W_x is the set of random strings that lead M to a correct result. By hypothesis,

$$\frac{\|W_x\|}{\|\{0,1\}^m\|} \geq \frac{2}{3}.$$

We claim that there are less than $2^{\gamma n}$ strings $z \in \{0,1\}^n$ such that

$$\left| \text{Prob}_{y \in \{0,1\}^d}(E(z,y) \in W_x) - \frac{\|W_x\|}{\|\{0,1\}^m\|} \right| > \frac{1}{100}.$$

[7] The extractor built in Theorem 5.10.5 only has $m = n^{\Omega(1)}$. There are known constructions of extractors with the required parameters [Zuc97].

The proof is by contradiction. Suppose that the set BAD of z satisfying the above relation has cardinality $\geq 2^{\gamma n}$. We consider the distribution Z on $\{0,1\}^n$ that assigns probability mass $2^{-\gamma n}$ to each element in some subset of BAD having $2^{\gamma n}$ elements, and 0 to all the other strings in $\{0,1\}^n$. The min-entropy of Z is γn and

$$\left|\text{Prob}_{z \in_Z \{0,1\}^n, y \in \{0,1\}^d}(E(z,y) \in W_x) - \frac{\|W_x\|}{\|\{0,1\}^m\|}\right| > \frac{1}{100}.$$

This contradicts the fact that E is a $(\gamma n, 1/100)$-extractor.

From our claim, it follows that, for $2^n - 2^{\gamma n}$ strings $z \in \{0,1\}^n$,

$$\text{Prob}_{y \in \{0,1\}^d}(E(z,y) \in W_x) \geq \frac{2}{3} - \frac{1}{100} > \frac{1}{2}.$$

Thus the algorithm has an error probability less than $2^{\gamma n}/2^n = 2^{-(1-\gamma)n}$.

The algorithm is using n random bits (i.e., the string z). To obtain error probability bounded by 2^{-t}, we need $(1-\gamma)n \geq t$. Since $n = O(m)$, it means that using $n = O(m+t)$ random bits suffices to reduce the error probability to 2^{-t}.

5.11 Comments and bibliographical notes

Shannon [Sha48] made the suggestion that cryptosystems that are breakable may still be considered satisfactory if the breaking procedure requires an unreasonable large amount of time. The emergence of computational complexity theory in the early '70s have allowed more precise formulations of similar ideas. The concept of a one-way function has been introduced by Diffie and Hellman [DH76]. Pseudo-random generators have a long history (see Knuth [Knu73]), however the classical algorithms (such as linear feedback shift registers, linear congruential sequences, etc.) are not suitable for cryptography. The notion of computational indistinguishability has been invented by Goldwasser and Micali [GM84], and Yao [Yao82] has defined formally the notion of a pseudo-random generator based on the concept of computational distance. The fact that weak one-way functions can be converted into strong one-way functions has been stated by Yao [Yao82]. Our proof of Theorem 5.2.1 is based on Cai's lecture notes [Cai01]. Blum and Micali [BM84] were the first to construct a pseudo-random generator based on an intractability assumption (namely, the intractability of the Discrete Log Problem). Blum and Micali [BM84] have also introduced the concept of a hidden bit (or a hard-core predicate) and have shown its essential role in the construction of a pseudo-random generator. Other pseudo-random generators have been built based on different hypothesis: Blum, Blum, and Shub [BBS87] have used the assumption that the quadratic residuosity problem is hard, and Alexi, Chor, Goldreich, and Schnorr [ACGS88] have used the weaker assumption that integer factoring is hard. The fact that any one-way function essentially has a hard-core predicate (see Theorem 5.3.9) has been established by Goldreich and Levin [GL89]. At that time, the operation of list decoding of error-correcting codes was not properly conceptualized and the realization that one important step in Goldreich and Levin's proof amounts to the list decoding of the Hadamard code came much later (see Sudan [Sud00]). The method for stretching the extension of a pseudo-random generator (see Theorem 5.4.1) has been presented by Goldreich, Goldwasser, and Micali [GGM86]. The same paper contains the construction of a pseudo-random function using as a building block a length-doubling pseudo-random generator (Theorem 5.5.1). Another construction of a pseudo-random function has been given by Naor and Reingold [NR99]. The fact that a one-way permutation can be converted into a pseudo-random generator has been shown by Yao [Yao82] using a more complicated method. The ultimate result in this line of research is that the existence of a one-way function is equivalent to the existence of a pseudo-random generator and has been demonstrated by Håstad, Impagliazzo, Levin and Luby [HILL99a].

Nisan and Wigderson [NW94] had observed that, in order to derandomize BPP computations, it is enough to have pseudo-random generators that are secure against adversaries whose running time is bounded by a fixed polynomial in the length of the output of the pseudo-random generator and which can be calculated in time that is polynomial in the same length. The same paper presents the construction of such a pseudo-random generator using as a building block an exponentially hard predicate (see Theorem 5.8.6). Subsequent papers have succeeded in relaxing the assumption regarding the hardness of the building block by showing that hardness can be amplified. Babai, Fortnow, Nisan, and Wigderson [BFNW93] have shown that a predicate that is worst-case hard can be converted into a predicate having the property that no adversary circuit of exponential size (i.e., of size 2^{cn}, for some constant $c > 0$) can calculate the predicate on more than a fraction of $(1 - 1/\text{poly}(n))$ of the inputs in Σ^n. Impagliazzo [Imp95] has shown that the latter type of predicate can be converted into a constant-rate hard predicate against adversary circuits of exponential size, and Impagliazzo and Wigderson [IW97] have shown how to transform a constant-hard predicate into an exponentially hard predicate against adversary circuits of exponential size, i.e., the kind of predicate that is needed in the Nisan-Wigderson construction. The last paper also presents Theorem 5.9.1, which shows that under a quite plausible hypothesis, P = BPP. The paper [BFNW93] utilizes the technique of polynomial encoding (see the proof of Theorem 5.6.3) developed in a series of earlier papers ([Lip89], [BF90], [GLR$^+$91], [GS92]). Our proof of Theorem 5.6.3 uses the method of polynomial encoding and the polynomial reconstruction algorithm (see Theorem 5.6.2) to push the hardness amplification from worst-case hard functions against superpolynomial (exponential) size adversary circuits to constant-rate hard functions against superpolynomial (exponential) size adversary circuits. The polynomial reconstruction algorithm was discovered by Sudan [Sud97] and was inspired by an algorithm of Berlekamp and Welch [BW86]. Sudan's algorithm needs to factor bivariate polynomials over a finite field. Different algorithms for this problem have been found by Kaltofen [Kal85], Lenstra [Len85], and Grigoriev [Gri84]. For a recent survey on the polynomial reconstruction problem (more commonly known as decoding or list decoding of polynomial error-correcting codes), we recommend the paper by Sudan [Sud01]. The proof of Theorem 5.6.12 that achieves hardness amplification from constant-rate hard functions to crypto-hard functions for adversary circuits of superpolynomial size is modeled after one of the proofs in the paper by Goldreich, Nisan, and Wigderson [GNW95]. This paper presents three proofs of the so-called XOR Lemma stated by Yao [Yao82], which, roughly speaking, shows that the hardness of a function f can be amplified by taking the "direct product" of f, which is the function f on multiple independent inputs. As we have seen in this chapter this method works fine for amplifying the hardness of functions against adversary circuits of superpolynomial size. However, in the case of adversary circuits of exponential size, the simple "direct product" method does not work because the input length is stretched too much and more refined variants of the XOR lemma are needed in which the inputs of the multiple copies are not independent. Such variants of the XOR lemma have been presented in the paper [Imp95] and in the paper [IW97], achieving the hardness amplification mentioned above. The proof of Theorem 5.6.12, (b), which we have skipped, can be found in the latter paper. A different approach to hardness amplification, based on list-decoding of error correcting codes, has been undertaken by Sudan, Trevisan, and Vadhan[STV01], who have succeeded to convert directly a worst-case predicate into an exponentially hard predicate. For references on list decoding of error-correcting codes, the reader can consult the survey paper of Sudan [Sud00]. The most efficient currently-known construction of a pseudo-random generator using as a building block a hard predicate has been given by Umans [Uma02].

The study of methods for repairing imperfect randomness has a long history. It starts probably with von Neumann's classical algorithm [vN51] for generating a sequence of unbiased bits from a source of of biased but independent and identically distributed bits. More and more general types of imperfect sources of randomness have been considered by Blum [Blu84], Santha and Vazirani [SV86], and Chor and Goldreich [CG88]. The general model for weak sources of randomness based on the notion of min-entropy has been introduced by Zuckerman [Zuc90]. Extractors have been first defined by Nisan and Zuckerman [NZ96]. There are numerous constructions of extractors and the reader is advised to consult the survey papers of Nisan and Ta-Shma [NTS99] and Shaltiel [Sha02], which contain a comprehensive coverage of extractors and their applications. The observation that the construction of a pseudo-random generator from a hard predicate can be used to build an extractor has been made by Trevisan [Tre01]. The reconstruction technique is at the origin of some of the best currently known constructions of extractions.

Chapter 6

Optimization problems

6.1 Chapter overview and basic definitions

Numerous NP-complete problems have originated as decision versions of optimization problems. For example, the CLIQUE problem asks whether a graph G has a clique of size at least k (on input a graph G and a natural number k). In fact, what we really want to know is the size of the maximum clique in G. Clearly, if P $\neq$ NP, then an optimization problem whose decision counterpart is NP-complete cannot be solved in polynomial time. In such situations, we should content with approximate solutions, i.e., solutions that are more or less close to the optimum. The quality of an approximate solution can be expressed in a precise numerical way by its closeness to the optimum, and this gives rise to an interesting and meaningful quantitative analysis of a large and important class of optimization problems.

Let us first define the object of our investigation.

Definition 6.1.1 (Optimization problem) *The elements that define an optimization problem A are:*

(1) *a set $\mathcal{I}_A$ of input instances; we assume that this set can be recognized in polynomial time; we also assume that an input instance $I \in \mathcal{I}_A$ is represented as a binary string and we let $|I|$ denote the length of this string;*

(2) *for each $I \in \mathcal{I}_A$, a set $\mathcal{F}_A(I)$ of feasible solutions associated to each input instance; we assume that each element in $\mathcal{F}_A(I)$ has size polynomially bounded in $|I|$, and*

(3) *an objective function f_A that assigns a real number to each pair (I, J) with $I \in \mathcal{I}_A$ and $J \in \mathcal{F}_A(I)$; we assume that this function is computable in polynomial time. There is also a default value for the cases when the set of feasible solutions is empty.*

If the objective function takes only non-negative values then A is called a positive optimization problem. *Given an instance $I \in \mathcal{I}_A$, the goal is to find*

$$\text{opt}_{J \in \mathcal{F}_A(I)} f_A(I, J)$$

or output the default value in case $\mathcal{F}_A(I)$ is empty, where opt *is* max *or* min *depending on what kind of an optimization problem we have. For convenience, we will often denote* $\text{opt}_{J \in \mathcal{F}_A(I)} f_A(I, J)$ *by* $\text{opt}_A(I)$.

We will restrict our attention to a class of optimization problems that are naturally associated to NP problems.

Definition 6.1.2 (NP optimization problem) *A max (min) optimization problem A is an* NP *optimization problem if the following associated decision problem B is in* NP*:*

Instance: An input instance $I \in \mathcal{I}_A$ and $k \in \mathbf{Z}$.

Question: Does there exists a feasible solution $J \in \mathcal{F}_A(I)$ such that $f_A(I, J) \geq k$ ($f_A(I, J) \leq k$, in the case of a min problem)?

Within the class of NP optimization problems we distinguish the subclass of *polynomially bounded* problems.

Definition 6.1.3 (Polynomially bounded optimization problem) *An* NP *optimization problem A is said to be polynomially bounded if there exists a polynomial p such that*

$$\text{opt}_A(I) \leq p(|I|),$$

for all $I \in \mathcal{I}_A$. The class of all polynomially bounded maximization (minimization) problems is denoted by MAX PB (MIN PB).

The quality of an approximation solution is measured by the *approximation ratio*, which is defined as follows.

Definition 6.1.4 (Approximation ratio) *Let A be an* NP *optimization problem with the objective function f_A. Let I be an input instance for A and J a feasible solution for I. The approximation ratio of J with respect to I is*

$$R(I, J) = \frac{f_A(I, J)}{\min_A(I)}, \text{ for a minimization problem, and}$$

$$R(I, J) = \frac{\max_A(I)}{f_A(I, J)}, \text{ for a maximization problem.}$$

The approximation ratio is always a number greater than or equal to 1 and the closer it is 1, the better the feasible solution J is.

Definition 6.1.5 (Approximation algorithm) *Let A be an optimization problem. An approximation algorithm B for A is a function that maps input instances $I \in \mathcal{I}_A$ to feasible solutions in $\mathcal{F}_A(I)$. As a technical convenience, we require that $f_A(I, B(I))$ and opt $A(I)$ have strictly positive value for all I. The approximation algorithm B has an approximation ratio $r_B : \mathbb{N} \to [1, +\infty)$ if, for all input instances I, $B(I)$ has approximation ratio at most $r_B(|I|)$ with respect to I.*

Given an optimization problem, we would like to know whether it admits a polynomial-time approximation algorithm that achieves a given approximation ratio. We will see that, under reasonable complexity-related assumptions, NP optimization problems can be very different from this point of view. The following are the basic classes of NP optimization problems defined in term of their approximation properties.

Definition 6.1.6 *The classes* PTAS, APX, log-APX, poly-APX *are defined as follows. Let A be an* NP *optimization problem.*

(a) *A has a polynomial-time approximation scheme (*PTAS*) (and A is in the class* PTAS*), if and only if, for any constant $\epsilon > 0$, there exists a polynomial-time approximation algorithm B_ϵ for A with an approximation ratio r_{B_ϵ} such that, for all input instances I,*

$$r_{B_\epsilon}(|I|) \leq 1 + \epsilon.$$

(b) *A is in the class* APX*, if and only if there exists a polynomial-time approximation algorithm B for A with an approximation ratio r_B and a constant c such that, for all input instances I,*

$$r_B(|I|) \leq c.$$

(c) *A is in the class* log-APX*, if and only if there exists a polynomial-time approximation algorithm B for A with an approximation ratio r_B and a constant $c > 0$ such that, for all input instances I,*

$$r_B(|I|) \leq c \log(|I|).$$

(d) *A is in the class* poly-APX*, if and only if there exists a polynomial-time approximation algorithm B for A with an approximation ratio r_B and a polynomial p such that, for all input instances I,*

$$r_B(|I|) \leq p(|I|).$$

Clearly, it holds that PTAS $\subseteq$ APX $\subseteq$ log-APX $\subseteq$ poly-APX, and, given a problem, it is desirable to find the smallest class in the above hierarchy that contains it.

In the context of approximation algorithms it is useful to consider a reduction mechanism that preserves approximability.

Definition 6.1.7 (L-reduction) *Let A and A' be two optimization problems. An* L*-reduction from A to A' (notation: $A \leq_{\mathrm{L}} A'$) consists of two polynomial-time functions f and g and the positive constants α and β that satisfy the following two properties:*

(a) *If I is an instance of A, f on input I produces an instance I' of A' and*

$$\mathrm{opt}_A(I) \leq \alpha \cdot \mathrm{opt}_{A'}(I');$$

(b) *Given a feasible solution S' of I' of cost c', g produces a feasible solution S of I of cost c and*

$$|c - \mathrm{opt}_A(I)| \leq \beta|c' - \mathrm{opt}_{A'}(I')|.$$

L-reductions are used to translate the approximation property of a problem to the approximation property of another problem.

Proposition 6.1.8 (a) *Let A and A' be two minimization problems such that $A \leq_{\mathrm{L}} A'$. If A' has a polynomial-time approximation algorithm with approximation ratio r', then A has a polynomial-time approximation algorithm with approximation ratio*

$$r(I) \leq 1 + \alpha\beta(r'(I') - 1),$$

where α and β are the constants from the definition of the L- reduction and I' is the input instance of A' that results from I by applying the reduction.

(b) *Let A and A' be two maximization problems such that $A \leq_{\mathrm{L}} A'$. If A' has a polynomial-time approximation algorithm with approximation ratio r' and if $1 - \alpha\beta(1 - 1/r'(I')) > 0$, then A has a polynomial-time approximation algorithm A' with approximation ratio*

$$r(I) \leq \frac{1}{1 - \alpha\beta(1 - 1/r'(I'))},$$

where α, β, and I' are as above.

Proof. (a) Let I be an instance of A. Using the function f given by the L-reduction, we calculate an instance I' of A' such that $\mathrm{opt}_{A'}(I') \leq \alpha \cdot \mathrm{opt}_A(I)$. Let *cost* be the objective function of problem A, and *cost'* be the objective function of problem A'. Using the polynomial-time approximation algorithm for A' we determine a feasible solution S' for I' such that $\mathrm{cost}'(S') \leq r'(I') \cdot \mathrm{opt}_{A'}(I')$. Using the function g of the L-reduction, we determine a solution S of I. We have

$$\begin{aligned}(\mathrm{cost}(S) - \mathrm{opt}_A(I)) &\leq \beta(\mathrm{cost}'(S') - \mathrm{opt}_{A'}(I')) \\ &\leq \beta(r'(I') - 1)\mathrm{opt}_{A'}(I') \\ &\leq \alpha\beta(r'(I') - 1)\mathrm{opt}_A(I).\end{aligned}$$

It follows that

$$\frac{\mathrm{cost}(S)}{\mathrm{opt}_A(I)} \leq 1 + \alpha\beta(r'(I') - 1).$$

(b) As above, from an instance I of A we determine in polynomial time an instance I' for A' with $\mathrm{opt}_{A'}(I') \leq \alpha \cdot \mathrm{opt}_A(I)$, and then a feasible solution S' for I' such that

$$\frac{\mathrm{opt}_{A'}(I')}{\mathrm{cost}'(S')} \leq r'(I').$$

Then using the polynomial-time function g, we determine a feasible solution S of I such that

$$\begin{aligned}(\text{opt}_A(I) - \text{cost}(S)) &\leq \beta(\text{opt}_{A'}(I') - \text{cost}'(S')) \\ &\leq \beta(1 - 1/r'(I'))\text{opt}_{A'}(I') \\ &\leq \beta(1 - 1/r'(I'))\alpha \cdot \text{opt}_A(I).\end{aligned}$$

Then, provided that $1 - \alpha\beta(1 - 1/r'(I')) > 0$,

$$\frac{\text{opt}_A(I)}{\text{cost}(S)} \leq \frac{1}{1 - \alpha\beta(1 - 1/r'(I'))}.$$ ∎

The classification of NP optimization problems in the hierarchy PTAS $\subset$ APX $\subset$ log-APX $\subset$ poly-APX requires the determination of upper and lower bounds on the approximation ratio achievable in polynomial time.

Upper bounds can be obtained in the standard way by designing approximation algorithms but also, perhaps surprisingly, by exploiting the relationship that exists between complexity classes and logic. A computational decision problem A can be characterized by a well-formed formula ϕ in a certain logic in the following sense: An input instance is in A if and only if that input instance, viewed as an interpretation, satisfies ϕ. In this way, one can establish connections between complexity classes and different logics. In particular, as we show in Section 6.2, a problem A is in NP if and only if A is characterized in the above sense by some formula ϕ in second-order logic with one existential second-order quantifier. By imposing further syntactical restrictions regarding the way formulas can be constructed, one can define subclasses of optimization problems. In Section 6.3, we introduce such classes and study the inclusion relations between them. The syntax that defines some of these classes also dictates the approximation properties of the class members. Thus, for many important NP optimization problems, one can establish an upper bound of the approximation ratio achievable in polynomial time by just giving a certain syntactical characterization of the problem. We undertake this approach in Section 6.4 for maximization problems, and in Section 6.5 for minimization problems.

As it is usually the case in complexity theory, given the current state of affairs, we cannot prove absolute non-trivial lower bounds for the approximation ratio. Note that if P $\neq$ NP, then any NP-optimization problem can be solved exactly in polynomial time. However, due mainly to another surprising characterization of NP, we can demonstrate good lower bounds under very reasonable assumptions such as P $\neq$ NP or NP $\subseteq$ DTIME$[2^{\log^{O(1)} n}]$. We show such lower bounds for some important NP-optimization problems in Section 6.6, Section 6.7, and Section 6.8. In particular, we consider all the main classes of NP optimization problems introduced earlier using the syntactical approach. For each such class $\mathcal{C}$, we pick a canonical $\mathcal{C}$-complete problem (under the L-reduction) and we establish a conditional but good lower bound of the approximation ratio that can be obtained in polynomial time.

The characterization of NP to which we have alluded above is given via an interactive protocol between two parties: A polynomial-time probabilistic machine V, called the *verifier*, and an all-mighty entity P, called the *prover*. Given a computational decision problem A, and an input string x, the prover wants to convince the verifier that $x \in A$. If $x \in A$, then the verifier should accept with probability one, or very close to one (this is called the *completeness property* of the protocol). If $x \notin A$, the verifier, regardless of what the prover has told him, does not accept but with very small probability (this is called the *soundness property* of the protocol). To illustrate, let us consider an NP problem A. There exist a verifier and a prover that behave as follows. For each $x \in A$, the prover can simply give the verifier a membership proof that $x \in A$, and the verifier checks the validity of the proof. If $x \notin A$, then there is no such membership proof and therefore the prover cannot fool the verifier to accept x. This is the standard characterization of NP (see Theorem 1.1.6). The new characterization is provided by the so-called PCP Theorem. It shows that for every $x \in A$, the prover can give the verifier a membership proof w, of length polynomial in $|x|$, with the following amazing property: The verifier needs to read only a constant number of w's bits in order to check its validity (such a membership proof is called a *holographic proof*). The PCP Theorem has a long and complicated proof that is beyond the scope of this book. However, in Section 6.9, we prove two weaker but still interesting variants. In the first one, the holographic proof is exponentially long. In the second one, the holographic proof has polynomial length (like in the PCP Theorem) but it is constructed by a polynomial-time prover that has access to a classical membership proof and the soundness property is weaker because it only holds that this type of restricted provers cannot fool the verifier to accept a fake holographic proof.

6.2 Logical characterization of NP

IN BRIEF: Any NP problem can be represented by a formula in second-order logic in which the second-order variables are quantified with existential quantifiers and the first order variables are quantified with universal quantifiers followed by existential quantifiers.

We start the exploration of NP optimization problems with a logical characterization of NP. We show that a set A is in NP if and only if there exists a logical formula of a quite particular form that is made true exactly by the input instances in A. Some clarifications are in order because apparently we are equating two unrelated things: The class NP is defined in terms of Turing machines that process strings, while a logical formula can be true or false with respect to an interpretation. However, an input instance for a problem can be encoded both by a string, which can be processed by Turing machines, and by a finite structure, (i.e., a finite set plus some relations defined on it), which can provide the interpretation of a logical sentence. Here we have in mind YES/NO problems, i.e., problems for which the goal is to determine whether a given input instance x has or not a certain property. We recall that such problems are called *decision problems*. For example, the satisfiability problem (SAT) is of this type because it consists of determining whether a boolean formula in conjunctive normal form is or not satisfiable. Input instances for a problem A can be encoded by strings over a fixed alphabet and we can consider the set of strings encoding instances for which the answer is YES. By a common abuse of notation this set is denoted by A as well. For example, boolean formulas in conjunctive normal form can be encoded somehow (the details are not important) by binary strings, and SAT also denotes the set of strings encoding formulas that are satisfiable.

On the other hand, an input instance can also be represented by a finite structure. For a rigorous treatment, we recall a few basic concepts from mathematical logic and introduce some notation. More details can be found in any standard textbook of mathematical logic, such as [End72].

Definition 6.2.1 (Signature) *A signature $\sigma = (\widetilde{R}_1, \ldots, \widetilde{R}_k)$ is a finite set of relation symbols (also called relation variables). Each relation symbol $\widetilde{R}_i$ has associated to it an integer $r_i \geq 0$ called the arity of $\widetilde{R}_i$. A relation symbol of arity* 0 *is called a constant symbol.*

Definition 6.2.2 (Relation) *For any natural number $n > 0$, a relation R of arity n over a set A is a subset of A^n. The fact that $\overline{x} = \{x_1, \ldots, x_n\}$ is in the subset R is denoted by $R(\overline{x})$. In this case, we also say that $\overline{x}$ is in the relation R or that $\overline{x}$ satisfies R. A relation symbol R of arity 0 over a set A denotes a fixed element of A.*

Definition 6.2.3 (Finite Structure) *Let $\sigma = (\widetilde{R}_1, \ldots, \widetilde{R}_k)$ be a signature with relation symbols of arities $r_1, \ldots, r_k$. A σ-structure $I = (D_I, R_1, \ldots, R_k)$ consists of a set D_I, called the domain of the structure I, and of relations $R_1, \ldots, R_k$ of arities $r_1, \ldots, r_k$, over D_I, $1 \leq i \leq k$. If the signature σ is clear from the context, or if it is not relevant, we will simply say that I is a structure. A structure I is finite if its domain D_I is a finite set. The size of a structure I, denoted $||I||$, is the cardinality of D_I.*

Abusing notation, we will use the same notation for relation symbols and for the relations themselves. We recall a few basic concepts of first-order logic. The formulas of first-order logic over a signature σ are built from the relation symbols of σ, a special binary symbol $=$, and variables $x_1, x_2, \ldots$, using the logical connectives $\wedge, \vee, \neg, \rightarrow$, and the quantifiers $\exists x_i$ and $\forall x_i$. Every formula ϕ of first-order logic can be given an *interpretation* under a structure I in the following way: The relation symbols of σ are interpreted by the corresponding relations of the structure, the special binary symbol $=$ is always interpreted as the equality relation on the domain of I, the connectors $\wedge, \vee, \neg, \rightarrow$ have their usual logical meaning (AND, OR, NOT, IMPLIES), and the variables in the quantifiers $(\exists x_i)$ and $(\forall x_i)$, $i \geq 1$, are ranging over the elements of the domain of I. A closed formula, i.e., a formula without free (i.e., not quantified) variables, is *true* or *false* under this interpretation and this is denoted by $I \models \phi$ (we say that the structure I is a *model* for the formula ϕ), and, respectively, $I \not\models \phi$. In case the formula ϕ has free variables, it becomes true or false under an assignment of elements from the domain of I to its free variables.

The logical description of the complexity class NP needs more than first-order logic (but not much more). To prepare the required additional logical concepts, let us recall the definition of the satisfiability problem (SAT).

Input: A set of boolean variables $\{x_1, \ldots, x_n\}$ and a boolean formula ϕ in CNF, i.e., ϕ is a conjunction $C_1 \wedge \ldots \wedge C_m$ and each clause C_i is the disjunction of some literals, where a literal is a variable or the negation of a variable.

Question: Is there a truth assignment of the variables under which ϕ is true?

Let $\phi = C_1 \wedge \ldots \wedge C_m$ be a boolean formula in CNF over the variables $\{x_1, \ldots, x_n\}$. To describe ϕ as a finite structure I, we take the domain

$$D_I = \{x_1, \ldots, x_n, C_1, \ldots, C_m\}$$

and consider over D_I two relations C and V of arity 1 and two relations P and N of arity 2 given by: $C(x)$ if and only if x is a clause, $V(x)$ if and only if x is a variable, $P(x, c)$ if and only if x is a variable that appears positively in the clause c, and $N(x, c)$ if and only if x is a variable that appears negatively in the clause c. The input instance for SAT given by the formula ϕ is fully described by the finite structure I with domain D_I of signature $\sigma = (C, V, P, N)$, where C and V are relation symbols of arity 1 and P and N are relation symbols of arity 2 (abusing notation we use the same letter for the relation symbol and for the relation itself). The formula ϕ is satisfiable if there is an assignment of truth values to the variables $x_1, \ldots, x_n$ under which ϕ is true. This fact can also be expressed in the logical framework that we have introduced. We use a new relation symbol S of arity 1 to represent an assignment of truth values. The intended meaning is that $S(x)$ if and only if the variable x has been assigned the value *true*. The formula ϕ is satisfiable if and only if there is a relation S such that

$$\forall x \, \exists y [C(x) \to (P(y, x) \wedge S(y)) \vee (N(y, x) \wedge \neg S(y))].$$

The above formula, let us call it ψ, is in first-order logic and has the particularity that in the prenex normal form (i.e., with the quantifiers pulled to the left) it has one alternation of quantifiers of the form $\forall \ldots \forall \exists \ldots \exists$. Such a first-order formula is said to be in the Π_2 *form*. If there is a relation S so that ψ is true, we can write

$$(I, S) \models \psi. \tag{6.1}$$

This means that the finite structure I together with the relation S defined on the domain of I is a *model* for the formula ψ. Note that S does not range over the elements of D_I but over relations over D_I. Such variables are called *second-order variables.* Also note that I represents a satisfiable formula if and only if there exists S such that Equation (6.1) holds. In a natural way, we write this as

$$I \in SAT \leftrightarrow I \models \exists S\psi.$$

The formula $\exists S\psi$ is an existential second-order formula. In general, if $S_1, \ldots, S_k$ are relation variables and ψ is a first-order formula, $\exists S_1 \ldots \exists S_k\psi$ is called an *existential second-order formula.* It is a second-order formula because it uses second-order variables and it is existential because all the second-order variables are quantified with the existential quantifier. Using the notation $\overline{S}$ for the tuple of relation symbols $(S_1, \ldots, S_k)$, we shall abbreviate $\exists S_1 \ldots \exists S_k\psi$ by $\exists\overline{S}\psi$. Sometimes, we will refer to a formula with a notation of the form $\phi(\overline{x}, \overline{y}, I, \overline{S})$ to emphasize that the formula has variables exclusively from the tuples $\overline{x}$ and $\overline{y}$, and that the relation symbols appearing in the formula are either from the tuple $\overline{S}$ or they correspond to relations of the finite structure I.

We have seen that a boolean formula ϕ is in SAT if and only if its representation as a finite structure I is a model for an existential second-order formula $\exists\overline{S}\psi$, with ψ a first-order formula in Π_2 form. This is not an accident: The next theorem states that this property characterizes all the sets in NP.

Before we pursue with the proof, we need a few preparations. Finite structures will be processed by Turing machines and consequently they need to be encoded by strings. We will assume that a finite structure $I = (D_I, P_1, \ldots, P_r)$, where $P_i, i = 1, \ldots, r$ are relations over the domain D_I, is encoded by a sequence of strings, called enc(I), as follows: The domain D_I, which can be considered to be the set $\{0, \ldots, n-1\}$ for some natural number n, will be encoded by the string 1^{n-1} (i.e., $n-1$ written in unary); for $i = 1, \ldots, r$, the relation P_i of arity m_i is encoded by the binary string π_i of length n^{m_i}, where the j-th bit of π_i is 1 if and only if, for all $j = 1, \ldots, n^{m_i}$, the j-th tuple in the lexicographical order of $\{0, \ldots, n-1\}^{m_i}$ is in the relation P_i.

As usual (for example we do the same for graph decision problems), we say that a set of structures is in NP, if their encodings are in NP.

Theorem 6.2.4 INFORMAL STATEMENT: *Any set in* NP *can be considered to be a set of finite structures satisfying a second-order formula in which the second-order variables are quantified with* $\forall$ *and the first-order variables are quantified with an alternation* $\forall\exists$.

FORMAL STATEMENT: *Let* σ *be a finite signature. A set* L *of finite* σ*-structures that is closed under isomorphism is in* NP *if and only if there exists an existential second-order sentence* ϕ *such that, for any* σ*-structure* I*, the sentences "*$I \in L$*" and "*$I \models \phi$*" are either both true or both false (i.e.,* $I \in L \leftrightarrow I \models \phi$*).*

Moreover, ϕ *can be chosen to be of the form* $\exists\overline{S}\,\forall\overline{x}\,\exists\overline{y}\,\psi(\overline{S}, \overline{x}, \overline{y})$*, with* ψ *a quantifier-free formula.*

Proof. One direction is easy to check but tedious to prove formally, so we shall content with a sketchy argument. If we are given a sentence ϕ of the form

$\exists \overline{S} \forall \overline{x} \exists \overline{y} \psi(\overline{S}, \overline{x}, \overline{y})$, we can build a nondeterministic polynomial-time Turing machine M that accepts an encoding enc(I) of a structure I if and only if $I \models \phi$. Assume that the domain of I is $\{0, \ldots, n-1\}$. First, M uses nondeterminism to guess the relations $\overline{S} = (S_1, \ldots, S_k)$. If the arity of S_i is q_i, then S_i is encoded by a binary string having length n^{q_i}. Therefore, the guessing of $\overline{S}$ takes time $n^{q_1} + \ldots + n^{q_t}$, i.e., polynomial time. Once $\overline{S}$ has been fixed and its encoding written on some tape of M, M has to check if

$$(I, \overline{S}) \models \forall \overline{x} \exists \overline{y} \psi(\overline{S}, \overline{x}, \overline{y}).$$

To this aim, M loops over all possible choices of assigning to $\overline{x}$ a value $\overline{a}$ from $\{0, \ldots, n-1\}^{m_1}$, where m_1 is the arity of $\overline{x}$. This loop has n^{m_1} iterations and each iteration consists of trying to find a value $\overline{b}$ for $\overline{y}$ so that $\psi(\overline{a}, \overline{b})$ holds. The formula $\psi(\overline{a}, \overline{b})$ is just a combinations of conjunctions, disjunctions, and negations of some determined terms and, therefore, its validity can be checked in polynomial time (actually constant time). The machine M accepts enc(I) if and only if each iteration of the first loop succeeds.

Conversely, let L be a set of finite σ-structures, for some fixed signature σ, that is in NP. This means that there is a polynomial-time nondeterministic Turing machine M that accepts exactly the encodings of the finite structures in L. Let us assume that $\sigma = (P_1, \ldots, P_r)$ and that the arity of each P_i is m_i, $1 \leq i \leq r$. We can assume that M has $r+1$ tapes with the tape symbols $0, 1$, and B (blank) and that the running time of M on an input enc(I) (where I is a finite structure) is bounded by $n^k - 1$, where $D_I = \{0, \ldots, n-1\}$. We can also assume that (1) the leftmost cell on the first tape contains a special symbol, $\diamondsuit$, which is never overwritten and which prevents the head on the first tape to move past its left end, (2) the tapes $2, \ldots, r+1$ are only scanned once from left to right (their content is perhaps copied on the first tape), and (3) the machine M makes exactly two nondeterministic choices at each computation step. Initially, tape 1 contains n written in unary, where the domain of I is $\{0, \ldots, n-1\}$, and for $j = 1, \ldots, r$, the $(j+1)$-th tape contains the encoding of the relation P_j. Clearly any type of nondeterministic polynomial-time Turing machine can be simulated by a machine with the above constraints in polynomial time. Because of the nondeterminism, M has many computation paths on an input enc(I). One such computation path can be described by a sequence of configurations $C_0, \ldots, C_t$, where t is the number of computation steps of M on enc(I). Each configuration C_i describes the content of the $r+1$ tapes, the cell on each tape where the tape head is placed, and the current state, all for the i-th step of the computation of M on enc(I). Each configuration C_i consists of an $(r+1)$-tuple $(C_{i,1}, \ldots, C_{i,r+1})$, each component of the tuple representing a tape. More precisely, if at step i M is in state q, the content of tape j is $b_0 \ldots b_{n^k-1}$ (because of the lack of time, the machine cannot reach a tape cell beyond the n^k-th one), and the j-th tape head is placed on cell h, then $C_{i,j} = b_0 \ldots b_{h-1}(b_h, q)b_{h+1} \ldots b_{n^k-1}$, where $(0, q), (1, q), (B, q)$, for all states q of M, are new symbols. We call the collection of all symbols that can be part of a configuration the *configuration symbols*. Let $\{\sigma_1, \ldots, \sigma_h\}$ be the set of all configuration symbols. From the sequence of configurations we build a sequence of computation tables $(T_j)_{j=1,\ldots,r+1}$, where T_j is

$$\begin{array}{c} C_{0,j} \\ C_{1,j} \\ \vdots \\ C_{n^k-1,j} \end{array}$$

Thus, T_j is an $n^k \times n^k$ table of configuration symbols and it represents the history of tape j. Since $t < n^k$, we have padded the sequence with the last configuration, C_t, so that we have a good control over the size of the table T.

Our next goal is to design an existential second-order formula ϕ that describes a sequence of valid computation tables $(T_j)_{j=1,\ldots,r+1}$. Indices in each T_j go from 0 to $n^k - 1$. Therefore we use k-tuples of variables $\overline{x} = (x_1, \ldots, x_k)$ and $\overline{y} = (y_1, \ldots, y_k)$ with x_i and y_i ranging from $\{0, \ldots, n-1\} = D_I$ to denote the rows and respectively the columns of T_j. In general, the variables that are overlined denote a k-tuple of simple variables.

To build the formula ϕ, we need to define some basic relations. First we introduce a new relation symbol L that will represent a linear ordering. To this aim, we consider the conjunction of the following formulas:

$$\begin{array}{l} \forall x \, \forall y \, (L(x,y) \vee L(y,x) \vee x = y) \\ \forall x \, \forall y \, \forall z \, ((L(x,y) \wedge L(y,z)) \rightarrow L(x,z)) \\ \forall x \, \neg L(x,x) \\ \forall x \, \forall y \, (L(x,y) \rightarrow \neg L(y,x)). \end{array}$$

The first formula states that every two distinct elements from the domain are comparable and the other formulas state that L is transitive, anti-reflexive, and anti-symmetric. A model I of the constructed formula (the conjunction of the four formulas from above) forces L to be a linear ordering on the domain of I. We will use the more common notation $x < y$ instead of $L(x,y)$ and also $x \leq y$, $y > x$, etc., expressions that can be build from L in the obvious way.

We introduce another binary relation symbol S that is meant to represent the successor relation on the domain of I. We define S by

$$\forall x \, \forall y \, (S(x,y) \leftrightarrow \forall z((x < y) \wedge ((x \leq z) \wedge (z \leq y)) \rightarrow (x = z \vee z = y))).$$

It is clear that S is forced to be the successor relation relative to the order defined by L.

Using S we can define some other useful relations. Thus, we introduce the Z and the N relations symbols and the formulas

$$\forall x \, (Z(x) \leftrightarrow \forall y \neg S(y,x))$$

and

$$\forall x\, (N(x) \leftrightarrow \forall y \neg S(x,y)).$$

Thus, $Z(x)$ states that x is the minimum element, which, by the closure under isomorphism, we identify with 0. Also, $N(x)$ states that x is the maximum element, which we identify with $n-1$. We introduce two symbols for constants, c_0 and c_{last}, and consider the formulas

$$Z(c_0) \text{ and } N(c_{\text{last}}).$$

Let $\overline{c_0} = (c_0, \ldots, c_0)$ and $\overline{c_{\text{last}}} = (c_{\text{last}}, \ldots, c_{\text{last}})$, where the tuples have k components. Next, we define the successor relation S_k on strings of k digits over the alphabet $\{0, \ldots, n-1\}$. In other words, $S_k(\overline{x}, \overline{y})$ holds if and only if $\overline{y}$ is the successor of $\overline{x}$ in the lexicographical order. To this aim, we define inductively the relations $S_1, \ldots, S_k$. The relation S_1 is S. Assuming that $S_{j-1}(x_1, \ldots, x_{j-1}; y_1, \ldots, y_{j-1})$ defines the successor relation for strings of $j-1$ characters, S_j is defined by the following expression universally quantified over all variables:

$$\begin{aligned}&[(x_1 = y_1) \wedge \ldots \wedge (x_{j-1} = y_{j-1}) \wedge S(x_j, y_j)] \vee \\ &\quad [N(x_j) \wedge Z(y_j) \wedge S_{j-1}(x_1, \ldots, x_{j-1}; y_1, \ldots, y_{j-1})].\end{aligned}$$

In other words, either x_j, the last digit of $\overline{x}$, is $n-1$ and then y_j must be 0 and $y_1 \ldots y_{j-1}$ must be the successor of $x_1 \ldots x_{j-1}$, or the last digit of $\overline{x}$ is not $n-1$ and then, y_j is the successor of x_j and the other digits of $\overline{x}$ and $\overline{y}$ are equal.

After these preparatory steps, we pursue with the part that describes a sequence of valid computation tables. For each configuration symbol σ and each computation table T_j, we introduce a relation symbol $T_{j,\sigma}$ of arity $2k$. $T_{j,\sigma}(\overline{x}, \overline{y})$ means that the entry in T_j at position (s,t) is σ, where s is the number encoded by $\overline{x}$ and t is the number encoded by $\overline{y}$. We need to say that each entry in each computation table contains a symbol and only one symbol. This fact is expressed by the conjunction of the following $r+1$ formulas:

$$\forall \overline{x}\, \forall \overline{y}\, [T_{j,\sigma_1}(\overline{x}, \overline{y}) \vee \ldots \vee T_{j,\sigma_h}(\overline{x}, \overline{y})] \qquad j = 1, \ldots, r+1,$$

where $\sigma_1, \ldots, \sigma_h$ are all the configuration symbols. To express the fact that each entry has at most one symbol, we introduce for each pair $k \neq \ell$, $k, \ell \in \{1, \ldots, h\}$, and for each $j \in \{1, \ldots, r+1\}$, the formula

$$\forall \overline{x}\, \forall \overline{y}\, [\neg T_{j,\sigma_k}(\overline{x}, \overline{y}) \vee \neg T_{j,\sigma_\ell}(\overline{x}, \overline{y})],$$

which says that it is not possible to have simultaneously $T_{i,\sigma_k}(\overline{x}, \overline{y})$ and $T_{i,\sigma_\ell}(\overline{x}, \overline{y})$.

We also introduce two relation symbols D_0 and D_1 to express that, at each step, M is either taking the first nondeterministic choice or the second nondeterministic choice. More precisely $D_0(\overline{x})(D_1(\overline{x}))$ states that the first (respectively, the second) choice has been selected at the step whose number is encoded by $\overline{x}$. We need to say that at each step exactly one of the nondeterministic choices is selected. This is expressed by the formula:

$$\forall \overline{x}\, [(D_0(\overline{x}) \vee D_1(\overline{x})) \wedge (\neg D_0(\overline{x}) \vee \neg D_1(\overline{x}))].$$

The second-order formula ϕ that we are buiding will be of the form

$$\exists L\, \exists S\, \exists S_1 \ldots \exists S_k\, \exists T_{1,\sigma_1} \ldots \exists T_{r+1,\sigma_h}\, \exists D_0\, \exists D_1\, \phi'. \tag{6.2}$$

The formula ϕ' is first-order and it will be a conjunction of formulas containing the formulas described above for the correctness of $L, S, S_1, \ldots, S_k,\ T_{1,\sigma_1}, \ldots, T_{r+1,\sigma_h}$, D_0, D_1, c_0, and c_{last}, and some additional formulas to be described below. These new formulas will state that the sequence of computation tables described by the relations $T_{1,\sigma_1}, \ldots, T_{r+1,\sigma_h}$ expresses a valid sequence of configurations of M on input enc(I), consistent with the nondeterministic choices stipulated by D_0 and D_1, and that leads to acceptance. For this, we need to express the following conditions:

(1) The top rows in each computation table contain the initial configuration for input enc(I);

(2) In each row, the tape head is scanning exactly one tape cell;

(3) For each $i \in \{0, \ldots, n^k - 2\}$, configuration C_{i+1} follows from C_i, according to the transition moves of M;

(4) The machine M reaches an accepting state;

Let us consider these conditions in turn. For (1), we must first say that the leftmost cell on row 0 of tape 1 contains $\diamondsuit$, after which row 0 of tape 1 contains $n-1$ written in unary, that the remaining symbols in row 0 of tape 1 are blank, and that the tape head is placed on the leftmost cell. All these conditions can be expressed by taking the conjunction of the following formulas:

$$\forall \overline{x}\, \forall \overline{y}\, [\overline{x} = \overline{c_0} \wedge \overline{y} = \overline{c_0} \rightarrow T_{1,(\diamondsuit,q_{start})}(\overline{x}, \overline{y})]$$
$$\forall \overline{x}\, \forall \overline{y}\, [\overline{x} = \overline{c_0} \wedge \overline{c_0} < \overline{y} \le \overline{c_{\text{last}}} \rightarrow T_{1,1}(\overline{x}, \overline{y})]$$
$$\forall \overline{x}\, \forall \overline{y}\, [\overline{x} = 0 \wedge \overline{y} > c_{\text{last}} \rightarrow T_{1,B}(\overline{x}, \overline{y})].$$

The notations $\overline{x} = \overline{c_0}$, $\overline{y} = \overline{c_0}$, $\overline{c_0} < \overline{y} \le \overline{c_{\text{last}}}$, and $\overline{y} > \overline{c_{\text{last}}}$ are notational abbreviations for the obvious corresponding logical expressions which can easily be written down using the relation L.

We must also say that the tape $j+1$, with $1 \le j \le r$, contains the encoding of relation P_j of I. This follows from the conjunction of the following formulas:

$$\forall y_1 \ldots \forall y_{m_j}\, [P_j(y_1, \ldots, y_{m_j}) \rightarrow T_{j+1,1}(\underbrace{c_0, \ldots, c_0}_{k}; \underbrace{c_0, \ldots, c_0}_{k-m_j}, y_1, \ldots, y_{m_j})]$$

$$\forall y_1 \ldots \forall y_{m_j}[\neg P_j(y_1,, \ldots, y_{m_j}) \rightarrow T_{j+1,0}(\underbrace{c_0, \ldots, c_0}_{k}; \underbrace{c_0, \ldots, c_0}_{k-m_j}, y_1, \ldots, y_{m_j})]$$

$$\forall y_1 \ldots \forall y_k[(y_1 > 0) \vee \ldots \vee (y_{k-m_j} > 0) \rightarrow T_{j+1,B}(\underbrace{c_0, \ldots, c_0}_{k}; y_1, \ldots, y_k)].$$

For condition (2), for tape 1 (and similarly for the other tapes), we insert in ψ the conjunction of the following two formulas:

$$\forall \overline{x}\, \exists \overline{y}[\vee T_{1,(\sigma,q)}(\overline{x}, \overline{y})],$$

where the disjunction is taken over all tape symbols σ and all states q, and

$$\forall \overline{x}\, \forall \overline{y_1}\, \forall \overline{y_2}[\wedge(T_{1,(\sigma_1,q_1)}(\overline{x}, \overline{y_1}) \wedge T_{1,(\sigma_2,q_2)}(\overline{x}, \overline{y_2}) \rightarrow \overline{y_1} = \overline{y_2})],$$

where the conjunction is taken over all tape symbols σ_1 and σ_2 and over all states q_1 and q_2.

Let us focus now on condition (3). The transition table of M induces a set of $(r+1)$-tuples $(\delta_1, \ldots, \delta_{r+1})$ as follows. Each δ_j is a 5-tuple $(\alpha_j, \beta_j, \gamma_j, c_j, \sigma_j)$, where $\alpha_j, \beta_j, \gamma_j$ and σ_j are configuration symbols and $c_j \in \{0, 1\}$ represents the nondeterministic choice. A 5-tuple $(\alpha_j, \beta_j, \gamma_j, c_j, \sigma_j)$ represents the fact that if in table T_j we have for some indices u and v

$$\begin{aligned} T_j(u-1, v-1) &= \alpha_j, \\ T_j(u-1, v) &= \beta_j, \\ T_j(u-1, v+1) &= \gamma_j \end{aligned}$$

and the nondeterministic choice at step $u-1$ is c_j, then (according to the transition function)

$$T_j(u, v) = \sigma_j.$$

For each $(r+1)$-tuple $(\delta_1, \ldots, \delta_{r+1})$ as above, we consider the conjunction of the following $r+1$ formulas:

$$\begin{aligned} &\forall \overline{x_1}\, \forall \overline{x}\, \forall \overline{y_1}\, \forall \overline{y}\, \forall \overline{y_2}[S_k(\overline{x_1}, \overline{x}) \wedge S_k(\overline{y_1}, \overline{y}) \wedge S_k(\overline{y}, \overline{y_2}) \\ &\qquad \wedge T_{j,\alpha_j}(\overline{x_1}, \overline{y_1}) \wedge T_{j,\beta_j}(\overline{x_1}, \overline{y}) \wedge T_{j,\gamma_j}(\overline{x_1}, \overline{y_2}) \wedge D_{c_j}(\overline{x_1}) \rightarrow T_{j,\sigma_j}(\overline{x}, \overline{y})], \end{aligned}$$

(one formula for each $j = 1, \ldots, r+1$).

The conjunction of all the formulas for all the $(r+1)$-tuples $(\delta_1, \ldots, \delta_{r+1})$ expresses condition (3).

Condition (4) is expressed by the following formula:

$$\exists \overline{x}\, \exists \overline{y}\, [\vee T_{1,(\sigma,q_{accept})}(\overline{x}, \overline{y})],$$

where the disjunction is taken over all configuration symbols σ.

The formula ϕ' from Equation (6.2) is now completely described (recall that it is the concatenation of all the formulas above) and it can be immediately checked that it is in form required in the statement of Theorem 6.2.4 (i.e., in the Π_2 form). From the construction, it follows that

$$I \models \exists L\, \exists S\, \exists S_1 \ldots \exists S_k\, \exists T_{1,\sigma_1} \ldots \exists T_{r+1,\sigma_h}\, \exists D_0\, \exists D_1\, \phi'$$

if and only if M has an accepting computation on the input enc(I). ∎

6.3 Logical characterization of NP optimization problems

IN BRIEF: Every NP optimization problem admits a logical formulation that involves a formula in first-order logic. By varying the allowed syntax for the logical formulas, one obtains several classes of NP optimization problems. The relations between these classes is investigated.

We turn to the main objective of this chapter, the investigation of optimization problems.

Let us have a look at a few problems in MAX PB and MIN PB. For each one of them, we give a standard description followed by a logical representation using the concepts introduced in the previous section.

Problem 6.3.1 MAX SAT problem:
Input: A set of clauses $C_1, \ldots, C_m$ each of them being the disjunction of some literals.
Goal: Find the maximum number of clauses that can be simultaneously satisfied by a truth assignment.

With the notations that were introduced in Section 6.2 for the SAT problem, MAX SAT can be formulated as follows:

$$\max_{\text{MAX SAT}}(I) = \max_{S,T} \{\|S\| \mid (I, S, T) \models \forall c\, \exists x\, [S(c) \rightarrow C(c) \wedge [(P(x,c) \wedge T(x)) \vee (N(x,c) \wedge \neg T(x))]\},$$

where I is the finite structure that represents the formula ϕ. The relation T represents the subset of variables that are assigned the value true and the relation S represents the subset of clauses that are satisfied. ∎

Problem 6.3.2 MAX CLIQUE (MC) Problem:
Input: A graph $G = (V, E)$ (as usual, V is the set of vertices, and $E \subseteq V \times V$ is the set of edges).
Goal: Find the size of the largest clique of G, where a clique is a set of nodes with the property that any two of them are connected by an edge.

MAX CLIQUE can be formulated as follows:

$$\max_{\text{MC}}(G) = \max_{S}\{\|S\| \mid (G,S) \models \forall x \, \forall y \, [S(x) \wedge S(y) \wedge x \neq y \rightarrow E(x,y)]\}.$$

Above, G represents the graph viewed as the finite structure with domain V and the binary relation E. The relation S represents the subset of nodes that form a clique. ∎

Problem 6.3.3 VERTEX COVER (VC) Problem:
Input: A graph $G = (V, E)$.
Goal: Find the size of the smallest set of nodes $V' \subseteq V$ such that each edge of E is incident upon some vertex of V'.

VERTEX COVER can be formulated as follows:

$$\min_{\text{VC}}(G) = \min_{S}\{\|S\| \mid (G,S) \models \forall x \, \forall y \, [E(x,y) \rightarrow (S(x) \vee S(y))]\}.$$

We have used the same notation as in MAX CLIQUE. ∎

All the above three problems are polynomially bounded; they all admit a logical characterization and syntactically these characterizations look quite similar. Recalling Theorem 6.2.4, this is not surprising and, indeed, the next theorem shows that all polynomially bounded NP optimization problems admit such a characterization.

Theorem 6.3.4 INFORMAL STATEMENT: *Every polynomially bounded* NP *optimization problem admits a logical formulation. In this formulation, the goal is to maximize or minimize the cardinality of a certain relation that is a part of a finite structure which satisfies a first-order formula in* Π_2 *form.*

FORMAL STATEMENT: *Let A be a polynomially bounded* NP *optimization problem whose inputs are represented as finite structures. There exists a finite type $\mathcal{S}$ and a closed first-order formula ϕ such that, for each input structure I,*

$$\text{opt}_A(I) = \text{opt}_S \; \{\|S_1\| \mid (I,S) \models \phi(\bar{x}, I, S)\},$$

where S is a finite structure of type $\mathcal{S}$ with the same domain as I and relations $S_1, S_2, \ldots, S_k$, $\bar{x}$ is a tuple of variables ranging over I's domain and opt is max or min. Moreover, formula ϕ has the form $\forall \bar{x} \; \exists \bar{y} \; \psi(\bar{x}, \bar{y}, I, S)$ with ψ quantifier-free.

Proof. (a) Let A be a problem in MAX PB. Let us assume that an input instance I has type σ and let d be an integer such that for any input instance I,

$$\max_{A}(I) \leq \|D_I\|^d,$$

where D_I is the domain of I. There must be such a constant d because the optimal value is polynomially bounded in the size of the input instance. Let us consider the following decision problem B associated to A.

Input: A structure I of type σ and a relation U over D_I of arity d.
Question: Is there a feasible solution $J \in \mathcal{F}_A(I)$ such that $f_A(I,J) \geq \|U\|$?

Since B is in NP, by the characterization of NP given in Theorem 6.2.4, there is a first-order formula ϕ in Π_2 form such that (I,U) is a Yes instance of B if and only if

$$(I,U) \models (\exists \overline{S})\phi(I,U,\overline{S}).$$

We claim that

$$\max_A(I) = \max_{\overline{S},U}\{\|U\| \mid (I,U,\overline{S}) \models \phi(I,U,\overline{S}).\}$$

Indeed, let $m^* = \max_A(I)$. Since $m^* \leq \|D_1\|^d$, there is a relation U over D_1^d so that $m^* = \|U\|$. It follows that (I,U) is a Yes instance of problem B and consequently

$$m^* \leq \max_{\overline{S},U}\{\|U\| \mid (I,U,\overline{S}) \models \phi(I,U,\overline{S})\}.$$

Conversely, let $\overline{S}, U$ be relations such that

$$(I,U,\overline{S}) \models \phi(I,U,\overline{S})$$

and such that $\|U\|$ is maximized with this property. Then (I,U) is a Yes instance for problem B. Hence, there is a feasible solution $J \in \mathcal{F}_A(I)$ such that $f_A(I,J) \geq \|U\|$. Thus

$$m^* \geq \max_{\overline{S},U}\{\|U\| \mid (I,U,\overline{S}) \models \phi(I,U,\overline{S})\}.$$

(b) Let A be a problem in MIN PB and I an input instance for A. In a similar way to (a), we have

$$\min_A(I) = \min_{\overline{S},U}\{\|U\| \mid (I,U,\overline{S}) \models \phi(I,U,\overline{S})\},$$

with ϕ a Π_2 formula. ∎

What about NP optimization problems that are not polynomially bounded? There are many examples of such problems. For example, each clause in an instance of MAX SAT can have a numerical weight attached to it and the goal is to find an assignment that satisfies a set of clauses with the maximum collective weight. Since the numerical weights can be represented by binary strings of length logarithmical in their value, the modified problem, weight MAX SAT, is no longer polynomially bounded. Can we find a syntactical characterization for such problems? The answer is yes. The price for removing the polynomial bounding restriction is the introduction of weights for all n_1-tuples over the I's domain, where n_1 is the arity of S_1. Note that the domain of I, being finite, can be identified with a set of the form $\{1,2,\ldots,n\}$.

Definition 6.3.5 (Weight assignment)

(1) *Let* $k \in \mathbb{N}$. *A* k*-weight assignment is a sequence of computable functions* $\{w_i\}_{i\geq 2}$*, where each* $w_i\colon \{1,2,\ldots,i\}^k \to \mathbf{R}$. *For each* k*-tuplet* $\bar{x}$ *and all* i *and* j*, if* $w_i(\bar{x})$ *and* $w_j(\bar{x})$ *are defined, then* $w_i(\bar{x}) = w_j(\bar{x})$.

(2) *A* k*-positive weight assignment is a sequence of computable functions* $\{w_i\}_{i\geq 2}$*, where each* $w_i\colon \{1,2,\ldots,i\}^k \to \mathbf{R}_+$. *For each* k*-tuplet* $\bar{x}$ *and all* i *and* j*, if* $w_i(\bar{x})$ *and* $w_j(\bar{x})$ *are defined, then* $w_i(\bar{x}) = w_j(\bar{x})$.

(3) *If* S *is a relation of arity* k *over* $\{1,2,\ldots,i\}$ *and* w *is a* k*-weight assignment, the weight of* S *is* $w(S) = \sum_{\bar{x}\in S} w_i(\bar{x})$.

Theorem 6.3.6 INFORMAL STATEMENT: *Every* NP *optimization problem admits a logical formulation. There exists a fixed way of assigning numerical weights to tuples such that, in the logical formulation, the goal is to maximize or minimize the total weight of tuples in a certain relation that is a part of a finite structure which satisfies a first-order formula in* Π_2 *form.*

FORMAL STATEMENT: *Let* A *be a (positive)* NP *optimization problem. There exists a signature* $\mathcal{S} = (\widetilde{S_1}, \widetilde{S_2}, \ldots, \widetilde{S_k})$ *with arities, respectively,* $n_1, \ldots, n_k$*, a closed first-order formula* ϕ *and a* n_1*-weight (positive) assignment* w *such that, for each input instance* I *whose set of feasible solutions is not empty,*

$$\mathrm{opt}_A(I) = \mathrm{opt}_S\ \{w(S_1) \mid (I, \overline{S}) \models \phi(\bar{x}, I, \overline{S})\}, \tag{6.3}$$

where $\overline{S}$ *is a finite structure of signature* $\mathcal{S}$ *with the same domain as* I *and relations* $S_1, S_2, \ldots, S_k$*,* $\bar{x}$ *is a tuple of variables ranging over* I*'s domain and opt is max or min. Moreover, formula* ϕ *has the form* $\forall\bar{x}\ \exists\bar{y}\ \psi(\bar{x}, \bar{y}, I, \overline{S})$ *with* ψ *quantifier-free.*

Proof. We consider the case of a maximization problem with arbitrary weights. The other cases are similar. Let A be such an NP optimization problem. We assume that the objective function f_A is integer-valued (the general case can be handled similarly). Let I be an input structure with domain $\{1,2,\ldots,n\}$. The structure I is encoded by a string whose length is bounded by a polynomial in n. Since the objective function f_A is polynomial-time computable, there exists a constant d such that, for all I,

$$|\mathrm{opt}_A(I)| \leq 2^{n^d} - 1.$$

We define inductively the following $(d+1)$-weight assignment w (at step j, we define w_j). Initially (this is step 2) we order in the lexicographical order the $(d+1)$ tuples over $\{1,2\}$ and we assign to them, in this order, the weights

$$-2^{2^d-1}, \ldots, -2^0, 2^0, 2^1, \ldots, 2^{2^d-1}.$$

At the end of step j, we have assigned the values

$$-2^{j^d-1}, \ldots, -2^0, 2^0, 2^1, \ldots, 2^{j^d-1}$$

to all $(d+1)$ tuples over $\{1,\ldots,j\}$. At stage $j+1$, we let w_{j+1} be equal to w_j on the $(d+1)$ tuples of $\{1,\ldots,j+1\}$ that do not contain $j+1$. Next, we order lexicographically all $(d+1)$ tuples over $\{1,\ldots,j+1\}$ that contain $j+1$ and assign to them the values

$$-2^{(j+1)^d-1},\ldots,-2^{j^d},2^{j^d},\ldots,2^{(j+1)^d-1},2^0,\ldots,2^0.$$

There are $(j+1)^{d+1}-j^{d+1}$ such tuples and $2((j+1)^d-j^d)$ values of the form $\pm 2^k (k \neq 0)$ to assign and, thus, all the values can be assigned.

The bottom line is that, for each $n \geq 2$ and for each integer m in the interval $[-(2^{n^d}-1), 2^{n^d}-1]$, there exists a set of $(d+1)$-tuples over $\{1,\ldots,n\}$ whose w weights sums to m (this follows from the binary representation of m). Let us suppose that A is a maximization problem (the case of a minimization problem is similar). We consider the following decision problem B:

Instance: A finite input structure $I \in \mathcal{I}_A$ with domain $\{1,2,\ldots,n\}$, a relation U over $\{1,2,\ldots,n\}$ of arity $(d+1)$, the $(d+1)$-weight assignment w defined above.

Question: Is there a feasible solution $J \in \mathcal{F}_A(I)$ such that $f_A(I,J) \geq w(U)$?

This is the decision problem associated to the NP optimization problem A and therefore is in NP. Consequently, by Theorem 6.2.4, (I,U) is a Yes instance of B if and only if there exists a quantifier-free first-order formula ψ and a finite structure R such that

$$(I,U,R) \models \forall \bar{x}\, \exists \bar{y}\, \psi(\bar{x},\bar{y},I,U,R).$$

Now, as in Theorem 6.3.4, it is easy to see that if ϕ is the formula $\forall \bar{x}\, \exists \bar{y}\; \psi(\bar{x},\bar{y},I,U,R)$, then

$$\max_A(I) = \max_{U,R}\{w(U) \mid (I,U,R) \models \phi(\bar{x},\bar{y},I,U,R)\}.$$

The proof for an arbitrary objective function (i.e., not necessarily integer-valued) is similar but more tedious. The key observation is that in polynomial time f_A can compute only rational values with a number of digits that is polynomial in n and all these possible values can be covered by weights assigned to tuples of constant arity. ∎

It should be remarked that, except for the arity, the weight assignment built in the proof of the above theorem is independent of the problem and, thus, we have proved a stronger fact. More precisely, there exists a canonical family of weight assignments $(w_1,\ldots,w_k,\ldots\}$, where, for each k, w_k is a k-assignment, such that for any NP optimization problem the weight assignment w in the expression (6.3) belongs to the family. Thus, the situation is completely similar to the polynomially bounded case from Theorem 6.3.4 (where the canonical weight assignment assigns value 1 to each tuplet). We can now classify all NP optimization problems with respect to the quantifier structure of the formulae describing them.

Definition 6.3.7 (Σ_n and Π_n formulas) *For $n \geq 1$, a $\Sigma_n(\Pi_n)$ formula is a first-order closed formula in prenex normal form that has n alternations of blocks of* *existential quantifiers and blocks of universal quantifiers starting with a block of existential (universal) quantifiers. A Σ_0 or Π_0 formula is a first-order quantifier-free closed formula.*

Let us observe that the problems from Example 6.3.1, Example 6.3.2, and from Example 6.3.3 admit also an alternative logical characterization. Thus, for MAX SAT we have

$$\max_{\text{MAX SAT}}(I) = \max_T \|\{c \mid (I,T) \models \exists x[C(c) \wedge ((P(x,c) \wedge T(x)) \vee (N(x,c) \wedge \neg T(x)))]\}\|,$$

and for MAX CLIQUE,

$$\max_{\text{MC}}(G) = \max_S \|\{x \mid (G,S) \models \forall y[S(x) \wedge S(y) \wedge x \neq y \rightarrow E(x,y)]\}\|.$$

These formulas use less quantifiers and are perhaps more natural. A similar formulation can be written for VERTEX COVER:

$$\min_{\text{VC}}(G) = \min_S \|\{x \mid (G,S) \models \forall x_1 \forall x_2[E(x_1,x_2) \rightarrow (S(x_1) \vee S(x_2))] \rightarrow S(x)\}\|.$$

Corresponding to these two types of syntax we define the following classes of NP optimization problems.

Definition 6.3.8 (Classes of optimization problems) *In all the statements below, $\overline{S}$ represents a tuple of relations defined over the domain of the input I (viewed as a finite structure). Also, ϕ is a first-order formula that only depends on the optimization problem A but not on the input instance I.*

(a) *A maximization problem A is in* MAX Σ_i (MAX Π_i), *where $i \geq 0$, if for all inputs I,*

$$\max_A(I) = \max_{\overline{S}} \|\{\overline{x} \mid (I,\overline{S}) \models \phi(\overline{x}, I, \overline{S})\}\|,$$

where ϕ is a first-order formula in Σ_i-form (Π_i-form) having $\overline{x}$ as its free variables.

(b) *A maximization problem A is in* MAX FΣ_i (MAX FΠ_i), *where $i \geq 1$, if for all inputs I,*

$$\max_A(I) = \max_{\overline{S}}\{\|S_1\| \mid (I,\overline{S}) \models \phi(\overline{x}, I, \overline{S})\},$$

where ϕ is a closed first-order formula in Σ_i-form (Π_i-form) and $\overline{S} = (S_1, \ldots)$.

(c) *The minimization classes* MIN Σ_i, MIN FΣ_i, MIN Π_i, *and* MIN FΠ_i *are defined in an analogous manner with* min *replacing* max.

(d) *In a similar way, we define the weight and the weight(+) analogues of the above classes. For example, a problem A is in* weight-MAX Σ_i (weight(+)-MAX Σ_i) *if there is a weight assignment (a positive weight assignment) so that, for all inputs I,*

$$\max_A(I) = \max_{\overline{S}} w(\{\overline{x} \mid (I, \overline{S}) \models \phi(\overline{x}, I, \overline{S})\}),$$

and ϕ is a Σ_i formula having $\overline{x}$ as its free variables.

Also, a problem A is in weight-MAX FΣ_i (weight(+)-MAX FΣ_i) *if there is a weight assignment (a positive weight assignment) so that, for all inputs I,*

$$\max_A(I) = \max_{\overline{S}}\{w(S_1) \mid (I, \overline{S}) \models \phi(\overline{x}, I, \overline{S})\}),$$

and ϕ is a closed Σ_i formula and $\overline{S} = (S_1, \ldots)$.

There are some general relations between these classes.

Proposition 6.3.9 (a) MAX FΠ_i = MAX Π_i, *for* $i \geq 1$.

(b) MAX F$\Sigma_i \subseteq$ MAX Σ_i, *for* $i \geq 1$.

(c) MAX $\Sigma_i \subseteq$ MAX FΠ_{i+1}, *for* $i \geq 1$.

(d) MIN FΠ_i = MIN Σ_i, *for* $i \geq 1$.

(e) MIN F$\Sigma_i \subseteq$ MIN Π_i, *for* $i \geq 1$.

(f) MIN $\Pi_i \subseteq$ MIN FΠ_{i+1}, *for* $i \geq 1$.

(g) *The same relations hold for the weight(+) variants of these classes.*

Proof. Let A be an optimization problem that is expressible as

$$\text{opt}_A(I) = \text{opt}_{\overline{S}}\{\|S_1\| \mid (I, \overline{S}) \models \phi(\overline{S})\},$$

where opt is max or min, ϕ is a closed first-order sentence and S_1 is the first relation symbol in the tuple of relations $\overline{S}$. It can be observed that

$$\max_{\overline{S}}\{\|S_1\| \mid (I, \overline{S}) \models \phi(\overline{S})\} = \max_{\overline{S}}\|\{\overline{x} \mid (I, \overline{S}) \models \phi(\overline{S}) \wedge S_1(\overline{x})\}\| \tag{6.4}$$

and

$$\begin{aligned}\min_{\overline{S}}\{\|S_1\| \mid (I, \overline{S}) \models \phi(\overline{S})\} &= \min_{\overline{S}}\|\{\overline{x} \mid (I, \overline{S}) \models \phi(\overline{S}) \rightarrow S_1(\overline{x})\}\| \\ &= \min_{\overline{S}}\|\{\overline{x} \mid (I, \overline{S}) \models \neg\phi(\overline{S}) \vee S_1(\overline{x})\}\|.\end{aligned} \tag{6.5}$$

Indeed, if $(I, \overline{S}) \models \phi(\overline{S})$ then

$$\{\overline{x} \mid (I, \overline{S}) \models \phi(\overline{S}) \wedge S_1(\overline{x})\} = \{\overline{x} \mid S_1(\overline{x}) \text{ is true }\}$$

and

$$\{\bar{x} \mid (I,\overline{S}) \models \phi(\overline{S}) \rightarrow S_1(\bar{x})\} = \{\bar{x} \mid S_1(\bar{x}) \text{ is true }\}$$

On the other hand, if $(I,\overline{S}) \models \neg\phi(\overline{S})$, then

$$\{\bar{x} \mid (I,\overline{S}) \models \phi(\overline{S}) \wedge S_1(\bar{x})\} = \emptyset$$

and

$$\{\bar{x} \mid (I,\overline{S}) \models \phi(\overline{S}) \rightarrow S_1(\bar{x})\} = D_I^m,$$

where D_I is the domain of the structure I and m is the arity of S_1.

These relations imply (6.4) and (6.5). It follows that

$$\mathrm{MAX\,F\Pi}_i \subseteq \mathrm{MAX\,\Pi}_i,$$

$$\mathrm{MAX\,F\Sigma}_i \subseteq \mathrm{MAX\,\Sigma}_i,$$

$$\mathrm{MIN\,F\Pi}_i \subseteq \mathrm{MIN\,\Sigma}_i,$$

and

$$\mathrm{MIN\,F\Sigma}_i \subseteq \mathrm{MIN\,\Pi}_i.$$

For the opposite direction, assume that A is an optimization problem that is expressible as

$$\mathrm{opt}_A(I) = \mathrm{opt}_{\overline{S}} \|\{\bar{x} \mid (I,\overline{S}) \models \phi(\bar{x},\overline{S})\}\|,$$

where opt is max or min and $\phi(\bar{x},\overline{S})$ is a first-order formula having $\bar{x}$ as its free variables. By introducing a new relation symbol T having the arity of $\bar{x}$, we can write

$$\begin{aligned}&\mathrm{opt}_{\overline{S}} \|\{\bar{x} \mid (I,\overline{S}) \models \phi(\bar{x},\overline{S})\}\| \\ &\qquad = \mathrm{opt}_{T,\overline{S}} \{\|T\| \mid (I,T,\overline{S}) \models \forall\bar{x}\,(T(\bar{x}) \Leftrightarrow \phi(\bar{x},\overline{S})\}. \end{aligned} \tag{6.6}$$

For a maximization problem, the right side of (6.6) is equal to

$$\max_{T,\overline{S}} \{\|T\| \mid (I,T,\overline{S}) \models \forall\bar{x}\,(T(\bar{x}) \rightarrow \phi(\bar{x},\overline{S}))\},$$

and, for a minimization problem, the right side of (6.6) is equal to

$$\min_{T,\overline{S}} \{\|T\| \mid (I,T,\overline{S}) \models \forall\bar{x}\,(\phi(\bar{x},\overline{S}) \rightarrow T(\bar{x}))\}.$$

It follows that for $i \geq 1$

$$\mathrm{MAX\,\Pi}_i \subseteq \mathrm{MAX\,F\Pi}_i,$$

$$\mathrm{MAX\,\Sigma}_i \subseteq \mathrm{MAX\,F\Pi}_{i+1}$$

and

$$\mathrm{MIN\,\Pi}_i \subseteq \mathrm{MIN\,F\Pi}_{i+1},$$

$$\text{MIN}\,\Sigma_i \subseteq \text{MIN}\,\text{F}\Pi_i.$$

The proof for the weight(+) version of the NP optimization classes is similar. ∎

From Theorem 6.3.4 we know that MAX PB $=$ MAX FΠ_2 and MIN PB $=$ MIN FΠ_2. In fact, we have seen in Theorem 6.3.6 that all NP optimization problems are in weight-MAX FΠ_2 or in weight-MIN FΠ_2. Can we characterize NP optimization problems with less quantifiers? Does the quantifier structure of formulae induce a proper hierarchy of the classes defined above? These are the questions which we study next. We focus first on maximization problems and start with the following observation.

Proposition 6.3.10 (a) MAX $\Sigma_2 =$ MAX Π_1.

(b) MAX F$\Sigma_2 =$ MAX FΠ_1.

Proof. We only have to show that MAX $\Sigma_2 \subseteq$ MAX Π_1 and MAX F$\Sigma_2 \subseteq$ MAX FΠ_1 (the converse inclusions are immediate from the definitions). Let A be a problem in MAX Σ_2. This means that if I is an input instance of A,

$$\max_A(I) = \max_{\overline{S}} \|\{\bar{x} \mid (I, \overline{S}) \models \exists \bar{y}\, \forall \bar{z}\, \psi(\overline{S}, \bar{x}, \bar{y}, \bar{z})\}\|.$$

For a tuple of relations $\overline{S}$ and a tuple $\bar{x}$, we say that a tuple $\bar{y}$ is a witness of $\bar{x}$ relative to $\overline{S}$ if

$$(I, \overline{S}) \models \forall \bar{z}\, \psi(\overline{S}, \bar{x}, \bar{y}, \bar{z}).$$

Observe that $\max_A(I)$ is equal to the number of tuples $\bar{x}$ that have a witness relative to $\overline{S}$, maximized over the choices of $\overline{S}$. We introduce a new relation symbol T and consider the set

$$\begin{aligned} C(\overline{S}, T) = \{(\bar{x}, \bar{y}) \mid (I, \overline{S}, T) \models \forall \bar{z}\, \psi(\overline{S}, \bar{x}, \bar{y}, \bar{z}) \wedge \\ T(\bar{x}, \bar{y}) \wedge \forall \bar{y_1}\, \forall \bar{y_2}\, (T(\bar{x}, \bar{y_1}) \wedge T(\bar{x}, \bar{y_2}) \rightarrow (y_1 = y_2))\}. \end{aligned} \tag{6.7}$$

The set $C(\overline{S}, T)$ consists of pairs of the form ($\bar{x}$, witness-of-$\bar{x}$) with the constraint that for each $\bar{x}$ at most one such pair is allowed if $\bar{x}$ has a witness (of course, no such pair is in $C(\overline{S}, T)$ if $\bar{x}$ has no witness). For each $\overline{S}$ there is a relation T such that

$$\|C(\overline{S}, T)\| = \text{ number of tuples } \bar{x} \text{ that have a witness relative to } \overline{S}.$$

Hence,

$$\max_A(I) = \max_{\overline{S}, T} \|C(\overline{S}, T)\|$$

and taking into account the syntactical description of $C(\overline{S}, T)$ given in (6.7), it follows that MAX $\Sigma_2 \subseteq$ MAX Π_1. Also, taking into consideration Proposition 6.3.9,

$$\text{MAX}\,\text{F}\Sigma_2 \subseteq \text{MAX}\,\Sigma_2 = \text{MAX}\,\Pi_1 = \text{MAX}\,\text{F}\Pi_1.$$

■

The next theorem clarifies the structure of the classes of NP optimization problems.

Theorem 6.3.11 (a) $\mathrm{MAX\,F\Sigma_1} \subsetneq \mathrm{MAX\,F\Pi_1} = \mathrm{MAX\,F\Sigma_2} \subsetneq \mathrm{MAX\,F\Pi_2} =$ MAX PB.

(b) weight(+)-$\mathrm{MAX\,F\Sigma_1} \subsetneq$ weight(+)-$\mathrm{MAX\,F\Pi_1} =$ weight(+)-$\mathrm{MAX\,F\Sigma_2} \subsetneq$ weight(+)-$\mathrm{MAX\,F\Pi_2}$.

Proof. From Theorem 6.3.4 we know that MAX PB $= \mathrm{MAX\,F\Pi_2}$ and from Proposition 6.3.10, it follows that

$$\mathrm{MAX\,F\Sigma_1} \subseteq \mathrm{MAX\,F\Sigma_2} = \mathrm{MAX\,F\Pi_1}.$$

We will show that MAX CLIQUE (MC) is in $\mathrm{MAX\,F\Pi_1} - \mathrm{MAX\,F\Sigma_1}$ and MAX CONNECTED COMPONENTS (MCC) is not in $\mathrm{MAX\,F\Pi_1}$. (MCC will be defined below.) Similar relations hold for the weight(+) version variants of MAX CLIQUE and of MAX CONNECTED COMPONENTS. This will prove (a) and (b).

By inspecting Example 6.3.2, we see that MAX CLIQUE is in $\mathrm{MAX\,F\Pi_1}$. Let us suppose that

$$\max_{\mathrm{MC}}(G) = \max_{\overline{S}}\{\|S_1\| \mid (G, \overline{S}) \models \exists \bar{x}\, \phi(\bar{x}, \overline{S})\}, \tag{6.8}$$

for some quantifier-free first-order formula ϕ. Consider two isomorphic disjoint graphs G_1 and G_2 and let G be the graph obtained by taking the union of G_1 and G_2. Let $\bar{S^1} = (S_1^1, \ldots, S_k^1)$ and $\bar{S^2} = (S_1^2, \ldots, S_k^2)$ be two tuples of relations that achieve the maximum in (6.8) for G_1 and, respectively, G_2. Define

$$\overline{S} = \bar{S^1} \cup \bar{S^2} = (S_1^1 \cup S_1^2, \ldots, S_k^1 \cup S_k^2).$$

The structures $(G_1, \bar{S^1})$ and $(G_2, \bar{S^2})$ are substructures of $(G, \overline{S})$ and since $(G_1, \bar{S^1}) \models \exists \bar{x} \phi(\bar{x}, \bar{S^1})$ and $(G_2, \bar{S^2}) \models \exists \bar{x} \phi(\bar{x}, \bar{S^2})$, it follows immediately that

$$(G, \overline{S}) \models \exists \bar{x}\, \phi(\bar{x}, \overline{S}), \tag{6.9}$$

because existential formulas, such as (6.6) are closed under extensions of structures. Consequently,

$$\max_{\mathrm{MC}}(G) \geq \|S^1\| + \|S^2\| = \max_{\mathrm{MC}}(G_1) + \max_{\mathrm{MC}}(G_2).$$

This is false because, from the definition of a clique,

$$\max_{\mathrm{MC}}(G) = \max_{\mathrm{MC}}(G_1) = \max_{\mathrm{MC}}(G_2).$$

The same argument shows that the weight(+) version of MAX CLIQUE is not in weight(+)-$\mathrm{MAX\,F\Sigma_1}$.

For separating MAX FΠ_2 from MAX FΠ_1 and weight(+)-MAX FΠ_2 from weight(+)-MAX FΠ_1, we introduce the problem MAX CONNECTED COMPONENT (MCC) (this is not a very interesting problem but serves well our purpose). For a change, we consider directly weight(+)-MCC (the unweighted case is similar).

Input: A graph $G = (V, E)$, a weight function $w\colon V \to \mathbf{R}_+$.

Goal: Determine the connected component of G of maximum weight. (The weight of a component is the sum of the weights of the vertices in the component.)

From Theorem 6.3.6, weight(+)-MCC is in weight(+)-MAX FΠ_2. Let us suppose that this problem is in weight(+)-MAX FΠ_1. Thus,

$$\max_{\text{MCC}}(G) = \max_{\overline{S}}\{w(S_1) \mid (G, \overline{S}) \models \forall \bar{x}\, \phi(\bar{x}, \overline{S})\}, \tag{6.10}$$

with ϕ a first-order, quantifier-free formula and w a weight function. Let m be the arity of S_1 and let W_n be the weight (according to the weight function w) of all tuples of arity m over $\{1, \ldots, n\}$. We consider a specific input instance for MCC: the graph G is the star graph, i.e., G has vertices $a_1, \ldots, a_n$, a_1 is connected to all the other nodes, and there is no other edge; the weight function is defined as follows:

$$\begin{aligned} w(a_1) &= W_n \\ w(a_i) &= 1, \quad i = 2, \ldots, n. \end{aligned}$$

Let $\bar{S^*} = (\bar{S_1^*}, \ldots)$ be the tuple of relations achieving the optimum in (6.10) for this graph G. Then

$$w(\bar{S_1^*}) = W_n + n - 1. \tag{6.11}$$

Let G^1 be the graph obtained from G by deleting a_1 and the edges adjacent to it. Let $\bar{S^1} = (\bar{S_1^1}, \ldots)$ be the restriction of S^* to $\{a_2, \ldots, a_n\}$. The structure $(G^1, \bar{S^1})$ is a substructure of $(G, \bar{S^*})$ and, since formulas in Π_1-form are closed under substructures,

$$(G^1, \overline{S}^1) \models \forall \bar{x}\, \phi(\bar{x}, \bar{S^1}).$$

It follows that

$$w(S_1^1) \le \max_{\text{MCC}}(G^1) = 1. \tag{6.12}$$

By a quick inspection of (6.11) and (6.12), it follows that the weight of all tuples of arity m that contain a_1 is greater than $W_n + (n-2)$. This is false, since the sum of all m-tuples is W_n. ∎

Let us turn now to classes of minimization problems.

Theorem 6.3.12 (1) MIN FΣ_1, MIN FΠ_1 $\subsetneq$ MIN FΣ_2 $\subsetneq$ MIN FΠ_2 = MIN PB *and* MIN FΣ_1 *and* MIN FΠ_1 *are incomparable.*

(2) *The same relations hold for the weight(+) versions of the above classes.*

Proof. *Part* (a). We first show that

$$\mathrm{MIN\,F\Pi_1} \not\subseteq \mathrm{MIN\,F\Sigma_1}$$

and the similar relation for the weight(+) analogue. From Example 6.3.3, the problem VERTEX COVER is in MIN FΠ_1. A problem A in MIN FΣ_1 is expressible as

$$\min_A(I) = \min_{\overline{S}}\{\|S_1\| \mid (I, \overline{S}) \models \exists\bar{x}\phi(\bar{x}, \overline{S})\}.$$

Observe that if we take two input structures I_1 and I_2 such that I_1 is a substructure of I_2 and if $\overline{S}$ is a tuple of relations such that $(I_1, \overline{S}) \models \exists\bar{x}\,\phi(\bar{x}, \overline{S})$, then $(I_2, \overline{S}) \models \exists\bar{x}\,\phi(\bar{x}, \overline{S})$. Consequently,

$$\min_A(I_1) \geq \min_A(I_2).$$

However, VERTEX COVER does not have this property. It is easy to find two graphs G_1 and G_2 such that G_1 is a subgraph of G_2 and

$$\min_{\mathrm{VC}}(G_1) < \min_{\mathrm{VC}}(G_2).$$

The same argument shows that the problem VERTEX COVER is neither in weight(+)-MIN FΣ_1 nor in weight-MIN FΣ_1.

Part (b). To show that

$$\mathrm{MIN\,F\Sigma_1} \not\subseteq \mathrm{MIN\,F\Pi_1}$$

and the similar relation for the weight(+) version, let A be an (artificial) optimization problem that on any graph G has $\min_A(G) = 1$. The problem is in MIN FΣ_1 because

$$\min_A(G) = \min_S\{\|S\| \mid (G, S) \models \exists\bar{x}\,S(\bar{x})\}.$$

Assume that A is in MIN FΠ_1 and thus in MIN Σ_1. This would mean that

$$\min_A(G) = \min_{\overline{S}}\|\{\bar{x} \mid (G, \overline{S}) \models \exists\bar{y}\,\phi(\bar{x}, \bar{y}, \overline{S})\}\|. \tag{6.13}$$

Let G_1 and G_2 be two disjoint graphs and let G be their union graph. Let $\overline{S}^*$ be a tuple of relations that achieves the optimum in (6.13) for G. Thus,

$$\|\{\bar{x} \mid (G, \overline{S}^*) \models \exists\bar{y}\,\phi(\bar{x}, \bar{y}, \overline{S})\}\| = 1. \tag{6.14}$$

Let $\overline{S}^{*,1}$ and $\overline{S}^{*,2}$ be the restriction of $\overline{S}^*$ to the vertices of G_1, respectively G_2. Since existential formulas are closed under extensions, for $i = 1, 2$,

$$\{\bar{x} \mid (G_i, \overline{S}^{*,i}) \models \exists\bar{y}\,\phi(\bar{x}, \bar{y}, \overline{S}^{*,i})\} \subseteq \{\bar{x} \mid (G, \overline{S}^*) \models \exists\bar{y}\,\phi(\bar{x}, \bar{y}, \overline{S}^*)\}.$$

Since $\min_A(G_i) = 1$, for $i = 1, 2$, it means that

$$\|\{\bar{x} \mid (G_i, \overline{S}^{*,i}) \models \exists \bar{y}\, \phi(\bar{x}, \bar{y}, \overline{S}^{*,i})\}\| \geq 1$$

and, since these sets for $i = 1, 2$ are disjoint,

$$\|\{\bar{x} \mid (G, \overline{S}^{*}) \models \exists \bar{y}\, \phi(\bar{x}, \bar{y}, \overline{S}^{*})\}\| \geq 2,$$

which contradicts (6.14).

The same argument works for the weight(+) case.

Part (c). Obviously, from the syntactical definition, $\mathrm{MIN\,F\Pi_1} \subseteq \mathrm{MIN\,F\Sigma_2}$ and $\mathrm{MIN\,F\Sigma_1} \subseteq \mathrm{MIN\,F\Sigma_2}$. From Part (a) and Part (b) it follows that these inclusions are in fact strict. Thus, we have shown that $\mathrm{MIN\,F\Pi_1} \subsetneqq \mathrm{MIN\,F\Sigma_2}$ and $\mathrm{MIN\,F\Sigma_1} \subsetneqq \mathrm{MIN\,F\Sigma_2}$.

Part (d). We next show that

$$\mathrm{MIN\,F\Sigma_2} \subsetneqq \mathrm{MIN\,F\Pi_2}$$

and the analogous relation for the weight(+) versions of these classes. To this aim, we introduce the problem MIN CYCLE (M-CYC):

Input: A graph $G = (V, E)$.

Goal: Determine the size of the shortest cycle in G.

By Theorem 6.3.4, M-CYC is in weight(+)-MIN $\mathrm{F\Pi_2}$. The assumption that this problem is in weight(+)-MIN $\mathrm{F\Sigma_2}$ would imply that we can write

$$\min_{\text{M-CYC}}(G) = \min_{\overline{S}}\{\|S_1\| \mid (G, \overline{S}) \models \exists \bar{x} \forall \bar{y}\, \phi(\bar{x}, \bar{y}, \overline{S})\} \tag{6.15}$$

Let the arity of $\bar{x}$ in the above expression be m. Consider a graph $G = (V, E)$ consisting of two disjoint cycles $(a_1, \ldots, a_n)$ and $(b_1, \ldots, b_{n+1})$, with $n > m$. Let $\bar{S^*}$ be the relation achieving the minimum in (6.15) for this graph G and let $\bar{c} = (c_1, \ldots, c_m)$ be an m-tuple of constants such that, for this graph G,

$$(G, \bar{S^*}) \models \forall \bar{y}\, \phi(\bar{c}, \bar{y}, \bar{S^*}).$$

Note also that

$$\|\bar{S_1^*}\| = \min_{\text{M-CYC}}(G) = n.$$

Since $n > m$, there is a node a_i in the first cycle that is distinct from any c_j, $j = 1, \ldots, m$. Let G^1 be the graph obtained by deleting in G the node a_i and the edges incident to it. Let $\bar{S^1}$ be the restriction of $\bar{S^*}$ to $V - \{a_i\}$. Since formulas in Π_1-form are closed under substructures,

$$(G^1, \bar{S^1}) \models \forall \bar{y}\, \phi(\bar{c}\bar{y}, \bar{S^1}).$$

It follows that $\min_{\text{M-CYC}}(G^1) \leq \|\bar{S_1^1}\| \leq \|\bar{S_1^*}\| = n$, which is false, because, clearly, $\min_{\text{M-CYC}}(G^1) = n + 1$. The same argument can be used for separating weight(+)-MIN $\mathrm{F\Sigma_2}$ from weight(+)-MIN $\mathrm{F\Pi_2}$. ∎

The exposition of the (known) relations between the syntactically defined classes of NP optimization problems is now complete and the results that we have established are depicted in Figure 6.1.

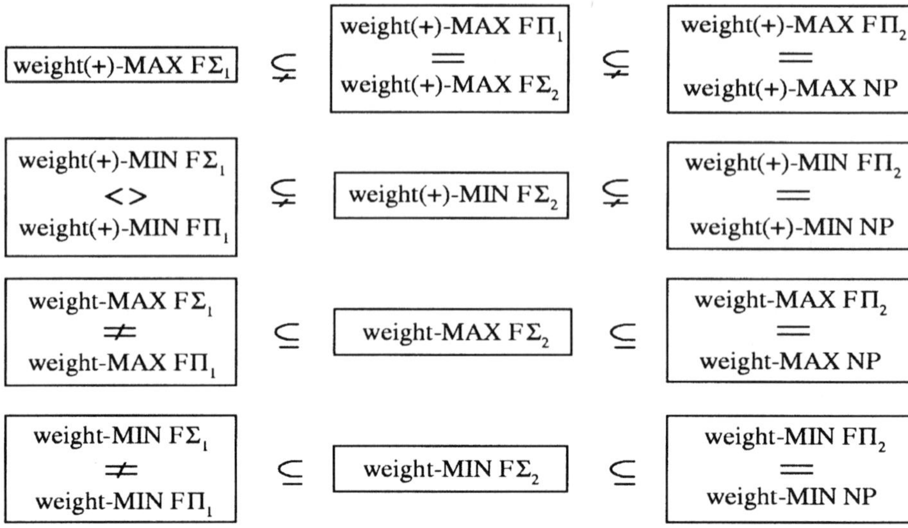

Figure 6.1: The relations between the syntactically defined NP optimization classes. (<> denotes incomparability.)

6.4 Approximation properties of maximization problems

IN BRIEF: A maximization problem whose logical representation involves a first-order formula with only existential quantifiers is constant approximable. Some natural problems, such as MAX 3-SAT, MAX CUT, MAX SUBDAG, admit such a logical characterization.

It is perhaps surprising that the approximation properties of many natural and important NP optimization problems are dictated by the problems' logical characterizations that we have seen in the previous section (or by some simple refinements of them). We start the exploration of this issue with MAX Σ_1.

Theorem 6.4.1 *weight*(+)-MAX $\Sigma_1 \subseteq$ APX.

Proof. We first illustrate the idea of the proof by sketching an algorithm for the MAX 3-SAT problem (given a boolean formula ϕ in 3-CNF form, find an assignment that satisfies as many clauses as possible). The crucial observation is that a random assignment satisfies any given clause C of the formula ϕ with probability $7/8$. Indeed, there are 8 possible truth assignments for the literals in C, and only one of these assignments makes the clause false. By the linearity of expectation, the expected number of clauses satisfied by a random assignment is $7/8 \cdot m$, where

m is the number of clauses in the formula ϕ. This, of course, is at least $7/8\cdot$ (maximum number of clauses that can be simultaneously satisfied). We can find deterministically and in polynomial time an assignment that performs at least as good as the expected value of a random assignment as follows. For the first variable x_1, we set $x_1 = true$ and then $x_1 = false$, and we determine in both cases the expected number of satisfied clauses conditioned by our setting of x_1. We choose for x_1 the truth value that yields a larger expected value. Then we proceed in the same way for x_2, then for x_3, and so on.

This idea can be generalized to any problem in weight(+)-MAX Σ_1. Let A be a problem in weight(+)-MAX Σ_1. Then if I is an input instance for A (which is viewed as a finite structure),

$$\max_A(I) = \max_{\overline{S}} w(\{\overline{x} \mid (I, \overline{S}) \models \exists\overline{y}\, \psi(\overline{S}, \overline{x}, \overline{y})\}),$$

where w is a weight assignment, ψ is a quantifier-free first-order formula and $\overline{S} = (S_1, \ldots, S_k)$ is a tuple of relation symbols. The arities of $\overline{x}$ and $\overline{y}$ are denoted by $n_{\overline{x}}$ and respectively $n_{\overline{y}}$. We will first show that, if we assign the values of the relations in $\overline{S}$ randomly, the expected value of $w(\{\overline{x} \mid (I, \overline{S}) \models \exists\overline{y}\, \psi(\overline{S}, \overline{x}, \overline{y})\})$ is within a multiplicative constant from the optimum. After that, we present an algorithm that outputs a value that is at least as good as the expected value.

Let I be a fixed structure over the universe $\{0, \ldots, n-1\}$. Each of the relations S_i, $i = 1, \ldots k$, can be viewed as a function $S_i\colon \{0, \ldots, (n-1)\}^{s_i} \to \{0, 1\}$, where s_i is the arity of S_i. According to our plan, we assign the values of $S_1, \ldots, S_k$ independently and uniformly at random and we introduce the random variable

$$X(\overline{S}) = w(\{\overline{u} \mid (I, \overline{S}) \models \exists\overline{y}\, \psi(\overline{S}, \overline{u}, \overline{y})\}).$$

We have that

$$\max_A(I) = \max_{\overline{S}}(X(\overline{S})) \geq E_{\overline{S}}(X(\overline{S})).$$

Let

$$\text{FEASIBLE} = \{\overline{u} \in \{0, \ldots, n-1\}^{n_{\overline{x}}} \mid \exists\overline{S} \text{ so that } (I, \overline{S}) \models \exists\overline{y}\, \psi(\overline{S}, \overline{u}, \overline{y})\}.$$

For each tuplet $\overline{u} \in$ FEASIBLE, we define the random variable

$$X_{\overline{u}}(\overline{S}) = \begin{cases} w(\overline{u}), & \text{if } (I, \overline{S}) \models \exists\overline{y}\, \psi(\overline{S}, \overline{u}, \overline{y}), \\ 0, & \text{otherwise.} \end{cases}$$

Clearly, $X(\overline{S}) = \sum_{\overline{u} \in \text{FEASIBLE}} X_{\overline{u}}(\overline{S})$ and

$$E(X(\overline{S})) = \sum_{\overline{u} \in \text{FEASIBLE}} E(X_{\overline{u}}(\overline{S})) \leq w(\text{FEASIBLE}).$$

Let us fix now $\overline{u} \in$ FEASIBLE. There exists a tuple of relations $\overline{S^*} = (S_1^*, \ldots, S_k^*\}$ and a tuple $\overline{v}$ in $\{0, \ldots, n-1\}^{n_{\overline{y}}}$ such that $(I, \overline{S^*}) \models \psi(\overline{S^*}, \overline{u}, \overline{v})$.

MAX CUT can be formulated as follows:

$$\max_{\text{MAX CUT}}(G) = \max_S \|\{(x,y) \mid (G,S) \models E(x,y) \wedge ((S(x) \wedge \neg S(y)) \vee (\neg S(x) \wedge S(y)))\}\|.$$

Problem 6.4.4 INDEPENDENT SET-B Problem:
Input: A graph $G = (V, E)$ with the degree of each node bounded by a constant b.
Goal: Find the size of the largest independent set (i.e., a set of nodes so that no two of them are adjacent).

A graph with bounded degree b can be represented by a relation A of arity $b+1$, where a tuple $(u, v_1, \ldots, v_b)$ is in A if the node u has the neighbors $v_1, \ldots, v_b$ (if there are less than b neighbors we repeat one of them). Then

$$\max_{\text{IND.SET-B}}(G) = \max_S \|\{(u, v_1, \ldots, v_b) \mid A(u, v_1, \ldots, v_b) \wedge S(u) \wedge \neg S(v_1) \wedge \ldots \wedge \neg S(v_k)\}\|$$

As we have mentioned, the class MAX Σ_0 is more commonly known as MAX SNP. A variant of MAX SNP is MAX SNP(π) in which the relation upon which the maximum is searched is required to be a permutation (i.e., a linear ordering) of the domain.

Definition 6.4.5 (MAX SNP(π)) *An optimization problem A is in* MAX SNP(π) *if there is a quantifier-free first-order formula ϕ such that, for any instance I,*

$$\max_A(I) = \max_\pi \|\{\overline{x} \mid (I, \pi) \models \phi(\overline{x}, \pi)\}\|,$$

where the maximum is taken over all permutations π of the domain of I. The weighted versions weight(+)-MAX SNP(π) *and weight*-MAX SNP(π) *are defined using the approach from Definition 6.3.8.*

This class contains interesting natural problems such as MAX SUBDAG and PRIORITY ORDERING.

Problem 6.4.6 MAX SUBDAG Problem:
Input: A graph $G = (V, E)$.
Goal: Find an acyclic subgraph $G' = (V, E')$ with E' as large as possible.

The following formulation shows that the problem is in MAX SNP(π).

$$\max_{\text{MAX SUBDAG}}(G) = \max_\pi \|\{(x,y) \mid (G,\pi) \models E(x,y) \wedge \pi(x) < \pi(y)\}\|$$

MAX SUBDAG can be approximated with an approximation ratio of 2 in the following way. One can take an arbitrary permutation π_1 and its inverse π_2 and calculate, for $i = 1, 2$,

$$A_i = \{(x, y) \mid E(x, y) \wedge \pi_i(x) < \pi_i(y)\}$$

and select the permutation π_i with a bigger A_i. Observe that A_1 and A_2 are disjoint and $A_1 \cup A_2 = E$. Thus, $\|A_1\| + \|A_2\| = \|E\|$. Consequently,

$$\max(\|A_1\|, \|A_2\|) \geq \frac{1}{2}\|E\| \geq \frac{1}{2} \max_{\text{MAX SUBDAG}}(G).$$

One can easily adapt the proof of Theorem 6.4.1 to show that all the problems in MAX SNP(π) (in fact, all problems in weight(+)-MAX SNP(π)) are in APX. The PRIORITY ORDERING problem, mentioned above, is defined as follows.

Problem 6.4.7 PRIORITY ORDERING (PO) Problem:
Input: A set V and real-valued weights for all pairs of distinct elements in V (the weight of a pair (x, y) represents the value of scheduling x before y).
Goal: Find a permutation π of V that maximizes

$$\sum_{\{(x,y) \mid \pi(x) < \pi(y)\}} w(x, y).$$

The problem is in weight-MAX SNP(π) because we can write

$$\max_{\text{PO}}(I) = \max_{\pi} w(\{(x, y) \mid (I, \pi) \models \pi(x) < \pi(y)\}).$$

6.5 Approximation properties of minimization problems

IN BRIEF: Two classes of minimization problems are introduced based on some syntactical restrictions in their logical representation. All the problems in the first class are constant approximable, and all the problems in the second class are log approximable. VERTEX COVER is a canonical representative of the first class, and SET COVER is a canonical representative of the second class.

We now turn to minimization problems. The syntactically-defined classes for minimization classes that we have seen in Section 6.3 do not have general approximation properties, but some refinements of them do. More precisely, two syntactically-defined classes, MIN $\mathrm{F}^+\Pi_1$ and MIN $\mathrm{F}^+\Pi_2(1)$, both containing important natural properties, have been identified to be in APX and, respectively, in log-APX.

Similarly,

$$\sum_{e \in U} \Big(\sum_{\{S | e \in S\}} x_S \Big) y_e \geq \sum_{e \in U} y_e.$$

Since

$$\sum_{S \in \mathcal{S}} \Big(\sum_{e \in S} y_e \Big) x_S = \sum_{e \in U} \Big(\sum_{\{S | e \in S\}} x_S \Big) y_e,$$

the conclusion follows. ∎

Thus a feasible solution for the dual problem provides a lower bound for the optimum solution of the original problem. The following algorithm utilizes this property: It builds in the set *Sol* a solution for the SET COVER problem whose value will be related to a feasible solution for the dual program. It is a greedy-type algorithm because, at each iteration, it selects the subset that covers new elements in the most cost-effective way, where the efficiency is measured by the ratio between the cost of the subset and the number of elements that are newly covered.

$Sol \leftarrow \emptyset, CoveredElems \leftarrow \emptyset$
while $CoveredElems \neq U$
  $\alpha = \min_{\{1,\ldots,k\}-Sol} \dfrac{c(S_i)}{\|S_i - CoveredElems\|}$;
  pick i for which the minimum is achieved;
  for all $x \in S_i - CoveredElems$, price$(x) \stackrel{\text{def}}{=} \alpha$;
  $Sol \leftarrow Sol \cup \{i\}$;
  $CoveredElems \leftarrow CoveredElems \cup S_i$;
end-while

The function price$(\cdot)$, defined for elements $x \in U$, is only needed for the analysis of the algorithm. For each $x \in U$, let

$$y_x = \frac{\text{price}(x)}{H_n}.$$

We show that the vector $y_{x_1}, \ldots, y_{x_n}$ is a feasible solution of the dual program. Indeed let $S \in Sol$ and let us suppose that S has ℓ elements. We number these elements in the order in which they enter the set *CoveredElems* in the greedy algorithm, obtaining $z_1, \ldots, z_\ell$. Consider the step at which z_i enters *CoveredElems*. Prior to this step, S contains at least $\ell - (i-1)$ uncovered elements. It follows that

$$\text{price}(z_i) \leq \frac{c(S)}{\ell - i + 1}.$$

Thus,

$$y_{z_i} \leq \frac{1}{H_n} \cdot \frac{c(S)}{\ell - i + 1}.$$

Therefore,

$$\sum_{i=1}^{\ell} y_{z_i} \leq c(S) \cdot \frac{1}{H_n}\left(\frac{1}{\ell} + \frac{1}{\ell - 1} + \ldots + 1\right) = \frac{H_\ell}{H_n} c(S) \leq c(S).$$

Thus, the vector $y_{x_1}, \ldots, y_{x_n}$ is a dual feasible solution and from Proposition 6.5.4 it follows that $\sum_{x \in U} y_x \leq OPT$, where OPT is the value of the optimum solution of (6.18) and thus it is, of course, the value of a feasible primal solution.

Observe now that at each iteration of the while loop in the greedy algorithm, the cost of the new set S that enters the solution Sol is distributed to the elements of S that are now covered for the first time. In other words, $c(S) = \sum \text{price}(x)$, where the sum is taken over the elements $x \in S$ that are now covered for the first time. Thus the cost of the solution constructed by the greedy algorithm is $\sum_{S \in Sol} c(S) = \sum_{x \in U} \text{price}(x)$. Now the conclusion follows easily, because $\sum_{S \in Sol} c(S) = \sum_{x \in U} \text{price}(x) = H_n(\sum_{x \in U} y_x) \leq H_n \cdot OPT$. ∎

Let us now consider the canonical problem for MIN $\text{F}^+\Pi_1$. VERTEX COVER can be approximated in polynomial time with approximation ratio 2 by building a maximal matching and taking the two endpoints of all the edges in the matching. The following algorithm does this:

```
Sol ← ∅
while there are edges left in G
        Pick an edge e = (u, v);
        Insert u and v in the covering, i.e., Sol ← Sol ∪ {u, v};
        Remove from G all the edges incident upon u or upon v;
end-while
```

The above algorithm clearly constructs a covering of G because an edge is not eliminated unless it has been covered, and, at the end, all the edges have been eliminated. On the other hand any vertex covering of G must cover the edges in the maximal matching by distinct nodes, because no two such edges have a common node. Since the number of edges picked in the maximal matching is $\|Sol\|/2$ it follows that $(\|Sol\|/2) \leq \min_{\text{VC}}(G)$.

This idea can be extended to the weight(+) version of VERTEX COVER. Observe first that VERTEX COVER is a particular case of SET COVER in which each element of the universe (in this case, the edges of G) belongs to exactly two subsets (in this case, a subset consists of the edges incident upon a node). It is easy to see from the proof of Theorem 6.5.3 that if each subset S has the cardinality bounded by a constant b, then the approximation ratio of the algorithm presented in the proof is bounded by a constant, namely H_b. We shall show that if each element in an instance of SET COVER belongs to at most b subsets, then again a constant approximation ratio can be achieved. This will prove that weight(+)-VERTEX COVER is in APX.

Theorem 6.5.5 *Let $b > 0$ be an integer constant and let* SET COVER(b) *be the* SET COVER *problem restricted to the instances in which each element belongs to at most b subsets. Then* SET COVER(b) *can be approximated in polynomial time with approximation ratio b.*

Proof. The proof utilizes again the framework of linear programming. Consider an instance with universe $U = \{1, \ldots, m\}$ and with subsets $S_j \subseteq U$, $j = 1, \ldots, n$, each subset S_j having the positive weight w_j. Let $A = (a_{ij})_{i=1,\ldots,m; j=1,\ldots,n}$ be the matrix defined by

$$a_{ij} = \begin{cases} 1, & \text{if the element } i \text{ is in subset } S_j, \\ 0, & \text{otherwise.} \end{cases}$$

Then the problem can be formulated as:

$$\begin{aligned} &\min \sum_{j=1}^{n} w_j x_j \\ &\text{subject to} \\ &\sum_{j=1}^{n} a_{ij} x_j \geq 1, i = 1, \ldots, m \\ &x_j \in \{0, 1\}, j = 1, \ldots, n. \end{aligned} \tag{6.20}$$

We relax the above problem by removing the integrality constraints and we observe that the constraints $x_j \leq 1$ are not needed. The dual of the relaxed version is:

$$\begin{aligned} &\min \sum_{i=1}^{m} y_i \\ &\text{subject to} \\ &\sum_{i=1}^{m} a_{ij} y_i \leq w_j, j = 1, \ldots, n \\ &y_i \geq 0, i = 1, \ldots, m. \end{aligned} \tag{6.21}$$

We will say that a dual feasible solution $\overline{y} = (y_1, \ldots, y_m)$ is maximal if there is no other dual feasible solution $\overline{y'} = (y'_1, \ldots, y'_m)$ with $y'_i \geq y_i$, for all $i = 1, \ldots, m$, and $\sum_{i=1}^{m} y'_i > \sum_{i=1}^{m} y_i$.

The algorithm that finds a solution for SET COVER(b) runs as follows:

Step 1: Find a maximal dual solution $\overline{y}$ for problem 6.21.

Step 2: Output the cover $C = \{j \mid \sum_{i=1}^{m} a_{ij} y_i = w_j\}$ (actually, C is the set of indices of subsets in the cover).

The first step can be done in polynomial time by solving the linear program (6.21), which clearly provides a maximal dual feasible solution (this step corresponds to finding a maximal matching in the algorithm that we have presented for VERTEX COVER). In the second step, we consider those j that saturate the constraints of (6.21).

Claim 6.5.6 *C is a feasible solution for* SET COVER(b).

Proof. Suppose it is not. Then there is an element $k \in \{1,\ldots,m\}$ that is not covered, i.e., $\sum_{i=1}^{m} a_{ij}y_i < w_j$, for all j with $k \in S_j$. Let

$$\delta = \min_{\{j|k\in S_j\}} \left\{ w_j - \sum_{j=1}^{m} a_{ij}y_i \right\} > 0.$$

Then if $e_k = (0,\ldots,1,\ldots,0)$ with the single 1 in position k, the vector $\overline{y'} = \overline{y} + \delta \cdot e_k$ is a dual feasible solution which contradicts the maximality of $\overline{y}$. ∎

Let $w(D) = \sum_{j\in D} w_j$ be the cost of a covering D, and let C^* be the optimal covering for the relaxed problem (6.20).

Claim 6.5.7 $w(C) \leq b \cdot \min_{\text{SET COVER}(b)}(G)$.

Proof. By the duality theorem (Proposition 6.5.4),

$$\sum_{i=1}^{m} y_i \leq \sum_{j=1}^{n} w_j x_j^*, \tag{6.22}$$

where $\overline{y} = (y_1,\ldots,y_m)$ is the maximal dual solution found in Step 1, and $\overline{x^*}$ is the vector corresponding to the optimal C^*. Then

$$\begin{aligned}
w(C) &= \sum_{j\in C} w_j && \text{by definition of } w(C)\\
&= \sum_{j\in C}\left(\sum_{i=1}^{m} a_{ij}y_i\right) && \text{by definition of } C\\
&\leq \max_i\left(\sum_{j\in C} a_{ij}\right)\cdot \sum_{i=1}^{m} y_i && \\
&\leq b\cdot \sum_{i=1}^{m} y_i && \text{because it is an instance of SET COVER}(b)\\
&\leq b\cdot \sum_{j=1}^{n} w_j x_j^* && \text{by Equation (6.22)}\\
&= b\cdot w(C^*) && \text{by definition of } w(C^*)\\
&\leq b\cdot \min_{\text{SET COVER}(b)}(G) && \text{because (6.20) is a relaxation.}
\end{aligned}$$

This finishes the proof of Claim 6.5.7 and of Theorem 6.5.5. ∎

Corollary 6.5.8 weight(+)-VERTEX COVER *can be approximated in polynomial time with approximation ratio* 2.

Proof. This follows immediately because as we have observed in the paragraph preceding Theorem 6.5.5, problem weight(+)-VERTEX COVER can be viewed as SET COVER(2). ∎

The constant approximability of SET COVER(b) can be transferred to all the problems in MIN $\mathrm{F}^+\Pi_1$ by using L-reductions.

Lemma 6.5.9 (a) *Let A be a problem in* MIN $\mathrm{F}^+\Pi_1$ *with b occurrences of the relation symbol S in formula (6.16). Then A is* L*-reducible to* SET COVER(b) *with the reduction parameters $\alpha = \beta = 1$.*

(b) *The same relation holds for* weight(+)-MIN $\mathrm{F}^+\Pi_1$ *and* weight(+)-SET COVER(b).

Proof. We prove the unweighted case, the other one being very similar. Let A be a problem in MIN $\mathrm{F}^+\Pi_1$. By definition, for every instance I we have

$$\min_A(I) = \min_S\{\|S\| \mid (I,S) \models \forall\overline{x}\,\phi(\overline{x},S)\},$$

where ϕ is quantifier-free and in conjunctive normal form, S is a single relation symbol with b occurrences in ϕ, all of them positive. Let m be the arity of S, t the arity of $\overline{x}$, and let n be the size of the domain of I. There are n^t possible values for $\overline{x}$; let $\{\overline{x}_1,\ldots,\overline{x}_{n_t}\}$ be the set of all these t-tuples. If ϕ_i denotes $\phi(\overline{x}_i,S)$, we can write

$$\min_A(I) = \min_S\{\|S\| \mid (I,S) \models \textstyle\bigwedge_i \phi_i\}.$$

Each ϕ_i is in conjunctive normal form and thus it is a conjunction of clauses $C_{i,1} \wedge \ldots C_{i,\ell}$ with each clause $C_{i,h}$ being of the form

$$\psi_{i,h} \vee \bigvee_{j=1}^{t_{i,h}} S(\overline{x}_{i_j}),$$

where $\psi_{i,h}$ is the part of the clause that does not contain the relation symbol S. Since we have instantiated the variables $\overline{x}$, using I, we can determine the truth value of each $\psi_{i,h}$ and simplify the formula. By repeating if needed some appearances of atoms of the form $S(\overline{x}_{i_j})$, we can assume that each simplified clause of ϕ_i contains b occurrences of the symbol S. In this way each ϕ_i has been transformed into a conjunction of clauses of the form $\phi'_i = \bigvee_{j=1}^{b} S(\overline{x}_{i_j})$, and

$$\min_A(I) = \min_S\{\|S\| \mid (I,S) \models \textstyle\bigwedge \phi'_i\}.$$

We construct an instance I' of SET COVER(b) as follows: The universe of I' consists of all b-tuples $\overline{y} = (\overline{x}_1,\ldots,\overline{x}_b)$, where each $\overline{x}_j$ is an m-tuple over the domain of I and there is some ϕ'_i with $\phi'_i = S(\overline{x}_1) \vee \ldots \vee S(\overline{x}_b)$ (thus, there is one $\overline{y}$ in the universe for each clause ϕ_i'); for each m-tuple $\overline{x}$ over the domain of I, I' will have in its collection of subsets the set $C_{\overline{x}}$ consisting of those $\overline{y}$ from the universe for which $\overline{x}$ is a component of $\overline{y}$. Clearly, each $\overline{y}$ is in at most b subsets.

It can be observed that $\mathcal{C} = \{C_{\overline{x}_{i_1}}, \ldots, C_{\overline{x}_{i_k}}\}$ is a covering for I' if and only if for $S = \{\overline{x}_{i_1}, \ldots, \overline{x}_{i_k}\}$ (viewing this time the relation S as a set of m-tuples) it holds that $(I, S) \models \forall \overline{x}\, \phi(\overline{x}, S)$. Since $\|\mathcal{C}\| = \|S\|$, it follows that $\min_A(I) = \min_{\text{SET COVER}(b)}(I')$. Consequently, we have L-reduced A to SET COVER(b) with the reduction parameters $\alpha = \beta = 1$. ∎

Thus we have established the following theorem.

Theorem 6.5.10 MIN $F^+\Pi_1$ *and* weight(+)-MIN $F^+\Pi_1$ *are in* APX.

Proof. This is an immediate consequence of the fact that SET COVER(b) is in APX and of Proposition 6.1.8. ∎

Using an L-reduction from SET COVER, we show that any problem in MIN $F^+\Pi_2(1)$ is in log-APX.

Lemma 6.5.11 (a) *Any problem in* MIN $F^+\Pi_2(1)$ *is* L*-reducible to* SET COVER *with the reduction parameters* $\alpha = \beta = 1$.

(b) *Any problem in* weight(+)-MIN $F^+\Pi_2(1)$ *is* L*-reducible to* weight(+)-SET COVER *with the reduction parameters* $\alpha = \beta = 1$.

Proof. Again, we prove only the non-weighted variant. Let A be a problem in MIN $F^+\Pi_2(1)$. This means that if I is an instance for A, then

$$\min_A(I) = \min_S\{\|S\| \mid (I, S) \models \forall \overline{x}\, \exists \overline{y} \phi(\overline{x}, \overline{y}, S)\},$$

where ϕ and S satisfy the requirements from the definition of MIN $F^+\Pi_2(1)$. Let m be the arity of S, p be the arity of $\overline{x}$, and let D be the domain of I. We say that a set $S \subseteq D^m$ covers $\overline{b} \in D^p$ if

$$(I, S) \models \exists \overline{y}\, \phi(\overline{b}, \overline{y}, S).$$

We next build an instance $I' = (U, C)$ of SET COVER. The universe of I' is D^p. For each subset $E \subseteq D^m$ of cardinality 1, we consider the set

$$C_E = \{\overline{x} \in U \mid \overline{x} \text{ is covered by } E\}.$$

We take C to be the collection of all these subsets C_E (for the weighted case, the weight of C_E is given by the weight of E).

Let R be a feasible solution for I. This means that

$$(I, R) \models \forall \overline{x}\, \exists \overline{y}\, \phi(\overline{x}, \overline{y}, R),$$

which, in other words, implies that R covers each element in D^p. Since ϕ is in DNF and each disjunct of ϕ has at most one occurrence of R, we deduce that each element in D^p is covered by a subset R' of R of cardinality 1. Therefore, the singleton subsets of R form a feasible solution for I'. Since there are $\|R\|$ such subsets, it follows that $\min_{SC}(I') \leq \min_A(I)$. Thus, the condition (a) in the definition of an L-reduction holds with $\alpha = 1$. Note that if $T = \{C_{E_1}, \ldots, C_{E_\ell}\}$ is a feasible solution for I', then $E = E_1 \cup \ldots \cup E_\ell$ is a feasible solution for I and $\|E\| = \|T\|$ (in the weighted case, the weight of E is equal to the weight of T). Consequently,

$$(\|E\| - \min_A(I)) = (\|T\| - \min_A(I)) \leq (\|T\| - \min_{SC}(I'))$$

which shows that condition (b) in the definition of an L-reduction holds as well with $\beta = 1$. ∎

As a corollary we obtain the following theorem.

Theorem 6.5.12 MIN $F^+\Pi_2(1)$ *and* weight(+)-MIN $F^+\Pi_2(1)$ *are in* log-APX.

Proof. Again this is derived immediately from Proposition 6.1.8 and the fact that SET COVER is in log-APX. ∎

6.6 Non-approximation properties

IN BRIEF: Lower bounds are demonstrated (under plausible hypothesis) for the approximation ratio that polynomial-time approximation algorithms can achieve for the arbitrary weighted versions of VERTEX COVER, MAX 3-SAT, and PRIORITY ORDERING. Some of the proofs are based on a certain type of interactive proof systems for NP, called probabilistically checkable proofs.

For NP optimization problems we cannot prove *absolute* results of the form "the problem A does not admit a polynomial-time algorithm achieving an approximation ratio $r = \ldots$." After all, if P = NP, then all NP optimization problems can be solved exactly in polynomial time. Therefore we shall content with proving *conditional* results of the form "under the hypothesis ..., the problem A does not admit a polynomial-time algorithm achieving an approximation ratio $r = \ldots$," where the hypothesis will be some very likely complexity-theoretic conjecture such as P $\neq$ NP.

One way to prove a result of this type is to build a polynomial-time reduction f from an NP complete problem, say 3-SAT, to the problem A such that for all boolean formulas ϕ there is a polynomial- time computable function g and some k with

$$\begin{aligned} \phi \in \text{3-SAT} &\rightarrow \max_A f(\phi) = g(|\phi|) \\ \phi \notin \text{3-SAT} &\rightarrow \max_A f(\phi) < \frac{1}{k} g(|\phi|) \end{aligned} \tag{6.23}$$

If we have such a reduction, then the existence of a polynomial-time approximation algorithm for A with approximation ratio k would imply a method for determining whether ϕ is in 3-SAT or not. The same idea can be applied to minimization problems. The crucial element is the gap between $g(|\phi|)$ and $(1/k)g(|\phi|)$. The following lemma displays such a reduction from 3-SAT to weight-VERTEX COVER (this problem has been defined in Example 6.3.3, and the case of positive weights has been treated in Corollary 6.5.8; here we consider the case in which negative numerical weights can also be assigned to the nodes of G).

Lemma 6.6.1 *If* P $\neq$ NP, *then, for every constant* q, *weight*-VERTEX COVER *is not approximable in polynomial time with approximation ratio* 2^{n^q}.

Proof. Let ϕ be a formula in 3-CNF having variables $x_1, \ldots, x_n$, and clauses $C_1, \ldots, C_m$. We build the following undirected graph $G = (V, E)$. The nodes in V are $x_1, \overline{x}_1, \ldots, x_n, \overline{x}_n$ and, for each clause $C_i = (\alpha \vee \beta \vee \gamma)$, we introduce the additional nodes $(\alpha, i), (\beta, i), (\gamma, i)$. The nodes described so far have each weight $W = 2^{n^q}$. There are two more special nodes, y and z, each of them having weight $(1 - nW - 2mW)/2$. For each variable x_i, there is the edge $(x_i, \overline{x}_i)$ and for each clause $C_i = (\alpha \vee \beta \vee \gamma)$ we introduce in E the edges $((\alpha, i), (\beta, i)), ((\beta, i), (\gamma, i))$ and $((\gamma, i), (\alpha, i))$ (forming a triangle). For all $(\alpha, i) \in V$, we introduce in E the edge $(\alpha, (\alpha, i))$. G also contains the edge (y, z). The nodes labeled with over-lined variables correspond to negated variables.

Observe that in order to cover an edge of the form $(x_i, \overline{x}_i)$ corresponding to the variable x_i, we need to select at least one of x_i or $\overline{x}_i$ in the vertex cover, and in order to cover the three edges of the triangle corresponding to a clause $C_i = (\alpha \vee \beta \vee \gamma)$, at least two of the nodes $(\alpha, i), (\beta, i), (\gamma, i)$ must be taken in the vertex cover. Also, one of y or z must be chosen in the vertex cover. It can be seen that if ϕ is satisfiable, there is a vertex cover of G having only the minimum number of nodes specified above and having weight $nW + 2mW + 1 - nW - 2mW = 1$. Indeed, if in the clause $C_i = (\alpha \vee \beta \vee \gamma)$, say, α is satisfied by the assignment, then we put in the covering the nodes α, (β, i) and (γ, i). On the other hand, if ϕ is not satisfiable, at least one more node, other than y or z, must be taken in any vertex cover and, thus, in this case, $\min_{VC}(G) \geq 1 + W$, because it can be easily seen that a vertex cover with the minimum number of nodes specified above would induce an assignment that satisfies the formula. Consequently, if weight-VERTEX COVER would be approximable in polynomial time with ratio less than W, then we could solve 3-SAT in polynomial time. ∎

For building reductions of the type (6.23), a surprising characterization of NP is extremely useful. Let us first review the characterization of NP given in Theorem 1.1.6. A problem A is in NP if there is a polynomial-time predicate Q and a polynomial p such that, for any string x,

$$x \in A \iff \exists w \, (|w| \leq p(|x|) \wedge Q(x, w)). \tag{6.24}$$

Indeed, if A is accepted by a polynomial-time nondeterministic Turing machine M, then for every string x accepted by M, we can take w to be an encoding of the nondeterministic choices made by M along one computation path that lead M to accept x. Clearly, the length of w is bounded by a polynomial in $|x|$. Given x and w, we can easily check in deterministic polynomial time whether M on input x and with the nondeterministic choices w accepts or not. The string w from Equation (6.24) is called a membership proof, or simply a proof, for x being in A.

By inspecting the Equation (6.24), we see that a problem A is in NP if there is a deterministic polynomial-time Turing machine, which is called a *verifier*, that evaluates $Q(x, w)$. It is convenient to consider that the verifier has in addition to its regular work tapes, two special read-only tapes (in the actual definition of a verifier, we will actually require three tapes, because we will allow a verifier to be probabilistic, and the third tape is needed for storing random bits): One tape that contains the input x, and another tape that contains the proof string, w. A string x is in A if and only if there is a proof w of length polynomial in $|x|$ such that the verifier starting with x on the input tape and with w on the special proof tape accepts. Implicitly, behind the scene, there is also a *prover*, an entity with unlimited computational power, that places the proof string on the proof tape before the verifier starts its work.

For natural problems, the proof w has usually a natural interpretation. For example for a boolean formula ϕ that is in 3-SAT, the proof w can be taken to be an assignment of truth values to the variables of ϕ. Clearly, in this example, the verifier needs to read the whole proof in order to see if it makes ϕ true. Can a more hurried verifier just browse a few bits of the proof and be convinced to accept or to reject its input?

Let us suppose that we have a set, called HASTY CNF, of boolean formulas in 3CNF (we recall that this means that ϕ is a conjunction of clauses where each clause is a disjunction of three literals) with the following property: For each formula in the set, either there is an assignment that satisfies it, or any assignment satisfies at most $9/10$ of its clauses. A verifier that is presented an assignment w that supposedly satisfies a given formula ϕ in HASTY CNF can randomly select one clause C of ϕ and check the truth values assigned to the variables appearing in C. If the assignment w satisfies ϕ, then C is satisfied as well. If the assignment w does not satisfy C, then C is satisfied by w with probability at most $9/10$. Consequently, the verifier will declare ϕ satisfiable if C is satisfied, and will say that ϕ is not satisfiable if C is not satisfied. This protocol has the following properties for any $\phi \in$ HASTY CNF:

- If ϕ is satisfiable, then there is an assignment w on which the verifier accepts with probability 1;
- If ϕ is not satisfiable, then for any assignment w, the verifier accepts with probability $\leq 9/10$.

By repeating the protocol k times, the error probability is decreased to at most $(9/10)^k$. Note that in this protocol the verifier uses at most $\log m$ random bits to select one clause among the m clauses of ϕ, and, quite remarkably, reads only 3 bits from the proof, the truth values of the three literals in the clause he has chosen.

The properties of HASTY CNF seem to be very special. Quite surprisingly, it is the case that every set in NP admits a similar protocol. Note that the above verifier is using random bits as well. Accordingly, we formally define a verifier as follows.

Definition 6.6.2 (a) *A verifier V is a polynomial-time probabilistic Turing machine that in addition to its working tapes has three read-only special tapes:*

- *the input tape, containing the input string x,*
- *the random tape, containing the random bits forming a string r that the verifier will use in its computation, and*
- *the proof tape, that contains the proof string w.*

The verifier based on the input x and on r, first decides in polynomial time what bits of the proof string w to read, reads these bits, and then performs some additional polynomial-time calculation at the end of which it accepts or it rejects the input.

(b) *Let $\mathrm{ACC}(V^w(x))$ be the probability over the random bits used by the verifier V, that x is accepted with w on the proof tape. Let*

$$\mathrm{ACC}(V(x)) = \max_w \mathrm{ACC}(V^w(x)).$$

(c) *The following parameters characterize the behavior of a verifier on a language L:*

- *the number of random bits $r(n)$, which is the maximum length of r over all input strings of length n;*
- *the number of queried bits $q(n)$, which is the maximum number of bits read from the proof tape over all input strings of length n;*
- *the completeness $c(n)$, which is a real number such that, for all $x \in L$ of length n,*
$$\mathrm{ACC}(V(x)) \geq c(n);$$
- *the soundness $s(n)$, which is a real number such that, for all $x \notin L$ of length n,*
$$\mathrm{ACC}(V(x)) \leq s(n).$$

Definition 6.6.3 *We say that a language L is in $\mathrm{PCP}_{c(n),s(n)}[r(n), q(n)]$ if there is a verifier V that accepts L with the following parameters: V uses $r(n)$ random bits, queries $q(n)$ bits of the proof string, has completeness $c(n)$, and soundness $s(n)$. We also say that V is a $\mathrm{PCP}_{c(n),s(n)}[r(n), q(n)]$ verifier.*

Note. PCP comes from *probabilistically checkable proof.*

From equation (6.24), we see that any language L in NP is in $\text{PCP}_{1,0}[0, \text{poly}(n)]$. The verifier that we have described for the hypothetical HASTY CNF is in $\text{PCP}_{1,1/2}[\log n, 3]$ (actually, the Definition 6.6.3 is used here abusively because the verifier achieves these parameters only for formulas ϕ in HASTY CNF). The following theorem, called the PCP Theorem, shows a new very surprising property of NP.

Theorem 6.6.4 *There exists positive constants c and q such that*

$$\text{NP} = \text{PCP}_{1,1/2}[c \log n, q].$$

In other words, for any problem in NP there is a "hasty" verifier that uses a logarithmic number of random bits and reads only a constant number bits from the proof. There is nothing special about the constant 1/2 in Theorem 6.6.4. If the verifier repeats its computations k times (with different random choices), the error probability becomes at most 2^{-k}. The proof of Theorem 6.6.4 is extremely complex and is beyond the scope of this book (we will present some weaker versions of it in Section 6.9).

Note that hidden in Theorem 6.6.4 there is a reduction to a maximization problem that has a gap similar to the one in the equations (6.23): If L is any language in NP, there is a verifier such that

$$x \in L \rightarrow \max_w \text{ACC}(V^w(x)) = 1,$$
$$x \notin L \rightarrow \max_w \text{ACC}(V^w(x)) \leq 1/2.$$

This fact can be exploited to show that, unless P = NP, some natural problems are hard to approximate. In such proofs, all the parameters of the PCP protocol have their role, and sometimes it is useful to have a smaller value for some parameters at the price of increasing some others. For example, a different combination of parameters that is able to handle any problem in NP is exhibited in Corollary 6.7.4.

In some situations, it is more convenient to use a different (but related to PCP) proof system, the *multi-prover interactive proof systems* (the standard abbreviation is MIP). As we have mentioned before, in a PCP protocol, a basic assumption is that the proof string is written on the proof tape before the verifier starts its computation and cannot be changed afterwards. It is as if a prover first provides the answers to all possible questions, writes them down on the proof tape and cannot later change his mind. By contrast, in a general multi-prover interactive systems, a prover answers to the questions as they are posed. Consequently, a cheating prover is more dangerous in a MIP protocol than in a PCP protocol, because he can now adapt his answers depending on the current question as well as on the previous questions and answers that have been emitted in the protocol. To counter this possibility, in a MIP protocol, there are more provers and they are not allowed to communicate with each other once the protocol has started. In this way, provers can be confronted and also no prover knows the entire history of the protocol. In view of our objectives, we also consider that each prover receives only one question of size q and returns an answer of size a (thus, a prover is regarded as a function from $\{0,1\}^q$ to $\{0,1\}^a$.) These systems are called *one-round multi-prover interactive proof systems*, abbreviated MIP_1. It can be shown that, with these constraints and with an appropriate choice of parameters, verifiers running PCP protocols and verifiers running MIP_1 protocols have the same power.

Definition 6.6.5 (a) *A* $\mathrm{MIP}_1(r,p,a,q,\epsilon)$ *is a one round multiprover interactive system in which the number of random bits is* $r(n)$*, the number of provers is* $p(n)$*, the size of each verifier's query is* $q(n)$*, the size of each prover's answer is* $a(n)$*, and the error probability is* $\epsilon(n)$*, where* n *is the size of the input (which, for conciseness, will often be omitted).*

(b) *A* $\mathrm{MIP}_1(r,p,a,q,\epsilon)$ *system involves* $p+1$ *parties: One verifier* V *and* p *provers* $P_1, P_2, \ldots, P_p$*. All these parties share a common input* x *and it is the joint goal of the provers to convince* V *to accept* x*. The interaction between* $V, P_1, \ldots, P_p$ *runs as follows. The verifier randomly selects a string* R *of length* r *and computes* $\pi(x,R) = (q_1, q_2, \ldots, q_p)$*, a* p*-tuplet of queries, all of length* q*, and sends, for all* i*, the query* q_i *to the prover* P_i*. The provers compute their answers* $a_i = P_i(x, q_i)$*, where* P_i *is the function defining the strategy of the prover* P_i *(note the overload in the notation* P_i*) and the length of all* a_i*'s is* a*. After the verifier receives the answers* $a_1, a_2, \ldots, a_p$*, it computes* $\rho(x, q_1, q_2, \ldots, q_p, a_1, a_2, \ldots, a_p)$ *which determines whether the input* x *is accepted or not. We assume that* π *and* ρ*, which define the verifier strategy, are polynomial-time computable functions, but there is no restriction on the functions* P_i*.*

(c) *Let* $\mathrm{ACC}_{V,(P_1,P_2,\ldots,P_p)}(x)$ *denote the probability that*

$$\rho(x, q_1, q_2, \ldots, q_p, a_1, a_2, \ldots, a_p) = \text{"accept,"}$$

when R *is chosen randomly of length* r *and the* q_i*'s and* a_i*'s are as above. The value of the verifier strategy* V *at* x *is the maximum of* $\mathrm{ACC}_{V,(P_1,P_2,\ldots,P_p)}(x)$ *over all* p*-tuples* $(P_1, P_2, \ldots, P_p)$ *of prover strategies. We denote this value by* $\mathrm{ACC}_V(x)$*.*

(d) *We say that* V *accepts a language* L *with error probability* ϵ *(where* $\epsilon\colon \mathbb{N} \to \mathbb{R}$ *and* $L \subseteq \Sigma^*$*) if:*

  (1) $x \in L$ *implies* $\mathrm{ACC}_V(x) = 1$*, and*

  (2) $x \notin L$ *implies* $\mathrm{ACC}_V(x) < \epsilon(|x|)$*.*

(e) *We say that* L *is in* $\mathrm{MIP}_1(r,p,a,q,\epsilon)$ *if there is a verifier* V *running the above protocol that accepts* L*.*

It is easy to see that a verifier V_1 running a PCP protocol can simulate a verifier V_2 runing a $\mathrm{MIP}_1(r,p,a,q,\epsilon)$ protocol. V_1 assumes that the proof tape contains p tracks, on track i being the answers to all possible questions of the i-th prover (this can be done by modifying the alphabet for the proof tape: A new symbol will be a p-tuple of old symbols). To simulate a query of V_2 to prover i, V_1 will prefix the query with i written in binary, read from the proof tape the position corresponding to the query, and then extract the content of track i of that position. We denote by $\mathrm{PCP}_{c,s}(r,p,a,q)$ a $\mathrm{PCP}_{c,s}(r,p)$ protocol where all queries have length q and the answers (i.e., the blocks of bits read by the verifier from the proof string) have length a. It follows that $\mathrm{MIP}_1(r,p,a,q,\epsilon) \subseteq \mathrm{PCP}_{1,\epsilon}(r,p,a,q+\log p)$. Simulating a PCP verifier V_1 with a MIP_1 verifier V_2 requires more care. The approach is to let V_2 ask Prover i the i-th query of V_1, take the answers, and then further simulate V_1 to get the verdict *accept* or *reject*. With this simulation strategy, we obtain the following relation.

Proposition 6.6.6 $\mathrm{PCP}_{1,\epsilon}(r,p,a,q) \subseteq \mathrm{MIP}_1(r,p,a,q,p^p \cdot \epsilon)$.

Proof. Let L be a language accepted by a verifier V_1 running a $\mathrm{PCP}_{1,\epsilon}(r,p,a,q)$ protocol. We consider the above simulation. If $x \in L$, there is a proof string w such that $\mathrm{ACC}(V_1^w(x)) = 1$. Then, if each prover P_i answers the questions as stipulated by w, we obtain $\mathrm{ACC}_{V_2,P_1,\ldots,P_p}(x) = 1$. Suppose now that $x \notin L$, and let $P_1,\ldots,P_p$ be provers such that $\mathrm{ACC}_{V_2,P_1,\ldots,P_p}(x) = \epsilon'$, for some $\epsilon' > 0$.

We define a random proof string as follows: For every position Q, randomly pick $i \in \{1,\ldots,p\}$, and set the Q-th bit of w to be equal to the answer of prover P_i to query Q. Then

$$
\begin{aligned}
&\mathrm{Prob}_{w,R}(V_1 \text{ accepts } x \text{ with } R \text{ and proof string } w)\\
&\quad\geq \mathrm{Prob}_R(V_2 \text{ accepts with } R \text{ and provers } P_1,\ldots,P_p) \times\\
&\qquad \mathrm{Prob}_{w,R}(\forall 1 \leq i \leq p, \text{ the answer of } P_i \text{ to the } i\text{-th query of } V_1 = \text{bit } i \text{ of } w)\\
&\quad\geq \epsilon'\left(\frac{1}{p}\right)^p.
\end{aligned}
$$

For the last inequality, we have used the fact that we can consider that the queries $Q_{R,1},\ldots,Q_{R,p}$ posed by V_1 are distinct, and thus, the events "the $Q_{R,i}$-th bit if w is the same as the answer of P_i to query $Q_{R,i}$" are independent.

It follows that there is some proof string w such that the error probability is at least $\epsilon'(1/p)^p$. However since V_1 can only err with probability ϵ, we conclude that $\epsilon' \leq \epsilon \cdot p^p$. ∎

Thus, there is a close connection between PCP and MIP_1 protocols. In the reductions that we will build, MIP_1 protocols having two provers are extremely useful. Such a protocol has been obtained by Feige and Lovász [FL92]. We state their result without proof.

Theorem 6.6.7 SAT *is in* $\mathrm{MIP}_1(O(\log^3 n), 2, O(\log^3 n), O(\log^3 n), 1/n)$.

Theorem 6.6.4 and Theorem 6.6.7 display two examples of maximization problems (namely, the problems implicitly appearing in the completeness and the soundness conditions), for which there is no feasible good approximation algorithm, unless $\mathrm{P} = \mathrm{NP}$, or $\mathrm{NP} \subseteq \mathrm{DTIME}[2^{(\log n)^{O(1)}}]$. As mentioned above, these problems can serve as starting points in reductions, and, thus, be used to show that other natural optimization problems do not admit feasible approximation algoritms (unless some unlikely facts hold true).

We start with MAX 3-SAT. As we have seen, this problem is in MAX Σ_0 and thus it is in APX. Could we strengthen this to show that the problem has a PTAS? The next result shows that if P $\neq$ NP, the answer is negative and the proof uses a reduction from a PCP verifier.

Lemma 6.6.8 *There is a polynomial-time reduction f from* SAT *to* MAX 3-SAT, *a polynomial-time computable function g, and a constant $\epsilon > 0$, such that, for any instance ϕ of* SAT,

$$\phi \in \text{SAT} \rightarrow \max_{\text{MAX 3-SAT}}(f(\phi)) = g(|\phi|)$$

$$\phi \notin \text{SAT} \rightarrow \max_{\text{MAX 3-SAT}}(f(\phi)) < \frac{1}{1+\epsilon} g(|\phi|).$$

Proof. We will consider first a problem similar to MAX 3-SAT but with fewer syntactical restrictions. This problem, let us call it FUNCTION SAT, is defined as follows. Let q be a fixed integer.

Input: A set of m boolean functions, $f_1, \ldots, f_m$, each of them being a function of q variables chosen from a common set of variables $y_1, \ldots, y_n$; the functions are given through their truth tables.

Goal: Find an assignment for $y_1, \ldots, y_n$ that maximizes the fraction of functions that evaluate to *true*.

We will first show a reduction of SAT to the above problem that exhibits a gap property similar to the one in the statement of the lemma. Since SAT $\in$ NP, by Theorem 6.6.4, SAT admits a $\text{PCP}_{1,1/2}(c \log n, q)$ verifier for some constants c and q. There are at most $2^{c \log n} = n^c$ choices for the random string r, and for each such choice at most q bits of the proof string are read. Thus, we can assume that the proof string is at most qn^c bits long. Let $N = qn^c$. We introduce the variables $y_1, \ldots, y_N$, one for each bit of the proof string. We can now view a proof string as an assignment of truth values to the variables $y_1, \ldots, y_N$ (i.e., $y_i = true$ if and only if the i-th bit of the proof string is 1). Let ϕ be a fixed instance for SAT. For each possible random string $r \in \{0,1\}^{c \log n}$, we will introduce a function f_r. To this aim, note that, with the input ϕ and the random string r being fixed, the positions of the q bits from the proof string that the verifier is reading are completely determined. Let $i_1(r), \ldots, i_q(r)$ be these positions. The function f_r has the variables $y_{i_1(r)}, \ldots, y_{i_q(r)}$ and is defined by: $f_r(b_1, \ldots, b_q) = true$ if and only if the verifier accepts the input ϕ when the bits of the proof strings in positions $i_1(r), \ldots, i_q(r)$ are $b_1, \ldots, b_q$ (or, in other words, if the verifier accepts in case the proof string assigns the values $b_1, \ldots, b_q$ to the variables $y_{i_1(r)}, \ldots, y_{i_q(r)}$). We have obtained an instance I of FUNCTION SAT that consists of the functions $(f_r)_{r \in \{0,1\}^{c \log n}}$. Note that I can be built in time polynomial in $|\phi|$.

Observe that for any proof string w,

$$\mathrm{ACC}(V^w(\phi)) = \frac{|\{f_r \mid f_r \text{ is true under the assignment } w\}|}{|(f_r)_{r\in\{0,1\}^{c\log n}}|}.$$

From Theorem 6.6.4, we have that

$$\phi \in SAT \to \max_w(\mathrm{ACC}(V^w(\phi)) = 1,$$
$$\phi \notin SAT \to \max_w(\mathrm{ACC}(V^w(\phi)) \leq 1/2.$$

It follows that

$$\phi \in SAT \to \max_{\text{FUNCTION SAT}}(I) = 1,$$
$$\phi \notin SAT \to \max_{\text{FUNCTION SAT}}(I) \leq 1/2.$$

It remains to reduce FUNCTION SAT to MAX 3-SAT. To this aim, we will make some transformations of the boolean functions $(f_r)_{r\in\{0,1\}^{c\log n}}$ that form the instance I. Let f_i be one of these functions. Since f_i has q variables, it can be written as a conjunction of at most 2^q clauses, each clause being a disjunction of at most q literals. Let $C_{i,1}, C_{i,2}, \ldots, C_{i,2^q}$ be these clauses. Then, if $\max_{\text{FUNCTION SAT}}(I) = 1$, it follows that the formula

$$\bigwedge_{i=1}^{n^c} \bigwedge_{j=1}^{2^q} C_{i,j} \tag{6.25}$$

is satisfiable.

If $\max_{\text{FUNCTION SAT}}(I) \leq 1/2$, then every assignment fails to satisfy at least half of the functions f_i, and thus at least one clause out of the 2^q clauses corresponding to each such unsatisfied f_i is unsatisfied as well. Thus, in this case, for any assignment, at least a fraction of $\frac{1}{2} \cdot \frac{1}{2^q}$ clauses are not satisfied.

For every clause $C_{i,j} = \ell_1 \vee \ell_2 \vee \ldots \ell_q$, if $q > 3$, we can write the following conjunction

$$(\ell_1 \vee \ell_2 \vee z_1) \wedge \left(\bigwedge_{t=1}^{q-4} (\overline{z_t} \vee \ell_{t+2} \vee z_{t+1}) \right) \wedge (\ell_{q-1} \vee \ell_q \vee \overline{z_{q-2}}), \tag{6.26}$$

where $z_1, \ldots, z_{k-2}$ are new variables. It can be checked that $C_{i,j}$ has a satisfying assignment if and only if the new formula has a satisfying assignment and that the two satisfying assignments, if they exist, can be taken to coincide on $\ell_1, \ldots, \ell_q$.

In this way, for each clause $C_{i,j}$ of (6.25), we get a set of $(q-2)$ clauses with 3 literals. The collection of all these clauses of size 3 form the instance J of MAX 3-SAT. The instance J has $n^c 2^q (q-2)$ clauses. If ϕ is satisfiable, then all the clauses of J can be satisfied simultaneously. Suppose now that ϕ is not satisfiable.

In this case, each assignment leaves at least a fraction of $2^{-(q+1)}$ clauses $C_{i,j}$ unsatisfied, and for each such unsatisfied $C_{i,j}$ there is at least one clause among the $q-2$ clauses in Equation (6.26) that is not satisfied. It follows that

$$\phi \in SAT \rightarrow \max_{\text{MAX 3-SAT}}(J) = n^c \cdot 2^q \cdot (q-2),$$

$$\phi \notin SAT \rightarrow \max_{\text{MAX 3-SAT}}(J) \leq \left(1 - \left(\frac{1}{q-2}\right)\frac{1}{2^{q+1}}\right) n^c \cdot 2^q \cdot (q-2).$$

Consequently, we have proved the lemma for ϵ verifying

$$\frac{1}{1+\epsilon} = 1 - \left(\frac{1}{q-2}\right)\frac{1}{2^{q+1}}. \qquad \blacksquare$$

An immediate corollary is the following.

Theorem 6.6.9 *If* P $\neq$ NP, *then* MAX 3-SAT *does not have a* PTAS. *Consequently, if* P $\neq$ NP, *there are problems in* MAX Σ_0 *that do not have a* PTAS.

Let us now turn to reductions that use MIP_1 protocols. If A is a maximization problem, the technique consists in reducing via a function Φ a $\text{MIP}_1(r,p,a,q,\epsilon)$ verifier strategy V to A in such a way that valid provers' strategies $(P_1, P_2, \ldots, P_p)$ correspond to feasible solutions, $\text{ACC}_{V,(P_1,P_2,\ldots,P_p)}(x) = f_A(J)$, where J is a feasible solution of $\Phi(x)$ and f_A is the objective function of problem A, and $\max_A \Phi(x) = 2^{r(|x|)} \cdot \text{ACC}_V(x)$.

To illustrate this method we consider the PRIORITY ORDERING problem (see Problem 6.4.7). The next theorem also shows that, very likely, weight-MAX SNP(π) does not have the good approximation property of weight(+)-MAX SNP(π) (namely, weight(+)-MAX SNP(π) is in APX).

Theorem 6.6.10 *For some constant* μ, *the problem* PRIORITY ORDERING *is not approximable in time* $O(2^{\log^\mu n})$ *with approximation ratio* $2^{\log^\mu n}$, *unless* NP $\subseteq$ DTIME$[2^{O(\log^{1/\mu} n)}]$.

Proof. Let V be a verifier executing a $\text{MIP}_1(d\log^3 n, 2, d\log^3 n, d\log^3 n, 1/n)$ protocol for SAT and let $n = |\phi|$, where ϕ is the common input of the protocol (we have used Theorem 6.6.7). Let $r(n) = 2^{d\log^e n}$. We build an instance I_ϕ for PRIORITY ORDERING (Pr. Ord) that has the following properties:

- $\max_{\text{Pr. Ord}}(I_\phi) = 2^{r(n)} \cdot \text{ACC}_V(\phi)$
- the construction of I_ϕ can be done in time $O(2^{\log^c n})$ for some constant c.

Let us show first that this is enough to establish the conclusion of the theorem. From the properties of the chosen MIP_1, we know that $\phi \in$ SAT $\rightarrow \text{ACC}_V(\phi) = 1$, and $\phi \notin$ SAT $\rightarrow \text{ACC}_V(\phi) \leq 1/n$.

From the second property of the construction it follows that $|I_\phi| \leq 2^{\log^c n}$. Note that

$$\phi \in \text{ SAT} \to \text{ ACC}_V(\phi) = 1 \to \max_{\text{Pr. Ord}} I_\phi = 2^{r(n)}$$

and

$$\phi \notin \text{ SAT} \to \text{ ACC}_V(\phi) \leq \frac{1}{n} \to \max_{\text{Pr. Ord}} I_\phi \leq \frac{1}{n} \cdot 2^{r(n)}.$$

Take $\mu = 1/c$ and suppose there is a DTIME$[2^{\log^\mu n}]$ algorithm B for A with approximation ratio $r_B(|I|) < 2^{\log^\mu |I|}$, which means $r_B(|I_\phi|) < n$. Consequently:

$$\phi \in \text{ SAT} \to B(I_\phi) \geq \frac{1}{r_B(|I_\phi|)} \cdot \max_{\text{Pr. Ord}} I_\phi > \frac{1}{n} \cdot 2^{r(n)}$$

and

$$\phi \notin \text{ SAT} \to B(I_\phi) \leq \max_{\text{Pr. Ord}} I_\phi \leq \frac{1}{n} \cdot 2^{r(n)}.$$

Since $B(I_\phi)$ can be computed in time $2^{O(\log^c n)}$, it follows that SAT can be solved in $2^{O(\log^{1/\mu} n)}$ time and, consequently, NP $\subseteq$ DTIME$[2^{O(\log^{1/\mu} n)}]$.

It remains to describe the construction of I_ϕ and to show that it has the purported properties. Let $N = 2^{d \log^3 n}$. The set of elements to be ordered is

$$X = \{s_1, s_2, s_3, s_4\} \cup X_1 \cup X_2 \cup X_3,$$

where s_1, s_2, s_3 and s_4 are four special elements called the "stakes", $X_1 = \{(u, v, a, 1) \mid u, v, a \in \Sigma^{d \cdot \log^3 n}\}$, $X_2 = \{(u, v, b, 2) \mid u, v, b \in \Sigma^{d \cdot \log^3 n}\}$ and $X_3 = \{(u, v, b, 3) \mid u, v, b \in \Sigma^{d \cdot \log^3 n}\}$. We first overview the construction. Intuitively, u and v should be interpreted as queries to the first and, respectively, to the second prover, and a and b, with or without subscripts, should be interpreted as answers provided by the first and respectively the second prover. By using large weights, we force that any arrangement π that candidates for being the optimal one achieves

(i) $s_1 <_\pi s_2 <_\pi s_3 <_\pi s_4$,

(ii) $s_2 <_\pi x <_\pi s_3$ for all $x \in X_2$,

(iii) $x <_\pi s_4$ for all $x \in X_1$, and

(iv) $s_1 <_\pi x$ for all $x \in X_3$.

For all pairs $(u, v) \in (\Sigma^{d \cdot \log^3 n})^2$ we order the set $X_{u,v,1} = \{(u, v, a, 1) \mid a \in \Sigma^{d \cdot \log^3 n}\}$ in the lexicographical order $<_l$, obtaining

$$X_{u,v,1} = \{t_1 <_l t_2 <_l \ldots <_l t_N\}.$$

We assign weights such that for a candidate ordering π, there exists $j \in \{0, \ldots, N\}$ such that

$$t_{j+1} <_\pi t_{j+2} <_\pi \ldots <_\pi t_N <_\pi s_1 <_\pi s_3 <_\pi t_1 <_\pi \ldots <_\pi t_j.$$

Similarly, for all pairs $(u,v) \in (\Sigma^{d \cdot \log^3 n})^2$ we order the set $X_{u,v,3} = \{(u,v,b,3) \mid a \in \Sigma^{d \cdot \log^3 n}\}$ in the lexicographical order $<_l$, obtaining

$$X_{u,v,3} = \{t_1 <_l t_2 <_l \ldots <_l t_N\}$$

and we force the existence of $k \in \{1, \ldots, N\}$ with

$$t_{k+1} <_\pi t_{k+2} <_\pi \ldots <_\pi t_N <_\pi s_2 <_\pi s_4 <_\pi t_1 <_\pi \ldots <_\pi t_k.$$

Using some other combinations of weights involving the "stakes" s_1, s_2, s_3, s_4, we guarantee that for all $u \in \Sigma^{d \cdot \log^3 n}$ there exists $j \in \{1, \ldots, N\}$ such that for all $v \in \Sigma^{d \cdot \log^3 n}$,

$$(u,v,a_{j+1},1) <_\pi \ldots <_\pi (u,v,a_N,1) <_\pi s_1 <_\pi s_3 \\ <_\pi (u,v,a_1,1) <_\pi \ldots <_\pi (u,v,a_j,1), \quad (6.27)$$

and for all $v \in \Sigma^{d \cdot \log^3 n}$ there exists $k \in \{1, \ldots, N\}$ such that for all $u \in \Sigma^{d \cdot \log^3 n}$,

$$(u,v,b_{k+1},3) <_\pi \ldots <_\pi (u,v,b_N,3) <_\pi s_2 <_\pi s_4 \\ <_\pi (u,v,b_1,3) <_\pi \ldots <_\pi (u,v,b_k,3). \quad (6.28)$$

These unique values of j and k for each u and v define a pair of provers' strategies (P_1, P_2). Moreover the weights are carefully defined in order to insure that

$$\max_\pi \sum_{\{(t,z) \mid \pi(t) < \pi(z)\}} w(t,z) = \mathrm{ACC}_V(x).$$

Now we proceed with the complete and formal description of the reduction. We order in the lexicographical order the sets of possible answers (i.e., the elements in $\Sigma^{d \cdot \log^3 n}$) obtaining: $a_1 < a_2 < \ldots < a_N$ and $b_1 < b_2 < \ldots < b_N$ (of course, $a_i = b_i$, but keep in mind the above intuitive interpretation). Let $S = 3N^3 + 4$ denote the number of elements in X. We define next the weights, $w(\cdot,\cdot)$, for some pairs in $X \times X$. Given an ordering π of X and a subset $Y \subseteq X$, the contribution of Y is defined to be

$$C_\pi(Y) = \sum_{x,y \in Y, \pi(x) < \pi(y)} w(x,y).$$

We call a permutation π *reasonable* if $C_\pi(X) \geq 0$. Since there exists an ordering σ (to be described below; it corresponds to a tuple of provers' strategies) such that $C_\sigma(X) \geq 0$, it follows that the optimal ordering π must be looked for among the reasonable ones.

For each 4-tuplet $(u,v,a,b) \in (\Sigma^{d \cdot \log^3 n})^4$, we denote

$$val_{u,v,a,b} = \begin{cases} \|\{R \in \Sigma^{d \cdot \log^3 n} \mid \pi(x,R) = (u,v)\}\|, & \text{if } \rho(x,u,v,a,b) = \text{"accept"} \\ 0, & \text{otherwise,} \end{cases}$$

and, by solving a linear system, we fix for each tuplet (u, v, a_i, b_h), a real value d_{u,v,a_i,b_h} such that for each $u, v \in \Sigma^{d\cdot\log^3 n}$, and each $j, k \in \{1, \ldots, N\}$,

$$\sum_{1\leq i\leq j}\sum_{1\leq h\leq k} d_{u,v,a_i,b_h} = val_{u,v,a_j,b_k}.$$

We state now the weights of pairs in $X \times X$ in six stages (1) -(6) (all the pairs that are not mentioned below have weight 0) .

(1) $w(s_2, s_1) = -M_1$, $w(s_3, s_2) = -M_1$, $w(s_4, s_3) = -M_1$,
$w(x, s_2) = -M_1$ and $w(s_3, x) = -M_1$ for all $x \in X_2$,
$w(s_4, x) = -M_1$ for all $x \in X_1$,
$w(x, s_1) = -M_1$ for all $x \in X_3$,
M_1 is defined as follows. Let k be the maximum weight for any pair defined in stages (2)-(6). Then $M_1 = \frac{S(S-1)}{2} \cdot k + 1$. The effect of these definitions is that if π is a reasonable ordering and $<_\pi$ denotes the order induced by π (i.e., $x <_\pi y$ if and only if $\pi(x) < \pi(y)$), then (because we need to avoid any pair with weight $-M$),

(i) $s_1 <_\pi s_2 <_\pi s_3 <_\pi s_4$,

(ii) $s_2 <_\pi x <_\pi s_3$ for all $x \in X_2$,

(iii) $x <_\pi s_4$ for all $x \in X_1$, and

(iv) $s_1 <_\pi x$ for all $x \in X_3$.

(2) For each pair $(u, a) \in (\Sigma^{d\cdot\log^3 n})^2$, we fix a linear order $<_o$ on the set $Y_{u,a,1} = \{(u, v, a, 1) \mid v \in \Sigma^{d\cdot\log^3 n}\}$, obtaining $Y_{u,a,1} = \{z_1 <_o z_2 <_o \ldots <_o z_N\}$. The order $<_o$ must have the following property: $\forall w\ \forall z$ [(w is the successor relative to $<_o$ of z) $\rightarrow$ (w is not the lexicographical successor of z) and (z is not the lexicographical successor of w)] and also z_1 and z_N are not the first and respectively the last elements from $Y_{u,a,1}$ with respect to the lexicographical ordering. We define the weights:

$$\begin{aligned} &w(s_3, z_1) = M_2, \qquad w(z_1, s_3) = -M_2,\\ &w(z_{i+1}, z_i) = -3M_2,\ i = 1, \ldots, N-1,\\ &w(z_N, s_1) = M_2, \qquad w(s_1, z_N) = -M_2. \end{aligned}$$

M_2 is defined by $M_2 = \frac{S(S-1)}{2} \cdot k + 1$, where k is now the maximum weight defined in the future stages (3)-(6). The effect of these definitions is that for all $u, a \in \Sigma^{d\cdot\log^3 n}$, if π is a reasonable ordering, then either all elements from $Y_{u,a,1}$ are smaller with respect to π than s_1 or all elements from $Y_{u,a,1}$ are larger with respect to π than s_3.

(3) For each pair $(v, b) \in (\Sigma^{d\cdot\log^3 n})^2$, we fix a linear order $<_o$ on the set $Y_{v,b,3} = \{(u, v, b, 3) \mid u \in \Sigma^{d\cdot\log^3 n}\}$, obtaining $Y_{v,b,3} = \{z_1 <_o z_2 <_o \ldots <_o z_N\}$. The

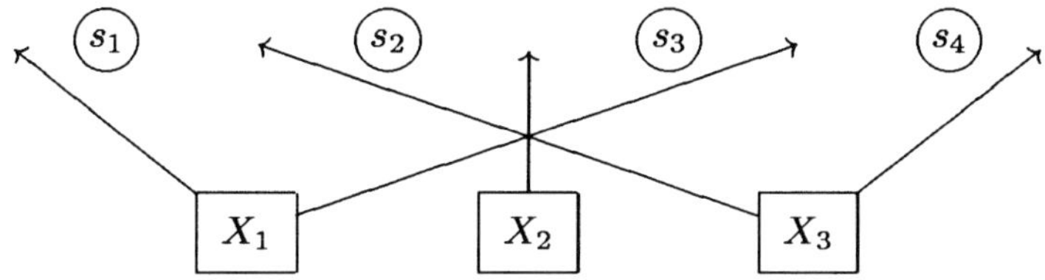

Figure 6.2: The effect of stages (1), (2), and (3): The elements of X_1, X_2 and X_3 must be positioned between the stakes s_1, s_2, s_3 and s_4 as in the figure.

order $<_o$ is as in (2). We define the weights:

$$\begin{aligned} &w(s_4, z_1) = M_3, \qquad w(z_1, s_4) = -M_3, \\ &w(z_{i+1}, z_i) = -3M_3,\ i = 1, \ldots, N-1, \\ &w(z_N, s_2) = M_3, \qquad w(s_2, z_N) = -M_3. \end{aligned}$$

M_3 is defined by $M_3 = \frac{S(S-1)}{2} \cdot k + 1$, where k is the maximum weight defined in the future stages (4)-(6). The effect of these definitions is that for all $v, b \in \Sigma^{d \cdot \log^3 n}$, if π is a reasonable ordering, then either all elements from $Y_{v,b,3}$ are smaller with respect to π than s_2 or all elements from $Y_{v,b,3}$ are larger with respect to π than s_4.

(4) For all pairs $(u, v) \in (\Sigma^{d \cdot \log^3 n})^2$, we order the set $X_{u,v,1} = \{(u, v, a, 1) \mid a \in \Sigma^{d \cdot \log^3 n}\}$ in the lexicographical order $<_l$, obtaining

$$X_{u,v,1} = \{t_1 <_l t_2 <_l \cdots <_l t_N\}.$$

We define the weights:

$$\begin{aligned} &w(s_3, t_1) = M_4, \qquad w(t_1, s_1) = -M_4, \\ &w(t_{i+1}, t_i) = -M_4,\ i = 1, \ldots, N-1, \\ &w(s_1, t_N) = -M_4, \end{aligned}$$

with $M_4 = \frac{S(S-1)}{2} \cdot k + 1$, where k is now the maximum weight defined in the future stages (5)–(6). It can be seen that, by the precautions we took when selecting the order $<_o$, there is no conflict between the definitions in stages (2) and (4). It is important to note that, relative to an ordering π, if there exists $j \in \{1, \ldots, N\}$ such that

$$t_{j+1} <_\pi t_{j+2} <_\pi \cdots <_\pi t_N <_\pi s_1 <_\pi s_3 <_\pi t_1 <_\pi \cdots <_\pi t_j, \tag{6.29}$$

then $C_\pi(X_{u,v,1} \cup \{s_1, s_3\}) = 0$ and $C_\pi(X_{u,v,1} \cup \{s_1, s_3\})$ is less than $-M_4$ in all other situations. Hence, a reasonable ordering π must satisfy the above inequalities.

(5) For all pairs $(u, v) \in (\Sigma^{d \cdot \log^3 n})^2$, we order the set $X_{u,v,3} = \{(u, v, b, 3) \mid a \in \Sigma^{d \cdot \log^3 n}\}$ in the lexicographical order $<_l$, obtaining $X_{u,v,3} = \{t_1 <_l$

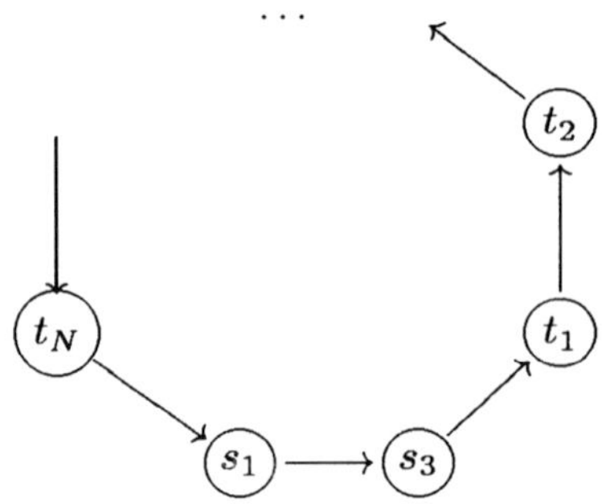

Figure 6.3: The effect of stage (4): Only one of the edges $(t_1, t_2), \ldots, (t_N, s_1)$ should be "broken" by a reasonable π.

$t_2 <_l \ldots <_l t_N\}$. We define the weights:

$$\begin{aligned} &w(s_4, t_1) = M_5, \quad w(t_1, s_4) = -M_5, \\ &w(t_{i+1}, t_i) = -M_5, i = 1, \ldots, N-1, \\ &w(s_2, t_N) = -M_5, \end{aligned}$$

with $M_5 = \frac{S(S-1)}{2} \cdot k + 1$, where k is now the maximum weight defined in stage (6). By the same arguments as in (4), a reasonable ordering π must satisfy

$$t_{k+1} <_\pi t_{k+2} <_\pi \ldots <_\pi t_N <_\pi s_2 <_\pi s_4 <_\pi t_1 <_\pi \ldots <_\pi t_k,$$

for some $k \in \{1, \ldots, N\}$.

(6) For all 4-tuples $(u, v, a, b) \in (\Sigma^{d \cdot \log^3 n})^4$

$$w((u, v, b, 2), (u, v, a, 1)) = d_{u,v,a,b}, \quad w((u, v, b, 3), (u, v, a, 1)) = -d_{u,v,a,b}.$$

Claim 6.6.11 *(i) Let $u \in \Sigma^{d \cdot \log^3 n}$ and let π be a reasonable ordering. There exists $j \in \{1, \ldots, N\}$ such that, for all $v \in \Sigma^{d \cdot \log^3 n}$,*

$$\begin{aligned} &(u, v, a_{j+1}, 1) <_\pi \ldots <_\pi (u, v, a_N, 1) <_\pi s_1 <_\pi s_3 \\ &\qquad <_\pi (u, v, a_1, 1) <_\pi \ldots <_\pi (u, v, a_j, 1). \end{aligned} \tag{6.30}$$

(ii) Let $v \in \Sigma^{d \cdot \log^3 n}$ and let π be a reasonable ordering. There exists $k \in \{1, \ldots, N\}$ such that, for all $u \in \Sigma^{d \cdot \log^3 n}$,

$$\begin{aligned} &(u, v, b_{k+1}, 3) <_\pi \ldots <_\pi (u, v, b_N, 3) <_\pi s_2 <_\pi s_4 \\ &\qquad <_\pi (u, v, b_1, 3) <_\pi \ldots <_\pi (u, v, b_k, 3). \end{aligned} \tag{6.31}$$

Proof. We prove (i). From the definitions in stage (4) and in particular the inequalities 6.29, for each $v \in \Sigma^{d \cdot \log^3 n}$ there exists a value $j(v)$ verifying the inequalities from (i). Suppose that for some pair $v \neq v'$, $j(v) < j(v')$. Then

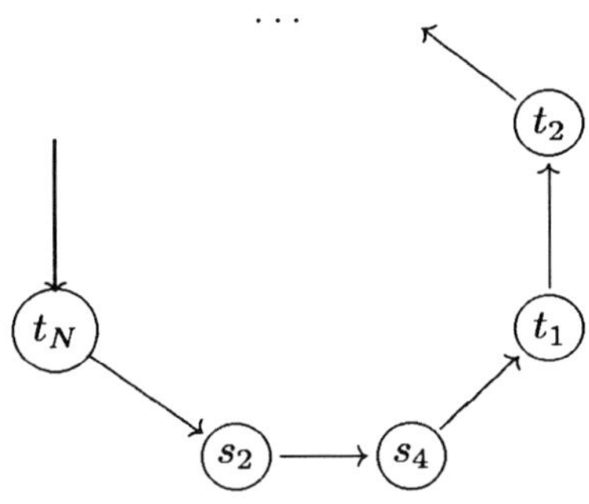

Figure 6.4: The effect of stage (5): Only one of the edges $(t_1, t_2), \ldots, (t_N, s_2)$ should be "broken" by a reasonable π.

$(u, v, a_{j(v')}, 1) <_\pi s_1 <_\pi s_3 <_\pi (u, v', a_{j(v')}, 1)$, which means that two elements from $Y_{u,a_{j(v')},1}$ are on distinct "sides" of $s_1 <_\pi s_3$. But then π cannot be a reasonable ordering by the discussion following the definitions in stage (2). It follows that for all $v \in \Sigma^{d \cdot \log^3 n}$, $j(v)$ has the same value. The proof of (ii) is similar. ∎

If π is a reasonable ordering and (u, a_j) are as in Claim 6.6.11, (i) and (v, b_k) are as in Claim 6.6.11, (ii), then we say that π *forces* answer a_j to u and answer b_k to v.

Claim 6.6.12 *Let π be a reasonable ordering. We define the provers' strategies (P_1, P_2) by $P_1(u) = a$ if π forces answer a to u and $P_2(v) = b$ if π forces answers b to v. Then $C_\pi(X) = 2^{d \log^3 n} \cdot \mathrm{ACC}_{V,(P_1,P_2)}(\phi)$.*

Proof. Observe that the contribution of all pairs whose weight is defined in stages (1) to (5) is 0. It remains to consider only the pairs whose weight is defined in stage (6). For each u, let $j(u)$ be defined by $a_{j(u)} = P_1(u)$, and for each v, let $k(v)$ be defined by $a_{k(v)} = P_2(v)$. We can see that only pairs (x, y) in which $x \in X_2$ and $y \in X_1$ or in which $x \in X_3$ and $y \in X_1$ matter. From the way the elements of X_1, X_2 and X_3 are arranged between the "stakes" s_1, s_2, s_3 and s_4 in the stages (1), (2) and (3), the first type of pairs occurs for (x, y) with $x \in X_2$ and $y \in X_1$, $y > s_3$, and the second type of pairs occurs for (x, y) with $x \in X_3$, $s_1 < x < s_2$ and $y \in X_1$, $s_3 < y < s_4$. Keeping in mind the definitions of $j(u)$, $k(v)$, and the "complementary" way in which weights are defined in stage (6), we get:

$$C_\pi(X) = \sum_u \sum_v \sum_{h=1}^{N} \sum_{i=1}^{j(u)} w((u, v, b_h, 2), (u, v, a_i, 1))$$

$$+ \sum_u \sum_v \sum_{h=k(v)+1}^{N} \sum_{i=1}^{j(u)} w((u, v, b_h, 3), (u, v, a_i, 1))$$

$$= \sum_u \sum_v \sum_{h=1}^{k(v)} \sum_{i=1}^{j(u)} d_{u,v,a_i,b_h} = \sum_u \sum_v val_{u,v,P_1(u),P_2(v)}$$
$$= 2^{d \log^3 n} \cdot \mathrm{ACC}_{V,(P_1,P_2)}(x).$$

In the above sums, u and v range over $\Sigma^{d \cdot \log^3 n}$. ∎

It follows from Claim 6.6.12 that $max_\pi C_\pi(X) \leq 2^{d \log^3 n} \cdot \mathrm{ACC}_V(\phi)$. On the other hand, from any pair of provers' strategies (P_1, P_2) one can easily build an ordering π such that $C_\pi(X) = 2^{d \log^3 n} \cdot \mathrm{ACC}_{V,(P_1,P_2)}(\phi)$. It is sufficient that π satisfies the requirements from the discussions following stages (1), (2) and (3) and that it *forces* to any queries u and v the real answers of the provers P_1 and P_2. In conclusion, $max_\pi C_\pi(X) = 2^{d \log^3 n} \cdot \mathrm{ACC}_V(\phi)$. It is not difficult to check that the whole construction can be made in DTIME$[2^{\log^c n}]$ for some constant c. ∎

6.7 Non-approximability of MAX CLIQUE and of some related problems

---

IN BRIEF: If P $\neq$ NP, then MAX CLIQUE does not admit a polynomial-time approximation algorithm with an approximation ratio of n^δ, for some constant δ. A similar result holds for the weighted version of MAX 2-SAT. If P $\neq$ NP, then VERTEX COVER cannot be approximated by polynomial-time algorithms with approximation ratios arbitrarily close to 1.

This section focuses primarily on the MAX CLIQUE problem defined in Problem 6.3.2. Besides being an important problem in its own, the analysis of this problem will lead to non-approximability results for VERTEX COVER and MAX 2-SAT, problems that are canonical for MIN $\mathrm{F}^+\Pi_1$ and respectively MAX Σ_0. Proving the non-approximability of MAX CLIQUE is particularly interesting since it uses a technique for the efficient utilization of random bits that can be applied elsewhere as well. We start by proving a weaker result, so that we can later get a better appreciation of this technique.

Proposition 6.7.1 *If* P $\neq$ NP, *then there is no polynomial-time approximation algorithm for* MAX CLIQUE *with approximation ratio less than 2.*

Proof. We use the standard notation $\omega(G)$ to denote the *clique number* of a graph G, i.e., the size of a maximum clique of G. Using Theorem 6.6.4, we will construct, for any input ϕ for SAT, in polynomial time in $|\phi|$, a graph G_ϕ with the following property

$$\phi \in \mathrm{SAT} \rightarrow \omega(G_\phi) = |\phi|^c \qquad (6.32)$$
$$\phi \notin \mathrm{SAT} \rightarrow \omega(G_\phi) \leq \frac{1}{2}|\phi|^c,$$

where c is a constant that will be specified shortly (the choice of SAT is not important; we can take any NP-complete language). As we have discussed at the beginning of Section 6.6, the conclusion follows easily from relations (6.32).

Let V be a $\mathrm{PCP}_{1,1/2}(c\log n, q)$ verifier for SAT as given by Theorem 6.6.4. On input ϕ, V can have many computations depending on the random string r and of the q bits of the proof string that are queried. Each such computation can be described by a tuple $(r, a_1, \ldots, a_q)$, where r, with $|r| = c\log n$ (where $n = |\phi|$), represents the random string and $a_1, \ldots, a_q$ are the q bits of the proof string that have been queried. The set of nodes of the graph G_ϕ will consist of the tuples $(r, a_1, \ldots, a_q)$ that make V accept ϕ. The number of such nodes is polynomial in $|\phi|$; also, determining whether a node $(r, a_1, \ldots, a_q)$ is a node of G_ϕ can be done in polynomial time by simulating V on input ϕ with the random string r and with the answers $a_1, \ldots, a_q$.

There is an edge between two nodes $(r, a_1, \ldots a_q)$ and $(r', a'_1, \ldots a'_q)$ if there is a proof string w that is consistent with both tuples; this means that for any position of w that is queried in both computations, the answers a_i and a'_j corresponding to this position coincide. For any two nodes of G_ϕ, one can check in polynomial time whether they are adjacent. Consequently, G_ϕ can be constructed in polynomial time in $|\phi|$.

For a fixed proof string w, the nodes that are consistent with w form a clique. Since these nodes represent all the computations of V that make the verifier accept ϕ with proof string w, the number of these nodes is $2^{c\log n}\mathrm{ACC}(V^w(\phi))$. Thus, for any proof string w, $\omega(G_\phi) \geq n^c \cdot \mathrm{ACC}(V^w(\phi))$. Conversely, if K is a clique of G_ϕ, then the nodes of K do not have any conflict with each other, and, thus, there is a proof w_0 that is consistent with all the nodes of K. It follows that $\omega(G_\phi) \leq n^c {\cdot} \mathrm{ACC}(V^{w_0}(\phi))$. Therefore, $\omega(G_\phi) = n^c {\cdot} \mathrm{ACC}(V(\phi))$. Since $\mathrm{ACC}(V(\phi))$ is 1 or at most 1/2 depending on whether ϕ is in SAT or not, the proof of the relations 6.32 is complete. ∎

As we have noted, the error probability in Theorem 6.6.4 can be made less than any constant ϵ and by using the above proof it can be shown that MAX CLIQUE does not have a PTAS.

In fact, we will show that, unless P = NP, MAX CLIQUE cannot be approximated in polynomial time with approximation ration n^δ, for some $\delta > 0$. This result can not be obtained directly from Proposition 6.7.1. Indeed, to apply the reduction from this theorem, we would need that SAT has a verifier that has probability of error $n^{-\epsilon}$, where ϵ depends on δ. To obtain such a probability error, the verifier from Theorem 6.6.4 would have to repeat its calculations k times, where $2^{-k} \leq n^{-\epsilon}$, which implies $k \geq \epsilon\log n$. If the verifier is using distinct random strings at each iteration, the total number of random bits would be $\Omega(\log^2 n)$ and the graph G_ϕ would have over-polynomial size.

To get around this difficulty, first observe that the reduction given in Proposition 6.7.1 works even if the number of queries is allowed to be $O(\log n)$. Besides that, the random bits must be used in a more efficient way. One approach is to use extractors and the method given at the end of Section 5.10. We will present another approach based on a special type of graph called *expander graph.* Intuitively speaking, an expander graph is a graph in which any "not-too-large" subset of nodes has a "large" set of neighbors. To be more precise, consider an undirected multigraph (i.e., there may be more edges between two vertices) $G = (V, E)$ with n vertices $v_1, \ldots, v_n$. For any subset $S \subseteq V$, we let $\Gamma(S)$ denote the set of all nodes adjacent to the nodes in S, i.e., $\Gamma(S) = \{v \mid (u, v) \in E \text{ for some } u \in S\}$. We would like that any subset of nodes S with $\|S\| \leq n/2$ has the property that $\|\Gamma(S)\| \geq c\|S\|$, for some constant c. We would also like to achieve this property for graphs of bounded degree (otherwise, it is easy, and, more importantly, not very useful). It turns out that there is an algebraic way to express these requirements, which is also helpful in the proofs to come. We associate to G its *adjacency matrix* $A(G)$; $A(G)$ is an $n \times n$ matrix where the (i, j) entry is equal to the number of edges between the nodes v_i and v_j. Let $\lambda_1 \geq \lambda_2 \geq \ldots \geq \lambda_n$ be the eigenvalues of $A(G)$. Since the graph is undirected, $A(G)$ is a symmetric matrix and we can fix corresponding eigenvectors $e_1, \ldots, e_n$ that form an orthonormal basis. A graph G is d-regular if the degree of each of its nodes is d. It can be shown that if G is d-regular then $\lambda_1 = d$ and $\lambda_2 < \lambda_1$ (for references see the notes at the end of the chapter).

Definition 6.7.2 *An (n, d, α)-expander graph is a bipartite graph $G = (X, Y, E)$ with $\|X\| = \|Y\| = n/2$ that is d-regular and has $\lambda_2 \leq \alpha$.*

Expanders graphs have the following property (which explains their name): There is a constant c which depends on α such that for any $S \subseteq X$,

$$\|\Gamma(S)\| \geq \left(1 + c\left(1 - \frac{2\|S\|}{n}\right)\right)\|S\|.$$

Several explicit construction of extractors of expanders are known. We describe one of them due to Gabber and Galil [GG81]. Fix a positive integer m, and let $n = 2m^2$. The vertices in X and Y are labeled with a pair of numbers (a, b) with a and b in $\mathbb{Z}_m$ (the ring of congruences modulo m). A vertex labeled (x, y) in X is connected to 7 nodes in Y whose labels are as follows: $(x, y), (x, 2x + y), (x, 2x + y + 1), (x, 2x + y + 2), (x + 2y, y), (x + 2y + 1, y)$ and $(x + 2y + 2, y)$. Clearly, this graph is 7-regular. It has been proved that, for the adjacency matrix A attached to this graph, $|\lambda_2| < \alpha = 7 - c^2/(1024 + c^2)$, where $c = (2 - \sqrt{3})/2$. Thus, the second eigenvalue of the matrix $(1/7)A$ is less than 1; this is the property that we need later. We will consider an expander graph $G = (V, E)$ having a node for each string r of length $c \log n$ (corresponding to a random string used by the verifier). Expander graphs have the following property: If we start a random walk from a given node, after traversing a constant number of edges we reach a node that is distributed uniformly among the nodes of G, and that is independent of the choice of the initial node.

In order to generate the $k = \epsilon \log n$ random strings r that the verifier needs, we will run the following algorithm that depends upon a parameter s.

Pick r_0 a uniformly at random chosen node of G (6.33)

for $i = 1$ to k do

  do a random walk in G starting from r_{i-1} that traverses s edges

  let r_i be the final node reached in the random walk

end for

We will show that if the verifier is using the strings $r_1, \ldots, r_k$ to repeat k times a $\mathrm{PCP}_{1,1/16}(c \log n, q)$ protocol for SAT, then the error probability is at most 2^{-k}.

To this aim, let ϕ be an unsatisfiable formula and let w be a proof string. We know that there exists a $\mathrm{PCP}_{1,1/16}(c \log n, q)$ verifier for SAT that accepts (erroneosly) ϕ with proof string w for at most a fraction of $1/16$ of the random strings r. Let C be the set of these misleading random strings r.

Lemma 6.7.3 *Let $G = (G_n)_{n \in \mathbb{N}}$ be a family of (n, d, α) expander graphs for some constant $\alpha < 1$. Let A_n be the adjacency matrix of G_n multiplied by $1/d$. We know that for all n, λ_1, the largest eigenvalue of A_n is 1 and that λ_2, the second largest eigenvalue of A_n, is less than 1. Let s be such that $(\lambda_2)^s \leq \frac{1}{4}$. Consider the sequence $r_1, \ldots, r_k$ produced by the algorithm (6.33) with the above parameter s. Then the probability that, for all $i \in \{1, \ldots, k\}$, $r_i \in C$ for all $i \in \{1, \ldots, k\}$ is at most 2^{-k}.*

Proof. We fix n and denote A_n more simply by A. Let p be the vector $p = (p_1, \ldots, p_n)$, where p_i denotes the probability of being at node v_i at a certain step in the random walk. Then Ap is the vector of probabilities at the next step (i.e., after traversing a randomly chosen edge from the d edges outgoing from a node). Similarly, $A^s p$ gives the probabilities after s steps in the random walk. We start from the vector

$$p_0 = \left(\frac{1}{n}, \ldots, \frac{1}{n} \right),$$

which represents the initial random choice of a node of G. Let B be the 0-1 matrix whose entries are all zero except for the diagonal where $B_{ii} = 1$ if and only if v_i belongs to C. Then BAp_0 gives the probabilities of being in a vertex of C after one iteration of the for loop in algorithm (6.33). Let $|\cdot|_1$ denotes the L_1 norm, i.e., $|p|_1 = |p_1 + p_2 + \ldots + p_n|$. Then, $|BA^s p_0|_1$ is the probability that r_1 is in C. The vector $A^s(BA^s p_0)$ gives the probabilities of the nodes reached at the second iteration of algorithm (6.33) when the second iteration started from a node in C. Thus $|(BA^s)^2 p_0|_1$ is the probability of having both r_1 and r_2 in C. Similarly, $|(BA^s)^k p_0|_1$ is the probability of having all $r_1, \ldots, r_k$ in C. To estimate this probability, we evaluate the L_2 norm $|\cdot|_2$ of $BA^s p$ for some arbitrary probability vector p (recall that for a vector $v = (v_1, \ldots, v_n)$, the L_2 norm is defined by $|v|_2 = \sqrt{v_1^2 + \ldots v_n^2}$). We will show that

$$|BA^s p|_2 \leq \frac{1}{2} |p|_2. \tag{6.34}$$

connected if the two tuples are not consistent. Let $\overline{G_\phi}$ be the graph that results from a boolean formula ϕ of size n. From Proposition 6.7.1 we have that

$$\phi \in \text{SAT} \rightarrow \max_{\text{Indep Set}}(\overline{G_\phi}) = n^c$$

$$\phi \notin \text{SAT} \rightarrow \max_{\text{Indep Set}}(\overline{G_\phi}) \leq n^c/2.$$

If N is the number of nodes in G_ϕ, from the above observations, it follows that

$$\phi \in \text{SAT} \rightarrow \min_{\text{VC}}(\overline{G_\phi}) = N - n^c$$

$$\phi \notin \text{SAT} \rightarrow \min_{\text{VC}}(\overline{G_\phi}) \geq N - n^c/2.$$

Thus, a polynomial time approximation algorithm with ratio less than $(N - n^c/2)/(N - n^c)$ would solve SAT.

Finally, by inspecting the construction of $\overline{G_\phi}$, it can be seen that $N = n^c 2^q$. This implies

$$\frac{N - n^c/2}{N - n^c} \leq \frac{n^c \cdot 2^q - n^c/2}{n^c \cdot 2^q - n^c} = \frac{2^q - 1/2}{2^q - 1} = 1 + \frac{1/2}{2^q - 1}.$$

We take $\epsilon = \frac{1/2}{2^q-1}$. This concludes the proof. ∎

Theorem 6.7.7 *If* P $\neq$ NP, *then there is some constant* δ *such that the problem* weight MAX 2-SAT *is not approximable in polynomial time with approximation ratio less than* n^δ.

Proof. The proof consists of an L-reduction of INDEPENDENT SET to weight MAX 2-SAT with parameters $\alpha = \beta = 1$. Since INDEPENDENT SET is just a disguised version of MAX CLIQUE, it is not approximable in polynomial time with a ratio less than n^δ (in the hypothesis P $\neq$ NP). The result follows from Proposition 6.1.8.

We only have to exhibit the L-reduction. Given $G = (V, E)$ an instance of INDEPENDENT SET, we build a formula ϕ in 2-CNF as follows: For each node i, we consider a clause x_i, and, for each edge (i, j), we consider a clause $(\overline{x}_i \vee \overline{x}_j)$. All these clauses have weight 1. We consider a new variable y and two more clauses, y and $\overline{y}$, each with weight $-\|E\|$. It is easy to observe the following fact.

Claim 6.7.8 *Given an assignment for* ϕ *with cost* c, *we can find in polynomial time an assignment with cost* $c' \geq c$ *such that all the clauses corresponding to edges are satisfied in the new assignment. Such an assignment is said to be in* normal form.

Now, it is easy to see that the nodes corresponding to the clauses that are made true by a normal form assignment form an independent set. Taking into account that one of the clauses y, $\neg y$ is satisfied by any assignment, it follows that:

$$\text{opt}_{\text{MAX 2-SAT}}(\phi) = \text{opt}_{\text{IND.SET}}(G) + \|E\| - \|E\| = \text{opt}_{\text{IND.SET}}(G).$$

Also, given any assignment with cost c', we can build in polynomial time a normal form assignment with cost $c \geq c'$ that corresponds to an independent set of size c and, thus, $\mathrm{opt}(G) - c \leq \mathrm{opt}(\phi) - c'$. ∎

Corollary 6.7.9 *There are problems in weight* MAX Σ_0 *that do not admit any polynomial-time approximation algorithms with an approximation ratio less than* n^δ *for some positive constant* δ, *unless* P = NP.

6.8 Non-approximability of SET COVER

IN BRIEF: If SAT cannot be solved in $2^{\mathrm{polylog} n}$ time, then SET COVER does not admit a polynomial-time approximation algorithm with an approximation ratio of $c \log n$, for some constant c.

We have seen in Theorem 6.5.3 that the SET COVER problem (see Problem 6.5.2) has a polynomial-time approximation algorithm with approximation ratio $\log n$, where n is the size of the universe. In this section, it will be demonstrated that, unless NP $\subseteq$ DTIME$[2^{\log^{O(1)} n}]$ (we will often abbreviate the latter class as DTIME$[2^{\mathrm{polylog}\, n}]$), this is basically the best one can achieve (if we ignore multiplicative constants in the approximation ratio). To this aim, we reduce a MIP_1 of the type displayed in Theorem 6.6.7 to the SET COVER problem.

The reduction uses as a basic building block a collection of finite sets called an *(m, ℓ)-system.*

Definition 6.8.1 *Let m and ℓ be positive integers. We say that a collection of sets $(B; C_1, \ldots, C_m)$ is an (m, ℓ)-system if*

(1) *B is a finite set, $C_1, \ldots, C_m$ are subsets of B, and*

(2) *if we take any ℓ subsets $D_{i_1}, \ldots, D_{i_\ell}$, $i_k \neq i_j$, where each D_{i_j} is either C_{i_j} or the complement of C_{i_j} (with respect to B), then $\bigcup_{j=1}^{\ell} D_{i_j} \neq B$.*

We will show later that, for any positive integers m and ℓ, there is an algorithm that (a) constructs an (m, ℓ)-system with $\|B\| = O(2^{2\ell} m^2)$, and (b) runs in time polynomial in $\|B\|$.

Theorem 6.8.2 *There is a constant c such that there is no polynomial-time approximation for* SET COVER *with approximation ratio $c \log n$, unless* NP $\subseteq$ DTIME$[2^{\mathrm{polylog}\, n}]$.

Proof. The starting point is a verifier V that performs the

$$\mathrm{MIP}_1(d \log^3 n, 2, d \log^3 n, d \log^3 n, 1/n)$$

protocol for SAT guaranteed by Theorem 6.6.7. We fix an input instance ϕ for SAT. It can be shown that V can be taken to also satisfy the following properties:

Claim 6.8.5 *If $\phi \notin$ SAT then*

$$\min_{\text{SET COVER}}(S_\phi) \geq \frac{\ell(\|Q_1\| + \|Q_2\|)}{16}.$$

Proof. If the relation does not hold then

$$\text{ACC}_V(\phi) \geq \frac{2}{\ell^2} = \frac{2}{(\log \phi)^6} > \frac{1}{|\phi|},$$

which contradicts the soundness of the verifier. ∎

We complete the proof of Theorem 6.8.2. The size of the instance S_ϕ is defined to be the number of elements in U, which is $\|\mathcal{R}\| \times \|B\|$, where $\mathcal{R}$ is the set of all possible random strings. We have $\|\mathcal{R}\| = 2^{d\log^3|\phi|}$ and $\|B\| = O(2^{2\ell}m^2) = O(2^{2\log^3|\phi|} \cdot 2^{2d\log^3|\phi|})$. It follows that $\ell = c \cdot \log\|S_\phi\|$, for some constant c. Thus, by inspecting Claim 6.8.3 and Claim 6.8.4, we have that

$$\phi \in \text{SAT} \rightarrow \min_{\text{SET COVER}}(S_\phi) = \|Q_1\| + \|Q_2\|$$

and

$$\phi \notin \text{SAT} \rightarrow \min_{\text{SET COVER}}(S_\phi) \geq \frac{c \cdot \log\|S_\phi\|}{16}(\|Q_1\| + \|Q_2\|).$$

Since the instance S_ϕ can be built in time polynomial in $\|S_\phi\|$, that is in DTIME$[2^{polylog\ n}]$, the conclusion follows. ∎

It remains to show how to construct an (m, ℓ)-system with the desired size.

Proposition 6.8.6 *For any positive integers m and ℓ, there is an (m, ℓ)-system $(B; C_1, \ldots, C_m)$ with $\|B\| = O(2^{2\ell}m^2)$. Such a set can be constructed in time polynomial in $\|B\|$.*

Proof. We will build a set B of binary vectors of length m with the following property: If we fix any ℓ coordinates $(i_1, \ldots, i_\ell)$, where each i_j is in $\{1, \ldots, m\}$, and if we take any string $y \in \{0,1\}^\ell$, there is a vector $x \in B$ such that $x(i_j) = y(j), j = 1, \ldots, \ell$. In other words, if we take the projection of all strings in B on any ℓ coordinates, we obtain all the binary strings of length ℓ. If we have such a set B, then we can take the sets C_i to be defined by

$$C_i = \{x \in B \mid x(i) = 1\},\ i = 1, \ldots, m.$$

Let us consider a collection of ℓ sets $D_{i_1}, D_{i_2}, \ldots, D_{i_\ell}$, where D_{i_j} is C_{i_j} or the complement of C_{i_j}. We take a vector $x \in B$ such that for any $j \in \{1, \ldots, \ell\}$,

$$x(i_j) = \begin{cases} 0, & \text{if } D_{i_j} = C_{i_j} \\ 1, & \text{if } D_{i_j} = \overline{C}_{i_j}. \end{cases}$$

Since the vector x is not in any of the sets $D_{i_j}, j = 1, \ldots, \ell$, it follows that

$$B \not\subset D_{i_1} \cup D_{i_2} \cup \ldots \cup D_{i_\ell}.$$

It remains to build the set B. Clearly, the set of all binary vectors of size m has the desired property, but its size, 2^m, is too big. The set B that we construct is a set of m-long vectors that represents the sample space of m random variables $x_1, \ldots, x_m$ that can take the values 0 or 1 and have the following property: If we consider any ℓ random variables $x_{i_1}, \ldots, x_{i_\ell}$ and any string $y \in \{0,1\}^\ell$, then the probability that $x_{i_j} = y_j$, $j = 1, \ldots, \ell$, satisfies

$$\left|\text{Prob}(x_{i_1} = y_1 \ \wedge \ldots \ \wedge \ x_{i_\ell} = y_\ell) - 2^{-\ell}\right| < 2^{-(\ell+1)}. \tag{6.38}$$

This will imply that the probability of having $x_{i_j} = y_j$, for all $j = 1, \ldots, \ell$, is positive, and so, there is an m-long vector in B having the values $y_1, \ldots, y_\ell$ in positions $i_1, \ldots, i_\ell$. The property that we desire for the random variables that yield B is important enough to deserve a definition.

Definition 6.8.7 *A set of random variables $x_1, \ldots, x_m$ that can take the values 0 and 1 is ℓ-wise ϵ-independent if for any subset of ℓ random variables $x_{i_1}, \ldots, x_{i_\ell}$ and for any ℓ-bit string $y \in \{0,1\}^\ell$,*

$$\left|\text{Prob}(x_{i_1} = y_1 \ \wedge \ldots \wedge \ x_{i_\ell} = y_\ell) - 2^{-\ell}\right| < \epsilon.$$

Observe that we need to produce a set of m random variables that is ℓ-wise, $2^{-(\ell+1)}$- independent over a sample space of size $O(2^{2\ell}m^2)$.

Definition 6.8.8 *Let $\delta > 0$. A $\{0,1\}$ random variable x is δ-biased, if*

$$|\text{Prob}(x = 0) - \text{Prob}(x = 1)| \leq \delta.$$

We view the m-long vectors as having their components in the finite field GF(2) and we remind that the notation $u \cdot v$ denotes the inner product of the vectors u and v, i.e., $u \cdot v = \sum_i u(i) \cdot v(i)$, where the arithmetical operations are performed in the field GF(2).

Definition 6.8.9 *A sample space S_m of binary vectors of length m is δ-biased with respect to linear tests if for any $\alpha \in \{0,1\}^m$ such that $\alpha \neq (0, \ldots, 0)$, $\alpha \cdot r$ is a δ-biased random variable, when r is chosen uniformly at random in S_m.*

The following lemma provides an interesting connection between sample spaces that are δ-biased with respect to linear tests and ℓ-wise ϵ-independent random variables.

Lemma 6.8.10 *Let m and ℓ be integer, positive numbers such that $\ell \leq m$. Let $S_m \in \{0,1\}^m$ be a sample space that is δ-biased with respect to linear tests. Then if we pick uniformly at random a vector $(x_1, \ldots, x_m)$ from S_m, the random variables $x_1, \ldots, x_m$ are ℓ-wise, $(1 - 2^{-\ell}) \cdot \delta$ - independent.*

Proof. Without loss of generality, let us look at the random variables $x_1, \ldots, x_\ell$. For a fixed $\alpha \in \{0,1\}^\ell$, let

$$p_\alpha = \text{Prob}(x_1 = \alpha(1) \ \wedge \ldots \ \wedge x_\ell = \alpha(\ell)).$$

For every $\beta \in \{0,1\}^\ell$, we define the application

$$\phi_\beta(\alpha) = (-1)^{\sum_{i=1}^{\ell} \alpha(i)\beta(i)}.$$

Let c_β be defined by $c_\beta = \sum_{\alpha\in\{0,1\}^\ell} \phi_\beta(\alpha)p_\alpha$ (c_β is the discrete Fourier transform of the sequence $(p_\alpha)_{\alpha\in\{0,1\}^\ell}$). We claim that, for any $\alpha \in \{0,1\}^\ell$,

$$p_\alpha = 2^{-\ell} \sum_{\beta\in\{0,1\}^\ell} \phi_\beta(\alpha)c_\beta. \tag{6.39}$$

(This is just the inverse Fourier transform.)

Let us prove this claim. It is convenient to regard β and α as the characteristic sequences of some subsets of $\{1,\ldots,\ell\}$. It is easy to check that for every β_1, β_2 and α in $\{0,1\}^\ell$, $\phi_{\beta_1}(\alpha)\cdot\phi_{\beta_2}(\alpha) = \phi_{\beta_1\triangle\beta_2}(\alpha)$ (where $\triangle$ denotes the symmetric difference of sets). Based on this relation, we derive that, for $\beta_1 \neq \beta_2$ and for any α,

$$\sum_{\alpha\in\{0,1\}^l} \phi_{\beta_1}(\alpha)\cdot\phi_{\beta_2}(\alpha) = 0$$

and

$$\sum_{\alpha\in\{0,1\}^\ell} \phi_{\beta_1}(\alpha)\cdot\phi_{\beta_1}(\alpha) = 2^\ell.$$

Then, for any fixed α_0,

$$\begin{aligned}
\sum_{\beta\in\{0,1\}^\ell} c_\beta\phi_\beta(\alpha_0) &= \sum_{\beta\in\{0,1\}^\ell}\Bigg(\sum_{\alpha\in\{0,1\}^\ell}\phi_\beta(\alpha)\cdot p_\alpha\cdot\phi_\beta(\alpha_0)\Bigg)\\
&= \sum_\alpha p_\alpha\Bigg(\sum_\beta \phi_\beta(\alpha)\phi_\beta(\alpha_0)\Bigg)\\
&= p_{\alpha_0}\sum_\beta\phi_\beta(\alpha_0)\cdot\phi_\beta(\alpha_0) + \sum_{\alpha\neq\alpha_0} p_\alpha\Bigg(\sum_\beta\phi_\beta(\alpha)\cdot\phi_\beta(\alpha_0)\Bigg)\\
&= p_{\alpha_0}\cdot 2^\ell + \sum_{\alpha\neq\alpha_0} p_\alpha\Bigg(\sum_\beta\phi_\alpha(\beta)\cdot\phi_{\alpha_0}(\beta)\Bigg)\\
&= p_{\alpha_0}\cdot 2^\ell.
\end{aligned}$$

(In the fourth equality, we have used the fact that $\phi_\beta(\alpha) = \phi_\alpha(\beta)$.) Note that if $\beta \neq (0,\ldots,0)$, then $|c_\beta|$ is the bias of the random variable $\beta\cdot x$, where x is the vector $(x_1,\ldots,x_\ell)$. Thus, $|c_\beta| \leq \delta$, for $\beta \neq (0,\ldots,0)$ (ℓ zeros). Also, for $\beta_0 = (0,\ldots,0)$, $c_{\beta_0} = 1$ and $\phi_{\beta_0}(\alpha)\cdot c_{\beta_0} = 1$.

Therefore, using Equation 6.39,

$$\begin{aligned}|p_\alpha - 2^{-\ell}| &= 2^{-\ell}\Big|\sum_{\beta\neq(0,\ldots,0)} \phi_\beta(\alpha)c_\beta\Big| \\ &\leq 2^{-\ell}\sum_{\beta\neq(0,\ldots,0)} |\phi_\beta(\alpha)|\cdot|c_\beta| \leq 2^{-\ell}(2^\ell - 1)\delta = (1-2^{-\ell})\delta.\end{aligned}$$

This ends the proof of Lemma 6.8.10 ∎

By inspecting Equation 6.38 and Lemma 6.8.10, it follows that it is enough to build a sample space S_m of m vectors that is δ-biased for a value of δ such that

$$(1-2^{-\ell})\cdot\delta \leq 2^{-(\ell+1)}. \tag{6.40}$$

The construction runs as follows. Let n be some parameter that will be specified later. We take two random values $u, v \in \mathrm{GF}(2^n)$, and we represent u and v in the usual way (i.e., as binary vectors giving the coefficients of two polynomials of degree at most $n-1$ over GF(2)). A random vector x of length m is built by taking the i-th bit of x to be $u^i \cdot v, i = 0, \ldots, m-1$, (i.e., we calculate u^i in $\mathrm{GF}(2^n)$ and then calculate the inner product $u^i \cdot v$ with u^i and v viewed as vectors over GF(2)). Note that a vector in the sample space S_m is specified by the $2n$ bits of u and v. Therefore the size of the sample space is 2^{2n}.

Lemma 6.8.11 *The sample space S_m described above is $((m-1)/2^n)$-biased with respect to linear tests.*

Proof. Let x be a sample point in S_m corresponding to $u, v \in \mathrm{GF}(2^n)$. Thus, for each $i \in \{0, \ldots, m-1\}$, $x(i) = u^i \cdot v$. If α is a binary vector of length m different from $(0, \ldots, 0)$, then

$$\begin{aligned}\alpha\cdot x &= \sum_{i=0}^{m-1}\alpha(i)x(i) = \sum_{i=0}^{m-1}\alpha(i)\cdot(u^i\cdot v) \\ &= \sum_{i=0}^{m-1}\alpha(i)\cdot\Big(\sum_{j=0}^{n-1}u^i(j)\cdot v(j)\Big) = \Big(\sum_{i=0}^{m-1}\alpha(i)\cdot u^i\Big)\cdot v.\end{aligned}$$

Let $q_\alpha(t) = \sum_{i=0}^{m-1}\alpha(i)\cdot t^i$. Let d be the number of zeros of the polynomial q_α over $\mathrm{GF}(2^n)$. Since the degree of q_α is $m-1$, it holds that $d \leq m-1$. If u is not a root of $q_\alpha(t)$, then $\sum_{i=0}^{n-1}\alpha(i)u^i \neq 0$, and it is easy to see that half of the v's yield $\big(\sum_{i=0}^{m-1}\alpha(i)\cdot u^i\big)\cdot v = 1$ and the other half of v's yield $\big(\sum_{i=0}^{m-1}\alpha(i)\cdot u^i\big)\cdot v = 0$. If u is a root of $q_\alpha(t)$, then for all $v \in \mathrm{GF}(2^n)$, $\big(\sum_{i=0}^{m-1}\alpha(i)\cdot u^i\big)\cdot v = 0$. Therefore, $\mathrm{Prob}(\alpha\cdot x = 0) - \mathrm{Prob}(\alpha\cdot x = 1)$ is

$$\left(\frac{1}{2}\frac{(2^n-d)\cdot 2^n}{2^{2n}} + \frac{d\cdot 2^n}{2^{2n}}\right) - \frac{1}{2}\frac{(2^n-d)2^n}{2^{2n}} = \frac{d}{2^n} \leq \frac{m-1}{2^n}.$$

∎

We can now complete the proof of Proposition 6.8.6. For our construction, from Equation 6.40, we must select n so that $(1 - 2^{-\ell})(m-1)/2^n \leq 2^{-(\ell+1)}$. It can be checked that $n = \lceil \log m \rceil + (\ell + 1)$ satisfies this requirement. The size of B is $2^{2n} = O(2^{2\ell} \cdot m^2)$ and it is easy to see that the construction can be done in time polynomial in $\|B\|$. ∎

6.9 Probabilistically checkable proofs

IN BRIEF: The PCP Theorem states that 3-SAT can be accepted by a polynomial-time verifier who reads a constant number of bits randomly chosen from a polynomially long proof string produced by an all-powerful adversary. If the tested formula ϕ is satisfiable, then the verifier always accepts, and if it is not satisfiable, the verifier can be fooled to accept with only small probability. The proof of the PCP Theorem is long and complex. Two weaker variants of this result are demonstrated. In the first variant, the proof string is exponentially long, and in the second variant, the prover has access to a satisfying assignment but is otherwise computationally bounded.

The utilization of Theorem 6.6.4, the *PCP Theorem* (and of the related Theorem 6.6.7) has been crucial in establishing most of the non-approximability results from the previous sections. The *PCP Theorem* is a stunning result but, unfortunately, its proof is extremely complex and is beyond the scope of this book. We will content to consider two related but weaker results whose proofs contain some of the main ideas used in the proof of Theorem 6.6.4, while avoiding the most intricate technicalities. We will first demonstrate the following theorem.

Theorem 6.9.1 3SAT *is in* $\mathrm{PCP}_{1,1/2}(cn^3, q)$ *for some constants* c *and* q.

Observe that the number of random bits used by the verifier in the above theorem is polynomial, which means that the prover should write down on the proof tape a proof string of exponential length (because it is possible that for each choice of random bits, the verifier queries distinct q locations in the proof string). Of course, this is much weaker than Theorem 6.6.4, in which the verifier utilizes only a logarithmical number of random bits and the proof string is of polynomial length. We will show later how to achieve this kind of parameters with a simple proof but at the price of weakening the soundness condition in the definition of a verifier that runs a PCP protocol.

Proof. We consider in the following a fixed formula in the 3-CNF format, i.e., a formula ϕ of the form $\phi = C_1 \wedge \ldots \wedge C_m$, where each clause C_i is of the form $C_i = \ell_{i_1} \vee \ell_{i_2} \vee \ell_{i_3}$ with each ℓ_j being a literal (i.e., a variable or the negation of a variable). The first step is to translate the formula ϕ into an algebraic expression. This step is called *arithmetization* and is done as follows. The boolean domain $\{true, false\}$ is identified with the finite field GF(2) of residues modulo 2 (0 is *false* and 1 is *true*). Let n be the number of variables in ϕ. We do the following transformations. For each clause $C_i = \ell_{i_1} \vee \ell_{i_2} \vee \ell_{i_3}$, we consider the "complementary" clause $\overline{C_i} = \overline{\ell_{i_1}} \wedge \overline{\ell_{i_2}} \wedge \overline{\ell_{i_3}}$. Next, every occurrence of a negated variable x is replaced by $1-x$, and every conjunction $\alpha \wedge \beta$ is replaced with $\alpha \cdot \beta$. For example, the clause $C = x_1 \vee \overline{x}_2 \vee x_3$ is transformed as follows:

$$\begin{aligned} x_1 \vee \overline{x}_2 \vee x_3 &\to \overline{x_1} \wedge x_2 \wedge \overline{x_3} \\ &\to (1-x_1)x_2(1-x_3) \\ &\to x_2 - x_1x_2 - x_2x_3 + x_1x_2x_3. \end{aligned}$$

In GF(2), the last expression is equal to $\widetilde{C} = x_2 + x_1x_2 + x_2x_3 + x_1x_2x_3$. In general, any clause C_i having the variables x_i, x_j and x_k will be transformed into an expression of the form

$$p_i = \alpha + \sum_{i \in D_1} x_i + \sum_{(i,j) \in D_2} x_i x_j + \sum_{(i,j,k) \in D_3} x_i x_j x_k, \tag{6.41}$$

where α is 0 or 1, and α, D_1, D_2, and D_3 depend on the form of C_i. In this way any clause C_i is transformed into a polynomial p_i over GF(2) and, for any truth assignment $a = (a_1, \ldots, a_k)$ (using the identification of $\{false, true\}$ with $\{0,1\}$), it can be easily checked that $C_i(a) = 1 - p_i(a)$. The formula $\phi = C_1 \wedge \ldots \wedge C_m$ will be represented by the vector of polynomials

$$(p_1, \ldots, p_m),$$

where each p_i is obtained by applying the above transformations to C_i. From our discussion, we derive the following useful fact.

Claim 6.9.2 *A vector $a \in (\text{GF}(2))^n$ corresponds to an assignment that satisfies $\phi = C_1 \wedge \ldots \wedge C_m$ if and only if $(p_1(a), \ldots, p_m(a))$ is the zero vector (i.e., all its entries are zero).*

In view of this fact, the prover, instead of displaying a truth assignment that satisfies ϕ, can equally well present a vector $a \in (\text{GF}(2))^n$ that is a zero for all the polynomials $p_1, \ldots, p_m$. This is not yet much of an improvement, because the prover still has to check all the entries of the vector. The next observations indicate how the verifiers can get through with only a constant number of queries. As before, for two vectors $x = (x_1, \ldots, x_m)$ and $y = (y_1, \ldots, y_m)$ over the field GF(2), let $x \cdot y = \sum_{i=1}^{m} x_i y_i$. Note that if $x \in (\text{GF}(2))^m$ is the zero vector, then for all $r \in (\text{GF}(2))^m$, $x \cdot r = 0$. On the other hand, if x is not zero, then if one picks a random $r \in (\text{GF}(2))^m$, the probability that $x \cdot r$ is not zero is 1/2. Thus, the computation of

$$(p_1(a), \ldots, p_m(a)) \cdot r$$

for a random string r, determines (probabilistically) whether $(p_1(a), \ldots, p_m(a))$ is the zero vector or not. By inspecting Equation (6.41), it can be seen that

$(p_1(a), \ldots, p_m(a)) \cdot r$ can be written as

$$\sum_{i=1}^{m} r_i p_i(a) = c(r) + \sum_{i \in S_1(r)} a_i + \sum_{(i,j) \in S_2(r)} a_i a_j + \sum_{(i,j,k) \in S_3(r)} a_i a_j a_k, \tag{6.42}$$

where the constant $c(r)$ and the sets $S_1(r)$, $S_2(r)$, and $S_3(r)$ depend only on ϕ and r but not on the assignment a. Thus, a honest prover in possession of a satisfying assignment $a = (a_1, \ldots, a_n)$ can provide in the proof string the sums

$$\sum_{i \in S_1(r)} a_i, \quad \sum_{(i,j) \in S_2(r)} a_i a_j, \quad \sum_{(i,j,k) \in S_3(r)} a_i a_j a_k$$

for all the possible r. The verifier calculates $c(r)$, $S_1(r)$, $S_2(r)$, and $S_3(r)$ for one random particular choice of r, selects from the proof string the entries corresponding to $S_1(r)$, $S_2(r)$, and $S_3(r)$, and, using Equation (6.42), checks if

$$\sum_{i=1}^{m} r_i p_i(a) = 0.$$

Thus, a constant number of queries are enough.

The verifier must rebut the "proofs" of dishonest provers as well. To this order, for an arbitrary vector $a = (a_1, \ldots, a_n)$, we define three linear functions over GF(2), $G_1\colon (\mathrm{GF}(2))^n \to \mathrm{GF}(2)$, $G_2\colon (\mathrm{GF}(2))^{n^2} \to \mathrm{GF}(2)$, $G_3\colon (\mathrm{GF}(2))^{n^3} \to \mathrm{GF}(2)$, given by

$$G_1(x_1, \ldots, x_n) = \sum_{i=1}^{n} a_i x_i$$

$$G_2(x_{11}, \ldots, x_{nn}) = \sum_{i=1}^{n} \sum_{j=1}^{n} a_i a_j x_{ij}$$

$$G_3(x_{111}, \ldots, x_{nnn}) = \sum_{i=1}^{n} \sum_{j=1}^{n} \sum_{k=1}^{n} a_i a_j a_k x_{ijk}.$$

The verifier expects the proof string to be of the form $g_1 g_2 g_3$ where g_1, g_2, and g_3 are binary strings of lengths 2^n, 2^{n^2} and 2^{n^3} that consist of all the values taken by G_1, G_2, and respectively G_3. More precisely, $g_1 = G_1(0\ldots0)\ldots G_1(1\ldots1)$ (the arguments of G_1 are n-bits long strings), and the strings g_2 and g_3 are defined similarly. The verifier must check two things:

(1) The strings g_1, g_2, and g_3 are encodings of three linear functions defined using the same vector of coefficients $a \in (\mathrm{GF}(2))^n$; this will ensure that the strings g_1, g_2, and g_3 define an assignment for the variables of ϕ.

(2) The vector a corresponds to a satisfying assignment.

We have seen earlier how to check for Condition (2). Testing Condition (1) is more complicated. In fact, it is not difficult to see that it cannot be done by reading only a constant number of bits of g_1, g_2, and g_3. However, we will show that by inspecting only a constant number of bits of g_i, $i = 1, 2, 3$, one can determine if g_i corresponds to a function that is equal to a linear function at all but a constant fraction of inputs from $(\mathrm{GF}(2))^n$.

Definition 6.9.3 *Let F and G be two finite sets. Two functions $f, g : F \to G$ are δ-close if*

$$\|\{x \in F \mid f(x) = g(x)\ \}\| \geq (1-\delta)\|F\|.$$

The following lemma shows how one can detect if a function is δ-close to a linear function.

Lemma 6.9.4 *Let $\delta < 1/3$ be a constant and let $\widetilde{G} : (\mathrm{GF}(2))^n \to \mathrm{GF}(2)$ be a function such that*

$$\mathrm{Prob}_{x,y\in(\mathrm{GF}(2))^n}(\widetilde{G}(x) + \widetilde{G}(y) = \widetilde{G}(x+y)) \geq 1 - \frac{\delta}{2}.$$

Then there exists a homogeneous linear function $G : (\mathrm{GF}(2))^n \to \mathrm{GF}(2)$ such that G and $\widetilde{G}$ are δ-close.

Proof. The function G is defined by taking $G(a)$ to be the majority element in the set $\{\widetilde{G}(a+y) - \widetilde{G}(y) \mid y \in (\mathrm{GF}(2))^n\}$. Observe that $G(0) = 0$. Let us first show that G and $\widetilde{G}$ are δ-close. Let us assume that this is not the case, i.e.,

$$\mathrm{Prob}_x(\widetilde{G}(x) \neq G(x)) > \delta.$$

By the definition of G, for each x, we also have

$$\mathrm{Prob}_y(G(x) = \widetilde{G}(x+y) - \widetilde{G}(y)) \geq \frac{1}{2}.$$

Since these events are independent, it follows that

$$\mathrm{Prob}_{x,y}(\widetilde{G}(x) \neq \widetilde{G}(x+y) - \widetilde{G}(y)) > \frac{1}{2}\delta,$$

and this contradicts the hypothesis of the lemma. We still need to demonstrate that G is a linear homogeneous function. First we show that, for all $a \in (\mathrm{GF}(2))^n$,

$$p_a \stackrel{\mathrm{def}}{=} \mathrm{Prob}_x(G(a) = \widetilde{G}(a+x) - \widetilde{G}(x)) \geq 1 - \delta. \tag{6.43}$$

Fix $a \in (\mathrm{GF}(2))^n$. By the hypothesis of the lemma,

$$\mathrm{Prob}_{x,y}(\widetilde{G}(x+a) + \widetilde{G}(y) \neq \widetilde{G}(x+a+y)) \leq \frac{1}{2}\delta, \text{ and}$$
$$\mathrm{Prob}_{x,y}(\widetilde{G}(x) + \widetilde{G}(y+a) \neq \widetilde{G}(x+y+a)) \leq \frac{1}{2}\delta.$$

To check for consistency, the verifier will perform the following test.

CONSISTENCY TEST
Test 1: Pick uniformly at random $x, x' \in (\mathrm{GF}(2))^n$.
Check if

$$(G_1(x+r) - G_1(r)) \cdot (G_1(x'+r') - G_1(r')) = (G_2(x \circ x' + r'') - G_2(r''))$$

for random r, r' and r''.
Test 2: Pick uniformly at random $x \in (\mathrm{GF}(2))^n$, $y \in (\mathrm{GF}(2))^{n^2}$.
Check if

$$(G_1(x+r) - G_1(r)) \cdot (G_2(y+r') - G_2(r')) = (G_3(x \circ y + r'') - G_3(r''))$$

for random r, r' and r''.

Claim 6.9.7 *Let $a = (a_i)_{1 \le i \le n}$, $b = (b_{ij})_{1 \le i,j \le n}$, $c = (c_{ijk})_{1 \le i,j,k \le n}$ be vectors over the field* GF*(2). For $x \in (\mathrm{GF}(2))^n$, let $H_1(x) = \sum_{i=1}^n a_i x_i$, for $y \in (\mathrm{GF}(2))^{n^2}$, let $H_2(y) = \sum_{i,j=1}^n b_{ij} y_{ij}$, and for $z \in (\mathrm{GF}(2))^{n^3}$, let $H_3(z) = \sum_{i,j,k=1}^n c_{ijk} z_{ijk}$. Suppose that G_1 is δ-close to H_1, G_2 is δ-close to H_2, and that G_3 is δ-close to H_3. Then, if $b \ne a \circ a$, Test 1 fails with probability (over the choices of x, x', r, r', and r'') at least $\frac{1}{4} - 6\delta$. Also, if $c \ne a \circ b$, Test 2 fails with probability at least $\frac{1}{4} - 6\delta$.*

Proof. We have observed that for vectors $\alpha \ne \alpha' \in (\mathrm{GF}(2))^n$

$$\mathrm{Prob}_{x \in (\mathrm{GF}(2))^n}(\alpha \cdot x \ne \alpha' \cdot x) = \frac{1}{2}. \tag{6.45}$$

If we regard in the natural way $a \circ a$ and b as matrices, the above relation implies

$$\mathrm{Prob}_{x \in (\mathrm{GF}(2))^n}((a \circ a)^T x \ne b^T x) \ge \frac{1}{2},$$

because $a \circ a$ and b have at least one pair of rows with the same index that are different. Using Equation (6.45) again, it follows that

$$\mathrm{Prob}_{x,x' \in (\mathrm{GF}(2))^n}(x'(a \circ a)^T x \ne x' b^T x) \ge \frac{1}{4}.$$

Note that $x'(a \circ a)^T x = H_1(x) \cdot H_1(x')$ and $x' b^T x = H_2(x \circ x')$. Therefore,

$$\mathrm{Prob}_{x,x' \in (\mathrm{GF}(2))^n}(H_1(x) \cdot H_1(x') \ne H_2(x \circ x')) \ge \frac{1}{4}. \tag{6.46}$$

Taking into account that G_1 is δ-close to H_1, that G_2 is δ-close to H_2, and using Claim 6.9.6, it follows that, for fixed x and x',

$$\mathrm{Prob}_r(G_1(x+r) - G_1(r) \ne H_1(x)) \le 2\delta$$
$$\mathrm{Prob}'_r(G_1(x'+r') - G_1(r') \ne H_1(x')) \le 2\delta$$
$$\mathrm{Prob}_r(G_2(x \circ x' + r'') - G_1(r'') \ne H_2(x \circ x')) \le 2\delta.$$

Combining these relations with Equation (6.46), the conclusion follows. The second part is proved in a similar way. ∎

By repeating the CONSISTENCY TEST a constant number of times, the probability that this test passes erroneously can be made arbitrarily small.

Claim 6.9.8 *Let $0 < \delta < 1/24$ be a fixed constant. Then there exists a constant $k_3(\delta)$ such that if there is no vector $a \in \mathrm{GF}(2)$ so that (a) G_1 is δ-close to $H_1(x) = a \cdot x$, (b) G_2 is δ-close to $H_2(x) = (a \circ a) \cdot x$, and (c) G_3 is δ-close to $H_3(x) = (a \circ a \circ a) \cdot x$, then with probability at least $1 - \delta$, at least one of $k_3(\delta)$ iterations of* LINEARITY TEST *and* CONSISTENCY TEST *fails.*

Proof. From Claim 6.9.5, we know that if the LINEARITY TEST passed $k_1(\delta/2)$ times, then we may assume with probability $1 - \delta/2$ that G_i is δ-close to some functions H_i, $i = 1, 2, 3$, where for some vectors $a \in (\mathrm{GF}(2))^n$, $b \in (\mathrm{GF}(2))^{n^2}$, and $c \in (\mathrm{GF}(2))^{n^3}$,

$$H_1(x) = \sum_{i=1}^{n} a_i x_i, \qquad H_2(y) = \sum_{i,j=1}^{n} b_{ij} y_{ij}, \qquad \text{and} \qquad H_3(y) = \sum_{i,j.k=1}^{n} c_{ijk} y_{ijk}.$$

If we take $k_2(\delta/2)$ such that $(1 - (\frac{1}{4} - 6\delta))^{k_2(\delta/2)} < \delta/2$, and if $b \neq a \circ a$ or $c \neq a \circ a \circ a$, then, with probability at least $1 - \delta/2$, one of the $k_2(\delta)$ iterations of CONSISTENCY TEST will fail. The conclusion follows with $k_3(\delta) = \max(k_1(\delta/2), k_2(\delta/2))$. ∎

If the proof string $g_1 g_2 g_3$ has passed $k_3(\delta)$ iterations of the LINEARITY TEST and $k_3(\delta)$ iterations of the CONSISTENCY TEST, the verifier will perform the following test that corresponds to checking condition (2), namely that the vector a that implicitly passed the tests corresponds to a satisfying assignment for the formula ϕ. The only ammendment to our earlier discussion is caused by the fact that the verifier only knows functions that are δ-close to $a \cdot x$, $(a \circ a) \cdot y$, and $(a \circ a \circ a) \cdot z$ and thus he has to use again self-correction.

SATISFIABILITY TEST

Input: Formula ϕ and functions G_1, G_2, G_3 given by their truth tables g_1, g_2, g_3.

Pick uniformly at random $r \in (\mathrm{GF}(2))^n$, compute $c(r), S_1(r), S_2(r)$ and $S_3(r)$ as in Equation (6.42).

Check if

$$\begin{aligned} c(r) + \; & (G_1(S_1(r) + r') - G_1(r')) \; + \\ & (G_2(S_2(r) + r'') - G_2(r'')) \; + \\ & (G_3(S_3(r) + r''') - G_3(r''')) = 0. \end{aligned}$$

for random r', r'' and r'''.

---

same as in Definition 6.6.2 except that it has an extra tape called the oracle tape. The oracle tape is shared by the verifier and the scribe. A PCPS$[r(n), q(n), t(n)]$ protocol on input a formula ϕ in 3CNF of length n runs as follows:

Round 1: A random string R of polynomial size is written on the oracle tape. The string R is called the public random string because the both the verifier and the scribe have access to it.

Round 2: The verifier produces a random string r of length $r(n)$ that it keeps private. Based on ϕ, R, and r, the verifier selects $q(n)$ addresses in the proof string (that will be provided by the scribe in Round 3). The bits of the proof strings at these addresses will be queried in Step 4.

Round 3: A scribe of complexity $t(n)$ is using an assignment a for ϕ and the string R as the oracle. The scribe produces bit by bit a proof string denoted $w^R(\phi, a)$ which is passed to the verifier.

Round 4: The verifier V reads from $w^R(\phi, a)$ the bits at the addresses selected at Round 2. At the end it accepts or it rejects the input.

We denote acceptance by 1, and rejection by 0, and we denote the output of the entire protocol by $V^R(r, w^R(\phi, a), \phi)$. The differences between a PCP and a PCPS protocol are: (a) the introduction of Round 1 in the PCPS protocol, which basically is used by the verifier to announce the prover the subset A of B as discussed in the Introduction, and (b) the fact that in the PCPS protocol, the prover (called a scribe), after being given an assignment, has limited resources to produce each bit of the proof string, while in the PCP protocol, the provers have unlimited computational power. We can consider that behind the scene there is an all-powerful prover that passes n bits of information (encoded as an assignment) to the scribe to help him make the verifier accept the formula ϕ.

Definition 6.9.10 *A language L is in* PCPS$[r(n), q(n), t(n)]$ *if there is a verifier running a* PCPS$[r(n), q(n), t(n)]$ *protocol with the following properties:*

(1) *(Completeness) If $x \in L$, then there is a scribe C of complexity $t(n)$ and a string a of length poly($|x|$) such that for any R*

$$\text{Prob}_r(V^R(r, w^R(x, a), x) = \textit{accept}) = 1,$$

where $w^R(x, a)$ is the proof string produced bit-by-bit by the scribe C on inputs x and a, and with oracle R.

(2) *(Soundness) If $x \notin L$, then with high probability of R (i.e., with probability at least $1 - 2^{-\Omega(n)}$), for any string a and for any scribe C of complexity $t(n)$, it holds that*

$$\text{Prob}_r(V^R(r, w^R(x, a), x) = \textit{accept}) \leq 1/2,$$

where $w^R(x, a)$ is the proof string produced bit-by-bit by the scribe C on inputs x and a, and with oracle R.

We will prove in Theorem 6.9.13, that, for any polynomial p, 3SAT is accepted by a verifier in a PCPS protocol utilizing $O(\log n)$ random bits and reading a constant number of bits from a polynomially long proof string, and such that the protocol is sound against scribes of complexity $p(n)$.

As indicated in the sketch, the proof relies on a certain sampling procedure. In general, for a function $a : \Sigma^n \to \{0, 1\}$, if we want to estimate the average value $2^{-n} \cdot \sum_{x \in \Sigma^n} a(x)$, it is enough to randomly select $O(n)$ points in Σ^n and calculate the average on these points. This sampling procedure utilizes a random string R to select the $O(n)$ sample points. With high probability of R, the average of a on the sample points is very close to the average of a on the entire Σ^n. We have however a different scenario: The string R is chosen first, and the function a can be chosen by an adversary that has access to R. The next theorem shows that if the adversary is an oracle circuit C of bounded size, then there exists a sampling procedure f that behaves as in the simple case presented above. The function f has two arguments, which we denote by R and r. In the notation $f(R, r)$, the string R is the random string used by the sampling procedure, and the string r indicates that f is producing the r-th sample point. We write C^R to indicate that the oracle circuit uses R as the oracle. The requirement for the sampling procedure is that $|r| = 8 \log(\text{size}(C))$ (a more careful proof shows that $|r| = (1 + \epsilon) \log(\text{size}(C))$ is in fact sufficient).

Theorem 6.9.11 (Sampling under adverse conditions) *Let $t \geq 3$. There is a function $f : \Sigma^* \times \Sigma^* \to \Sigma^*$ and a polynomial p such that for any natural number m,*

(a) *For R with $|R| = p(m)$ and for r with $|r| = 8 \cdot t \cdot \log m$, $f(R, r)$ has length m;*

(b) *With probability of R at least $1 - 2^{-poly(m)}$, for any oracle circuit C with inputs of length m, having size m^t, and that outputs 1 or 0,*

$$\left| \frac{1}{2^{|r|}} \sum_r C^R(f(R, r)) - \frac{1}{2^m} \sum_z C^R(z) \right| \leq m^{-1} \qquad (6.47)$$

where the first sum is taken over all the strings r of length $8 \cdot t \cdot \log m$, and the second sum is over all the strings of length m,

(c) *$f(R, r)$ can be computed in time poly(m).*

Proof. An inspection of statement (b) reveals that the desired function $f(R, \cdot)$ (with R fixed) is, with high probability of R, a pseudo-random generator secure against circuits that have oracle access to R (with security $(1/m, m^t)$, where m is the output length). Even though the existence of pseudo-random generators is an outstanding open problems (implying P $\neq$ NP), the construction of a family of functions $(f(R, \cdot))_R$ such that $f(R, \cdot)$ is with high probability a pseudo-random generator is doable (realize that for almost oracles R, $\text{P}^R \neq \text{NP}^R$, see Section 3.5). The construction follows the method indicated in Sections 5.3 and 5.4. More precisely, we first construct a permutation that depends on a string R and that, with

$\text{Prob}_{x\in\Sigma^n}(C^R(x) = R^{-1}(x)) \leq 2^{-0.2n}$. We fix a permutation $R : \Sigma^n \to \Sigma^n$ among the (overwhelming) fraction of permutations that satisfy the above relation. Thus, the function R is a one-way permutation with security $(2^{-0.2n}, 2^{0.7n})$ and moreover the security holds with respect to adversary oracle circuits that have oracle access to R.

Step 2. Theorem 5.3.11 and the prior results upon which it is based (presenting the construction of an extender starting from a one-way permutation) remain true (with basically the same proof) for the case of adversary oracle circuits. Recall that, starting with the one-way permutation R, the construction amounts to taking the function $h_R : \Sigma^{2n} \to \Sigma^{2n+1}$ given by $h_R(x \odot s) = R(x) \odot s \odot (x \cdot s)$, where $|x| = |s| = n$, $\odot$ denotes string concatenation, and $x \cdot s$ is the inner product of x and s viewed as vectors over GF(2). By Theorem 5.3.9 and Lemma 5.3.4, the function h_R is an extender with security $\left(\frac{4}{3}\cdot 2^{-0.2n}, O\left((2^{-0.4n}/n^2)\cdot 2^{0.7n}\right)\right)$, and the security holds with respect to oracle circuits that have access to R. This formally means that for any oracle circuit C with size$(C) \leq c(2^{-0.4n}/n^2)\cdot 2^{0.7n}$, for some constant c,

$$\left|\text{Prob}_{r\in\Sigma^{2n}}(C^R(h_R(r)) = 1) - \text{Prob}_{z\in\Sigma^{2n+1}}(C^R(z) = 1)\right| < 2^{-0.2n}.$$

Step 3. We now use Theorem 5.4.1, which presents the construction of pseudo-random generators with large extensions starting from an extender. This theorem holds with basically the same proof for adversary oracle circuits. We obtain a function $g_R : \Sigma^{2n} \to \Sigma^{2^{0.1n}}$ that is a pseudo-random generator with security $(2^{-0.1n}, 2^{0.28n})$ and the security holds with respect to oracle circuits that have access to R. This formally means that for any oracle circuit C with size$(C) \leq 2^{0.28n}$,

$$\left|\text{Prob}_{r\in\Sigma^{2n}}(C^R(g_R(r)) = 1) - \text{Prob}_{z\in\Sigma^{2^{0.1n}}}(C^R(z) = 1)\right| < 2^{-0.1n}.$$

We choose n such that (1) $m \leq 2^{0.1n}$ (so that the output length of g_R is at least m), (2) $2^{0.28n} \geq m^t$, (so that the security holds for circuits of size m^t), and (3) $2^{-0.1n} \leq (1/m)$ (so that the bias of the pseudo-random generator is less than $(1/m)$). The value $n = 4t\log m$ verifies all these requierements. We take $f(R,r)$ to be the first m bits of $g_R(r)$. The function $R : \Sigma^n \to \Sigma^n$ can be specified with $n2^n = \text{poly}(m)$ bits and the length of r is $|r| = 2n = 8t\log m$. Clearly, for any oracle circuit C with size$(C) \leq m^t \leq 2^{0.28n}$,

$$\left|\text{Prob}_{r\in\Sigma^{8t\log m}}(C^R(f(R,r)) = 1) - \text{Prob}_{z\in\Sigma^m}(C^R(z) = 1)\right| < 2^{-0.1n} < m^{-1}.$$

This relation holds with probability of R at least $1 - 2^{-2^{0.8n}} = 1 - 2^{-\text{poly}(m)}$, and the theorem is proved. ∎

We can now prove the promised result.

Theorem 6.9.13 (Light PCP theorem)
For any $k \geq 3$, 3SAT $\in$ PCPS$[O(\log n), O(1), n^k]$.

Proof. Let us fix a boolean formula ϕ in 3CNF, and let n be the length of ϕ. According to Theorem 6.9.1, there is a verifier V' running a $\mathrm{PCP}_{1,1/4}(cn^3, q)$ protocol for some constants c and q (in Theorem 6.9.1, the soundness value is 1/2, but clearly by doing two iterations of the protocol, we can bring it down to 1/4). The computation of V' depends on the formula ϕ, the random string ρ, and the proof string w.

We build a verifier V that simulates V', but runs a PCPS protocol with $O(\log n)$ private random bits. One obvious problem for the simulation is that V' is using cn^3 random bits that are not disclosed to the prover, while V can only use $O(\log n)$ private random bits. To solve this problem, we consider the function f promised by Theorem 6.9.11 with $m = cn^3$, and a constant value of t which will be specified later. In the first round, V writes a random string R of size $p(m)$ (p is the polynomial from Theorem 6.9.11), with the intention of using as the random string of length cn^3 needed in the simulation of V' only strings from the set

$$X_R = \{f(R, r) \mid |r| = 8 \cdot t \cdot \log m\}.$$

Let ℓ denote $8 \cdot t \cdot \log m$. In the second round, V selects uniformly at random a string r in $\{0,1\}^\ell$, calculates $f(R, r)$, simulates V' with the random string $f(R, r)$ having the desired length of $m = cn^3$ to determine the addresses in the proof string that V' is going to query later. In the third round, a scribe having ϕ and an assignment a for ϕ, and having access to R, calculates a proof string $w^R(\phi, a)$ spending no more than n^k time per bit of the proof string. It then passes $w^R(\phi, a)$ to the verifier. In the fourth round, V simulates V' with the queries established in round 2, and with the proof string $w^R(\phi, a)$. If some queried addresses are not in $w^R(\phi, a)$, then V rejects. Otherwise the simulation can be completed and V gives the verdict (1 for accept, 0 for reject) that the simulated V' gives.

Let us assume that ϕ is satisfiable and a is an assignment that makes ϕ to be true. According to Theorem 6.9.1, there is a proof string w such that

$$\mathrm{Prob}_\rho(V'(\rho, w, \phi) = 1) = 1.$$

An inspection of the proof of Theorem 6.9.1 shows that given a, each bit of w can be calculated in $O(n^3)$ time. For any R written by V in the first round, a scribe, having the assignment a, produces the bits of w that V' queries when the random string ρ is taken from X_R. It follows that for every R, and for every r of length ℓ,

$$V^R(r, w^R(\phi, a), \phi) = V'(f(R, r), w, \phi) = 1.$$

Thus, if ϕ is satisfiable, for any string R,

$$\mathrm{Prob}_{r \in \{0,1\}^\ell}(V^R(r, w^R(\phi, a), \phi) = 1) = 1.$$

This proves the completeness condition for the PCPS protocol.

Let us consider the case in which ϕ is not satisfiable. Then for any proof string w, $\mathrm{Prob}_\rho(V'(\rho, w, \phi) = 1) \leq 1/4$. Let us fix an assignment a for ϕ and a scribe S producing each bit of the proof string in time n^k. The scribe S, on input ϕ and a, and with some R on the oracle tape, tries to convince V to accept. We build next an oracle circuit C that simulates the whole protocol run by the verifier V' and the scribe S. The circuit C has ϕ and a embedded into its circuitry and has as input a string ρ of size cn^3. C first simulates V' and determines the addresses $i_1, \ldots, i_q$ in the proof string that are queried by V' on input formula ϕ and random string ρ. Next it simulates the scribe S to determine what are the bits at addresses $i_1, \ldots, i_q$ of the proof string $w^R(\phi, a)$ that is produced by S. Finally, the circuit C simulates the last round of the computation of V' and accepts or rejects accordingly. Thus the circuit C simulates V' on the following input: The boolean formula ϕ, the random string ρ, and the bits of the proof string obtained as specified above. C also simulates the scribe S to determine the q bits of the proof string queried by V'. The simulation of V' takes $p_1(n)$ time, for some polynomial p_1, and S produces every single bit in time n^t. Thus the size of the circuit C is bounded by $p_1(n) \log p_1(n) + qn^k \leq (cn^3)^t = m^t$ for some constant t. This is the value of t for which we use Theorem 6.9.11. We denote by $\mathcal{R}$ the set of strings R for which the equation (6.47) holds. Recall that the size of $\mathcal{R}$ represents a fraction of $1 - 2^{-poly(m)}$ from the set of strings of length $p(m)$.

Let $g^R(\rho)$ be the output of C on input ρ running with oracle R. From the simulation it holds that for every R,

$$g^R(\rho) = V'(\rho, w^R(\phi, a), \phi).$$

Also, g^R is calculated by the oracle circuit C of size m^t. Thus, by Theorem 6.9.11, if $R \in \mathcal{R}$,

$$\left|\text{Prob}_{r\in\{0,1\}^\ell}(g^R(f(R,r)) = 1) - \text{Prob}_{\rho\in\{0,1\}^m}(g^R(\rho) = 1)\right| \leq m^{-1} = \frac{1}{cn^3}.$$

Since

$$\text{Prob}_{\rho\in\{0,1\}^m}(g^R(\rho) = 1) = \text{Prob}_{\rho\in\{0,1\}^m}(V'(\rho, w^R(\phi, a), \phi) = 1) \leq \frac{1}{4},$$

it follows that for $R \in \mathcal{R}$,

$$\text{Prob}_{r\in\{0,1\}^\ell}(g^R(f(R,r)) = 1) \leq \frac{1}{4} + \frac{1}{cn^3} < \frac{1}{2}.$$

Note that $g^R(f(R,r))$ is exactly $V^R(r, w^R(\phi, a), \phi)$. Therefore, for any $R \in \mathcal{R}$, if $w^R(\phi, a)$ is produced by a scribe of complexity n^t, and a is any assignment for ϕ,

$$\text{Prob}_{r\in\{0,1\}^\ell}(V^R(r, w^R(\phi, a), \phi) < 1/2.$$

This proves the soundness condition for the PCPS protocol run by V. ∎

6.10 Comments and bibliographical notes

Theorem 6.2.4 has been proved by Fagin in [Fag74]. This result is the starting point for an active direction of research in computational complexity, called *descriptive complexity*, that investigates the logical resources needed to characterize various complexity classes. For an introduction in the area of descriptive complexity theory the reader could consult the survey articles by Immerman [Imm95, Imm89]. A thorough coverage is provided in Immerman's monograph [Imm99].

The logical approach to NP optimization problems started with the influential paper of Papadimitriou and Yannakakis [PY91] that introduced the classes MAX SNP, MAX NP, and the concept of L-reduction. A systematic study of the classification of NP optimization problems in terms of the syntax of their logical characterization has been undertaken by Kolaitis and Thakur in [KT94] and [KT95]. These two papers represent the main sources for Section 6.3. Zimand [Zim99] has investigated the classes of weighted NP optimization problems. Theorem 6.3.6 appeared in this paper.

The results relating the logical characterization of NP optimization problem to their approximation properties are summarized in Table 6.10.

The literature on approximation algorithms for NP optimization problems is huge, in consonance with the importance of the area. Most algorithms books such as the one by Cormen, Leiserson, and Rivest [CLR90] or the one by Papadimitriou and Steiglitz [PS82] contain chapters dedicated to this topic. The collection of articles [Hoc97] surveys the state of the art in this area. Theorem 6.4.1 is from the article of Papadimitriou and Yannakakis [PY91]. The greedy approximation algorithm for SET COVER and its analysis presented in Theorem 6.5.3 are due to Chvátal [Chv79]. The analysis is important since it introduces the idea of using linear programming duality in approximation. Johnson [Joh74] and Lovász [Lov75] have earlier demonstrated that the greedy algorithm is an approximation algorithm with $H(d)$ approximation ratio for the unweighted SET COVER problem, where $H(d) = \sum_{i=1}^{d}(1/i)$ and d is the size of the largest set. Johnson's paper has introduced the classes APX and PTAS presented here in Definition 6.1.6. The algorithm for SET COVER(b) and its analysis presented here in Theorem 6.5.5 are from a paper of Hochbaum [Hoc82]. The results from Lemma 6.5.9 and Lemma 6.5.11 are from Kolaitis and Thakur [KT95].

Until quite recently very few non-approximability results were known because, in principle, the traditional techniques for building reductions are not adequate for obtaining the kind of gaps exhibited in Equation 6.23 that are suitable for getting this type of results. The events that changed this state of affairs dramatically have been the development of the theory of probabilistic checking of proofs and the appearance of the paper by Feige, Goldwasser, Lovász, Safra, and Szegedy [FGL$^+$91] that displayed the reduction from the PCP protocol to the MAX CLIQUE problem that is presented here in Theorem 6.7.1. The connection between proof checking and approximation was also observed in a paper by Condon [Con91].

Table 6.1: Approximation properties of NP optimization problems. $\tilde{\mathrm{P}}$ denotes $\mathrm{DTIME}[2^{\log^{O(1)} n}]$.

Class		Unweighted	Weight(+)	Weight
MAX Σ_0	upper bound	APX	APX	
	lower bound	no PTAS unless P = NP	no PTAS unless P = NP	not approx. with ratio n^δ, for some $\delta > 0$ unless P = NP
MAX Σ_1	upper bound	APX	APX	
	lower bound	no PTAS unless P = NP	no PTAS unless P = NP	not approx. with ratio n^δ, for some $\delta > 0$ unless P = NP
MAX SNP(π)	upper bound	APX	APX	
	lower bound	no PTAS unless P = NP	no PTAS unless P = NP	not approx. with ratio $2^{\log^\mu n}$, for some $\mu > 0$ unless NP $\subseteq \tilde{\mathrm{P}}$
MIN $\mathrm{F}^+\Pi_1$	upper bound	APX	APX	
	lower bound	no PTAS unless P = NP	no PTAS unless P = NP	not approx. with ratio 2^{n^q}, for any $q > 0$ unless P = NP
MIN $\mathrm{F}^+\Pi_2(1)$	upper bound	log-APX	log-APX	
	lower bound	not approx. with ratio $c \log n$ for some $c > 0$ unless NP $\subseteq \tilde{\mathrm{P}}$	not approx. with ratio $c \log n$ for some $c > 0$ unless NP $\subseteq \tilde{\mathrm{P}}$	not approx. with ratio 2^{n^q}, for some $\delta > 0$ unless P = NP

Numerous papers have contributed to the techniques needed to prove Theorem 6.6.4, the *PCP Theorem*. In an attempt to generalize the definition of NP based on proof checking, Babai and Moran [Bab85, BM88] and independently Goldwasser, Micali, and Rackoff [GMR85] have introduced the idea of using randomness and interaction between a verifier and a prover; the class of languages accepted by a polynomial-time verifier running this type of protocol is called IP. Ben-Or, Kilian, and Wigderson [BOGKW88] considered the variant in which the verifier interacts with two (or more) provers (who are not allowed to communicate to each other). obtaining the class MIP. Theses classes turned out to be surprisingly large. Lund, Fortnow, Karloff, and Nisan [LFKN90] have shown that IP contains the polynomial hierarchy, a result shortly superseded by Shamir's proof that IP = PSPACE [Sha92]. Babai, Fortnow, and Lund [BFL91] proved that MIP equals NEXP, the class of problems solvable in nondeterministic exponential time. This latter result shows amazingly that an exponentially long proof string can be checked in randomized polynomial time with low error probability. Then in a sequence of papers, Babai, Fortnow, Levin, and Szegedy [BFLS91], Feige, Goldwasser, Lovász, Safra, and Szegedy [FGL+91], Arora and Safra [AS92], the MIP = NEXP result was scaled down to NP (replacing NEXP) with increasingly better values of the different parameters that affect the protocol. Finally, Arora, Lund, Motwani, Sudan, and Szegedy [ALM+92] have obtained the PCP Theorem. A complete self-contained proof of this important result is available in [Aro94] and also in [HPS94]. The chapter contains the complete proofs of two easier variants of the PCP Theorem. The first variant, Theorem 6.9.1, is part of the proof of the full PCP Theorem, and the second one, Theorem 6.9.13, was proven by Zimand [Zim01].

Following the ideas pioneered in [FGL+91], increasingly stronger non-approximability results have been obtained for many combinatorial problems. Theorem 6.6.9 is due to Arora, Lund, Motwani, Sudan, and Szegedy [ALM+92] and Theorem 6.6.10 has been demonstrated by Zimand [Zim99].

The general idea of the reductions from Theorem 6.7.1 and Theorem 6.7.5 is from [FGL+91]. The employment of random walks on expander graphs for an efficient utilization of random bits has been discovered by Impagliazzo and Zuckerman [IZ89] relying on a method of Ajtai, Komlós, and Szemerédy [AKS87]. The Gabber-Galil expander graphs used in the proof of Theorem 6.7.5 are from [GG81]. Another construction with better parameters is due to Lubotzky, Phillips and Sarnak [LPS86]. For an in-depth coverage of expander graphs, the reader could consult the monograph by Chung [Chu97]. The strongest currently known non-approximation result for MAX CLIQUE (namely that, unless ZPP = NP, there is no polynomial-time approximation algorithm for this problem with approximation ratio $n^{1-\epsilon}$) is due to Håstad [Hås99] (see also the paper by Engebretsen and Holmerin [EH00] containing a slightly stronger inapproximability result under a stronger assumption).

The non-approximability result for SET COVER (Theorem 6.8.2) has been obtained by Lund and Yannakakis [LY94]. The construction of the δ-biased sample space that is used in the proof of Theorem 6.8.2 is from a paper by Alon, Goldreich, Håstad, and Peralta [AGHR92] relying on some results of Naor and Naor [NN93].

A comprehensive compendium of NP optimization problems, containing the best approximation results known for each problem as well as the strongest known non-approximability property, is available on the world wide web at http://www.nada.kth.se/~viggo/problemlist.

Appendix A

Tail bounds

In general, a "well-behaved" random variable has most of its values concentrated around its expectation. This phenomenon is expressed in various precise ways by some well-known theorems in probability theory such as the law of large numbers, the central limit theorem, and the de Moivre-Laplace theorem. These theorems, typically, cannot be used directly in complexity theory because they have an asymptotic nature that does not say anything precise about the behaviour of a random process depending on a parameter n (for example, n is the input length of a probabilistic algorithm), even if n is large.

We present here a few standard non-asymptotical inequalities that bound the probability that a random variable deviates from its expected value by a certain amount. Such inequalities are called *tail bounds*.

If X is a random variable, then $\mathrm{E}(X)$ denotes its expected value, and $\mathrm{Var}(X)$ denotes its variance (i.e., $\mathrm{Var}(X) = \mathrm{E}((X - E(X))^2)$).

Theorem A.0.1 (Markov's Inequality) *Let X be a random variable that takes positive values. Then, for every $t > 0$,*

$$\mathrm{Prob}(X \geq t) \leq \frac{\mathrm{E}(X)}{t}.$$

Proof.

$$\begin{aligned}\mathrm{E}(X) &= \sum_x x \cdot \mathrm{Prob}(X = x) = \sum_{x<t} x \cdot \mathrm{Prob}(X = x) + \sum_{x \geq t} x \cdot \mathrm{Prob}(X = x) \\ &\geq \sum_{x \geq t} x \cdot \mathrm{Prob}(X = x) \geq t \cdot \sum_{x \geq t} \mathrm{Prob}(X = x) = t \cdot \mathrm{Prob}(X \geq t).\end{aligned}$$

Therefore, $\mathrm{Prob}(X \geq t) \leq \mathrm{E}(X)/t$. ∎

Proof. (a) We evaluate the expression $\mathrm{e}^d/(1+d)^{(1+d)}$ that occurs in inequality (A.1). We have $(1+d)^{(1+d)} = \mathrm{e}^{(1+d)\ln(1+d)}$ and, by Taylor expansion, $(1+d)\ln(1+d) = (1+d)(d - d^2/2 + O(d^3)) = d + d^2/2 + O(d^3)$. It follows (with some additional use of calculus) that $(1+d)\ln(1+d) \geq d + d^2/3$ for all $0 \leq d < 1$, and thus $\mathrm{e}^d/(1+d)^{(1+d)} \leq \mathrm{e}^d/\mathrm{e}^{d+d^2/3} = \mathrm{e}^{-d^2/3}$. Substituting in inequality (A.1),

$$\operatorname{Prob}(X > (1+d)\mu) \leq \left(\frac{\mathrm{e}^d}{(1+d)^{(1+d)}}\right)^\mu \leq \mathrm{e}^{-\frac{d^2\mu}{3}}.$$

(b) We evaluate the expression $\mathrm{e}^{-d}/(1-d)^{(1-d)}$ that occurs in inequality (A.2). We have $(1-d)^{(1-d)} = \mathrm{e}^{(1-d)\ln(1-d)}$ and, by Taylor expansion, $(1-d)\ln(1-d) = (1-d)(-d - d^2/2 - d^3/3 - O(d^4)) = -d + d^2/2 + O(d^3)$. It follows that $(1-d)\ln(1-d) \geq -d + d^2/3$ for all $0 \leq d < 1$, and thus $\mathrm{e}^{-d}/(1-d)^{(1-d)} \leq \mathrm{e}^{-d}/\mathrm{e}^{-d+d^2/2} = \mathrm{e}^{-d^2/2}$. Substituting in inequality (A.2),

$$\operatorname{Prob}(X < (1-d)\mu) \leq \left(\frac{\mathrm{e}^{-d}}{(1-d)^{(1-d)}}\right)^\mu \leq \mathrm{e}^{-\frac{d^2\mu}{2}}.$$

∎

In some situations, we need to bound the probability that $X > \mu + d$ or $X < \mu - d$, for some d.

Theorem A.0.5 (Chernoff's bounds; additive form) *Let $X_1, \ldots, X_n$ be independent random variables that take the values 0 and 1. Assume $\operatorname{Prob}(X_i = 1) = p$, $i = 1, \ldots, n$, for some $0 < p < 1$. Let $X = \sum_{i=1}^n X_i$ be a random variable and $\mu = \mathrm{E}(X) = np$. Then,*

(1) *for any $0 < d < p$,*

$$\operatorname{Prob}\Big(\frac{1}{n}(X_1 + \ldots + X_n) > p + d\Big) = \operatorname{Prob}(X > \mu - nd) \leq \mathrm{e}^{-\frac{1}{3p}d^2 n}, \quad \text{(A.5)}$$

(2) *for any $0 < d < p$,*

$$\operatorname{Prob}\Big(\frac{1}{n}(X_1 + \ldots + X_n) < p - d\Big) = \operatorname{Prob}(X < \mu - nd) \leq \mathrm{e}^{-\frac{1}{2p}d^2 n}, \quad \text{(A.6)}$$

Proof. We have

$$\operatorname{Prob}\Big(\frac{1}{n}\sum X_i > p + d\Big) = \operatorname{Prob}(X > np + nd) = \operatorname{Prob}\Big(X > \mu\Big(1 + \frac{d}{p}\Big)\Big)$$
$$\leq \mathrm{e}^{-\frac{1}{3}\cdot\frac{d^2}{p^2}\cdot pn} = \mathrm{e}^{-\frac{1}{3p}d^2 n}.$$

Similarly, for the second inequality

$$\operatorname{Prob}\Big(\frac{1}{n}\sum X_i < p - d\Big) = \operatorname{Prob}(X < np - nd) = \operatorname{Prob}\Big(X < \mu\Big(1 - \frac{d}{p}\Big)\Big)$$
$$\leq \mathrm{e}^{-\frac{1}{2}\cdot\frac{d^2}{p^2}\cdot pn} = \mathrm{e}^{-\frac{1}{2p}d^2 n}.$$

∎

Bibliography

[ACGS88] W. Alexi, B. Chor, O. Goldreich, and C. Schnorr. RSA and Rabin functions: Certain parts are as hard as the whole. *SIAM Journal on Computing*, 17(2):194–209, April 1988. Special issue on cryptography.

[AF98] A. Ambainis and R. Freivalds. 1-way quantum finite automata: strengths, weaknesses and generalizations. In *Proceedings of the 38th IEEE Symposium on Foundations of Computer Science*, pages 332–341, 1998.

[AGHR92] N. Alon, O. Goldreich, J. Håstad, and R. Peralta. Simple constructions of almost k-wise independent random variables. *Random Structures and Algorithms*, 3(3):289–304, 1992.

[AKS87] M. Ajtai, J. Komlós, and E. Szemerédi. Deterministic simulation in LOGSPACE. In *Proceedings of the 19th ACM Symposium on Theory of Computing*, pages 132–140, 1987.

[ALM+92] S. Arora, C. Lund, R. Motwani, M. Sudan, and M. Szegedy. Proof verification and intractability of approximation problems. In *Proceedings of the 32nd IEEE Symposium on Foundations of Computer Science*, pages 14–23, 1992.

[Aro94] S. Arora. *Probabilistic Checking of Proofs and Hardness of Approximation Problems.* PhD thesis, Princeton University, 1994.

[AS92] S. Arora and S. Safra. Probabilistic checkable proofs: A new characterization of NP. In *Proceedings of the 32nd IEEE Symposium on Foundations of Computer Science*, pages 1–13, 1992.

[AS96] K. Ambos-Spies. Resource-bounded genericity. In *Computability, Enumerability, Unsolvability*, London Mathematical Society. LNS 224, pages 1–59. Cambridge University Press, 1996.

[ASB97] K. Ambos-Spies and L. Bentzien. Separating NP-completeness under strong hypotheses. In *Proceedings of the 12th Structure in Complexity Theory Conference*, pages 121–127. IEEE Computer Society Press, 1997.

[ASM97] K. Ambos-Spies and E. Mayordomo. Resource-bounded measure and randomness. In A. Sorbi, editor, *Complexity, Logic, and Recursion Theory*, Lecture Notes in Pure and Applied Mathematics, pages 1–47. Marcel Dekker, 1997.

[Cal88] C. Calude. *Theories of Computational Complexity.* North-Holland, 1988.

[CIZ92] C. Calude, G. Istrate, and M. Zimand. Recursive Baire classification and speedable functions. *Z. Math. Logik Grundlag. Math.*, 38:169–178, 1992.

[CZ92] C. Calude and M. Zimand. On three theorems in abstract complexity theory: A topological glimpse. In *Abstracts of the 2nd International Colloquium on Words, Languages and Combinatorics, Kyoto, Japan*, pages 11–12, 1992.

[CZ96] C. Calude and M. Zimand. Effective category and measure in abstract complexity theory. *Theoretical Computer Science*, 154(2):307–327, 1996.

[CG88] B. Chor and O. Goldreich. Unbiased bits from sources of weak randomness and probabilistic communication complexity. *SIAM Journal on Computing*, 17:230–261, 1988.

[Chu97] Fan R. K. Chung. *Spectral Graph Theory.* Conference Board of the Mathematical Sciences. American Mathematical Society, 1997.

[Chv79] V. Chvátal. A greeedy heuristics for the Set-Covering Problem. *Mathematics of Operations Research*, 4:233–235, 1979.

[CLR90] T. Cormen, C. Leiserson, and R. Rivest. *Introduction to Algorithms.* MIT Press, 1990.

[Con72] R. L. Constable. The operator gap. *Journal of the ACM*, 19(3):175–183, 1972.

[Con91] A. Condon. The complexity of the max word problem and the power of one-way interactive proof systems. In *Proceedings of the 8th Annual Symposium on Theoretical Aspects of Computer Science*, pages 456–465. Springer-Verlag Lecture Notes in Computer Science #480, 1991.

[Deu85] D. Deutsch. Quantum theory, the Church-Turing principle and the universal quantum computer. *Proceedings of the Royal Society of London*, 1985.

[DH76] W. Diffie and M. E. Hellman. New directions in cryptography. *IEEE Trans. on Info. Theory*, IT-22:644–654, 1976.

[DJ92] D. Deutsch and R. Jozsa. Rapid solutions of problems by quantum computation. *Proceedings of the Royal Society of London*, 1992.

[DK00] D. Du and K. Ko. *Theory of Computational Complexity.* John Wiley and Sons, Inc., 2000.

[EH00] L. Engebretsen and J. Holmerin. Clique is hard to approximate within $n^{1-o(1)}$. In *Proceedings of the 27th International Colloquium on Automata, Languages, and Programming*, pages 2–12. Springer-Verlag Lecture Notes in Computer Science #1853, 2000.

[End72] H. B. Enderton. *A Mathematical Introduction to Logic.* Academic Press, New York, 1972.

[Fag74] R. Fagin. Generalized first-order spectra and polynomial-time recognizable sets. In R. Karp, editor, *Complexity of Computation*, pages 27–41. Proceedings of SIAM-AMS Symposium in Applied Mathematics, 1974.

[Fel68] W. Feller. *Introduction to Probability Theory and its Applications.* John Wiley & Sons, 3rd ed. edition, 1968.

[Fen95] S. Fenner. Resource-bounded Baire category: a stronger approach. In *Proceedings of the 10th Structure in Complexity Theory Conference*, pages 182–192. IEEE Computer Society Press, 1995.

[Fey82] R. Feynman. Simulating physics with computers. *International Journal of Theoretical Physics*, 21(6/7):467–488, 1982.

[FGL+91] U. Feige, S. Goldwasser, L. Lovász, S. Safra, and M. Szegedy. Approximating clique is almost NP complete. In *Proceedings of the 31st IEEE Symposium on Foundations of Computer Science*, pages 2–12. IEEE Computer Society Press, 1991.

[FL92] U. Feige and L. Lovász. Two-prover one round systems: Their power and their problems. In *Proceedings of the 24th ACM Symposium on Theory of Computing*, pages 733–744. ACM Press, 1992.

[GG81] O. Gabber and Z. Galil. Explicit constructions of linear-sized superconcentrators. *Journal of Computer and System Sciences*, 22(3):407–420, June 1981.

[GGM86] O. Goldreich, S. Goldwasser, and S. Micali. How to construct a random functions. *Journal of the ACM*, 33(4):792–807, 1986.

[GL89] O. Goldreich and L. Levin. A hard-core predicate for all one-way functions. In *Proceedings of the 21st ACM Symposium on Theory of Computing*, pages 25–32, 1989.

[GNW95] O. Goldreich, N. Nisan, and A. Wigderson. On Yao's XOR lemma. Technical Report TR95-050, Electronic Colloquium on Computational Complexity, March 1995. `http://www.eccc.uni-trier.de/eccc`.

[Gol97] O. Goldreich. Notes on Levin's theory of average complexity. Technical Report TR97-058, ECCCTR: Electronic Colloquium on Computational Complexity, Technical reports, November 1997.

[GHJY91] J. Goldsmith, L. Hemachandra, D. Joseph, and P. Young. Near-testable sets. *SIAM Journal on Computing*, 20(3):506–523, 1991.

[GHS91] J. Geske, D. Huynh, and J. Seiferas. A note on almost-everywhere complex sets and separating deterministic time complexity classes. *Information and Computation*, 92:97–104, 1991.

[GLR+91] P. Gemmell, R. Lipton, R. Rubinfeld, M. Sudan, and A. Wigderson. Self-testing/correcting for polynomials and for approximate functions. In ACM, editor, *Proceedings of the twenty third annual ACM Symposium on Theory of Computing, New Orleans, Louisiana, May 6–8, 1991*, pages 33–42, New York, NY, USA, 1991. ACM Press.

[GM84] S. Goldwasser and S. Micali. Probabilistic encryption. *Journal of Computer and System Sciences*, 28(2):270–299, 1984.

[GMR85] S. Goldwasser, S. Micali, and C. Rackoff. The knowledge complexity of interactive proof systems. *SIAM Journal on Computing*, 18(1):186–208, 1985.

[KW97] A. Kondacs and J. Watrous. On the power of quantum finite state automata. In *Proceedings of the 37th IEEE Symposium on Foundations of Computer Science*, pages 66–75, 1997.

[Lad75] R.. Ladner. On the structure of polynomial time reducibility. *Journal of the ACM*, 22(1):155–171, 1975.

[Lec63] Y. Lecerf. Machines de Turing réversibles. récursive insolubilité en $n \in \mathbb{N}$ de l'equation $u = \theta^n$ est un isomorphisme de codes. *Comptes rendus de l'Académie française des sciences*, 1963.

[Len85] A. Lenstra. Factoring multivariate polynomials over finite fields. Journal of Computer and System Sciences, 30(2):515–534, April 1985.

[Lev73] L. Levin. On storage capacity for algorithms. *Soviet Mathematics Doklady*, 14(5):1464–1466, 1973.

[Lev74] L. Levin. Complexity of computation of computable functions. In V. A. Kozmidiadi, N. A. Maslov, and N. V. Petri, editors, *Complexity of computations and algorithms*, pages 174–185. Mir, Moscow (in Russian), 1974.

[Lev86] L. Levin. Average case complete problems. *SIAM Journal on Computing*, 15:285–286, 1986.

[Lev95] L. Levin. Computational complexity of functions. *Theoretical Computer Science*, 157:268–271, 1995.

[LFKN90] C. Lund, L. Fortnow, H. Karloff, and N. Nisan. Algebraic methods for interactive proof systems. In *Proceedings of the 30th IEEE Symposium on Foundations of Computer Science*, pages 2–10. IEEE Computer Society Press, October 1990.

[Lip89] R.. Lipton. New directions in testing. In *Proceedings of DIMACS Workshop on Distributed Computing and Cryptography*, 1989.

[LM96] J. Lutz and E. Mayordomo. Cook vesus Karp-Levin: Separating completeness notions if NP is not small. *Theoretical Computer Science*, 164:141–163, 1996.

[Lov75] L. Lovász. On the ratio of optimal integral and fractional covers. *Discrete Mathematics*, 13:383–390, 1975.

[LPS86] A. Lubotzky, R.. Phillips, and P. Sarnak. Explicit expanders and the Ramanujan conjecture. In *Proceedings of the 18th Annual ACM Symposium on Theory of Computing (STOC-86)*, pages 240–246, Berkeley, California, May 1986.

[Lut90] J. Lutz. Category and measure in complexity theory. *SIAM Journal on Computing*, 19(6):1100–1131, 1990.

[Lut92] J. Lutz. Almost everywhere high nonuniform complexity. *Journal of Computer and System Sciences*, 44:220–258, 1992.

[Lut97] J. Lutz. The quantitative structure of exponential time. In L. Hemaspaandra and A. Selman, editors, *Complexity Theory Retrospective II*, pages 225–260. Springer-Verlag, 1997.

[LY94] C. Lund and M. Yannakakis. On the hardness of approximating minimization problems. *Journal of the ACM*, 41(5):960–981, 1994.

[Mü93] H. Müller. A note on balanced immunity. *Mathematical Systems Theory*, 26:157–167, 1993.

[Meh73] K. Mehlhorn. On the size of sets of computable functions. In *Proceedings of the 14th IEEE Symposium on Foundations of Computer Science*, pages 190–196, 1973.

[MF72] A. R. Meyer and P. C. Fischer. Computational speed-up by effective operators. *J. Symbolic Logic*, 37(1):55–68, 1972.

[MM65] E. McCreight and A. Meyer. Classes of computable functions defined by bounds on computation: preliminary report. In *Conf. Rec. ACM Symp. on Theory of Computing*, pages 79–88, 1965.

[MS98] T. Mihara and S. Sung. A quantum polynomial algorithm in the worst case for Simon's problem. In *Proceedings of the 9th International Symposium on Algorithms and Computation*, Lecture Notes in Computer Science # 1533, pages 229–236. Springer-Verlag, 1998.

[MW79] A. R. Meyer and K. Winklman. The fundamental theorem of complexity theory. In J. W. de Bakker and J. van Leeuwen, editors, *Found. Comput. Sci. III, Part 1: Automata, Data Structures, Complexity*, pages 97–112. Mathematical Centre Tracts, vol. 108, Amsterdam, 1979.

[MY78] M. Machtey and P. Young. *An Introduction to Automata Theory, Languages and Computation*. North-Holland, 1978.

[NN93] J. Naor and M. Naor. Small-bias probability spaces: Efficient constructions and applications. *SIAM Journal on Computing*, 22(4):838–856, August 1993.

[NR99] M. Naor and O. Reingold. Synthesizers and their application to the parallel construction of pseudo-random functions. *JCSS: Journal of Computer and System Sciences*, 58, 1999.

[NTS99] N. Nisan and A. Ta-Shma. Extracting randomness: A survey and new constructions. *Journal of Computer and System Sciences*, 58(1):148–173, 1999.

[NW94] N. Nisan and A. Wigderson. Hardness vs. randomness. *Journal of Computer and System Sciences*, 49:149–167, 1994.

[NZ96] N. Nisan and D. Zuckerman. Randomness is linear in space. *Journal of Computer and System Sciences*, 52:43–52, 1996.

[Ogi94] M. Ogihara. Polynomial-time membership comparable sets. In *Proceedings of the 9th Structure in Complexity Theory Conference*, pages 2–11. IEEE Computer Society Press, July 1994.

[PS82] C. H. Papadimitriou and K. Steiglitz. *Combinatorial Optimization, Algorithms and Complexity*. Prentice-Hall, Englewood Cliffs, NJ, 1982.

[PY91] C. H. Papadimitriou and M. Yannakakis. Optimization, approximation and complexity classes. *Journal of Computer and System Sciences*, 43:425–440, 1991.

[Rab59] M. Rabin. Speed of computation of functions and classification of recursive sets. In *Proc. Third Conven. of Scientific Societies*, pages 1–2. Israel, 1959.

[Zim01] M. Zimand. Probabilistically checkable proofs the easy way. Technical Report TR01-027, ECCCTR: Electronic Colloquium on Computational Complexity, Technical reports, 2001.

[Zuc90] D. Zuckerman. General weak random sources. In *Proceedings of the 30th IEEE Symposium on Foundations of Computer Science*, pages 534–543. IEEE Computer Society Press, 1990.

[Zuc97] D. Zuckerman. Randomness-optimal oblivious sampling. *Random Structures and Algorithms*, 11:345–367, 1997.

Index